Polynomials and
Linear Control Systems

MONOGRAPHS AND TEXTBOOKS IN
PURE AND APPLIED MATHEMATICS

1. *K. Yano,* Integral Formulas in Riemannian Geometry (1970) *(out of print)*
2. *S. Kobayashi,* Hyperbolic Manifolds and Holomorphic Mappings (1970) *(out of print)*
3. *V. S. Vladimirov,* Equations of Mathematical Physics (A. Jeffrey, editor; A. Littlewood, translator) (1970) *(out of print)*
4. *B. N. Pshenichnyi,* Necessary Conditions for an Extremum (L. Neustadt, translation editor; K. Makowski, translator) (1971)
5. *L. Narici, E. Beckenstein, and G. Bachman,* Functional Analysis and Valuation Theory (1971)
6. *D. S. Passman,* Infinite Group Rings (1971)
7. *L. Dornhoff,* Group Representation Theory (in two parts). Part A: Ordinary Representation Theory. Part B: Modular Representation Theory (1971, 1972)
8. *W. Boothby and G. L. Weiss (eds.),* Symmetric Spaces: Short Courses Presented at Washington University (1972)
9. *Y. Matsushima,* Differentiable Manifolds (E. T. Kobayashi, translator) (1972)
10. *L. E. Ward, Jr.,* Topology: An Outline for a First Course (1972) *(out of print)*
11. *A. Babakhanian,* Cohomological Methods in Group Theory (1972)
12. *R. Gilmer,* Multiplicative Ideal Theory (1972)
13. *J. Yeh,* Stochastic Processes and the Wiener Integral (1973) *(out of print)*
14. *J. Barros-Neto,* Introduction to the Theory of Distributions (1973) *(out of print)*
15. *R. Larsen,* Functional Analysis: An Introduction (1973) *(out of print)*
16. *K. Yano and S. Ishihara,* Tangent and Cotangent Bundles: Differential Geometry (1973) *(out of print)*
17. *C. Procesi,* Rings with Polynomial Identities (1973)
18. *R. Hermann,* Geometry, Physics, and Systems (1973) *(out of print)*
19. *N. R. Wallach,* Harmonic Analysis on Homogeneous Spaces (1973) *(out of print)*
20. *J. Dieudonné,* Introduction to the Theory of Formal Groups (1973)
21. *I. Vaisman,* Cohomology and Differential Forms (1973)
22. *B.-Y. Chen,* Geometry of Submanifolds (1973)
23. *M. Marcus,* Finite Dimensional Multilinear Algebra (in two parts) (1973, 1975)
24. *R. Larsen,* Banach Algebras: An Introduction (1973)
25. *R. O. Kujala and A. L. Vitter (eds),* Value Distribution Theory: Part A; Part B. Deficit and Bezout Estimates by Wilhelm Stoll (1973)
26. *K. B. Stolarsky,* Algebraic Numbers and Diophantine Approximation (1974)
27. *A. R. Magid,* The Separable Galois Theory of Commutative Rings (1974)
28. *B. R. McDonald,* Finite Rings with Identity (1974)
29. *J. Satake,* Linear Algebra (S. Koh, T. Akiba, and S. Ihara, translators) (1975)

Other Volumes in Preparation

Polynomials and Linear Control Systems

STEPHEN BARNETT

University of Bradford
Bradford, West Yorkshire, England

MARCEL DEKKER, INC.   New York and Basel

Library of Congress Cataloging in Publication Data

Barnett, Stephen, [date]
  Polynomials and linear control systems.

  Includes index.
  1. Polynomials. 2. System analysis. I. Title.
QA161.P59B37 1983   512.9'42   83-5309
ISBN 0-8247-1898-4

MARCEL DEKKER, INC.
270 Madison Avenue, New York, New York 10016

Current printing (last digit):
10 9 8 7 6 5 4 3 2 1

PRINTED IN THE UNITED STATES OF AMERICA

... for the Use of some Friends who have neither Leisure, Convenience, nor, perhaps, Patience, to search into so many different Authors, and turn over so many tedious Volumes, as is unavoidably required to make but tolerable progress in the Mathematics. [William Jones, *A New Introduction to the Mathematics*, London, 1706]

Preface

Problems involving polynomials possess a long history, but
have recently come under extensive reexamination because of their
importance in areas of contemporary applied mathematics, including
linear control systems, electrical networks, signal processing, and
coding theory. There is a need for new textbooks to replace out-
dated treatments of the "theory of equations," where the profusion
of manipulations is often tedious and uninspiring. The aims of
this book are twofold: first, to develop properties of polynomials
and polynomial matrices using matrix techniques; and second, to
describe related topics in the theory of linear control systems.

A glance at the Contents will reveal the scope of the book,
but a few comments can usefully be added. For reasons of space,
only polynomials having real or complex coefficients are considered. The emphasis is on qualitative aspects, which are begun in
Chapter 1 with determination of greatest common divisors of polynomials. This gives an opportunity to establish four primary
themes which recur throughout the book: the applications of
matrices having companion, Sylvester and bezoutian forms, and of
tabular arrays based on euclidean-type recursions. These are
developed in several directions: in Chapter 3, to a study of the
stability and location of roots of polynomials; in Chapter 4, to
problems of relative primeness and greatest common divisor of

polynomial matrices; and in Chapter 5, to the case when general
polynomial bases are used. Applications to control systems are
interspersed with this material: canonical forms, linear feedback,
and realization are presented for systems with a single input and
output in Chapter 2, and for multivariable systems in Chapter 4.

My objective has been to bring together in a coherent fashion
a large number of varied results which have appeared in the mathematical and engineering literature over the past few decades. A
good proportion of the material is not readily available in book
form: for example, the application of companion matrices to link
together various techniques for greatest common divisors (Chapter
1); the proofs of classical stability and root location criteria
using Liapunov-type matrix equations (Chapter 3); and the extraction of greatest common divisors of polynomial matrices (Chapter
4). It is my hope that the book will be useful for both study and
reference purposes, and will stimulate further research. To further these aims I have included a large bibliography and many
worked examples. I have also put special effort into devising a
considerable number of problems, which often extend some results in
the text. The book could therefore be used for advanced under-
graduate or graduate courses, the only prerequisites being a work-
ing knowledge of matrix algebra, and some acquaintance with transform methods for differential and difference equations. Using
straightforward mathematics, students can be brought close to
problems of current research interest.

There is no other existing single volume which contains a
similar collection of topics, but whether the influence of my
personal interests has been too great is for readers to judge.
Doubtless, some mathematicians would prefer a more abstract presen-
tation, and engineers might feel that the control content is rather
thin. However, to keep the book a reasonable length, I have been
forced to curtail the treatment of many subjects, and to omit
altogether some (e.g., polynomials in several variables) which I
had originally intended to include. There is no shortage of

material to fill future complementary works.

I have labored long and hard to remove both typographical and factual errors, but apologize in advance for those which must inevitably remain. It is a pleasure to acknowledge suggestions and comments from colleagues at Bradford University, and also from R. E. Hartwig, L. R. Fletcher, and A. T. Fuller; special thanks are due to E. I. Jury for his invaluable and extensive remarks on Chapter 3. I am also grateful to secretaries in the School of Mathematical Sciences for their skillful typing, especially R. H. Wilkinson and V. M. Hunter; the latter did a superb job on the camera ready typescript. Part of the book was written while I was a Visiting Professor at North Carolina State University, Raleigh, whose hospitality was much appreciated.

Stephen Barnett

Contents

Polynomials and
Linear Control Systems

1

Polynomials: Approaches to Greatest Common Divisor

1.0 INTRODUCTION

It is assumed that the reader has encountered the elementary
algebra of polynomials, including the concept of greatest common
divisor (g.c.d.), with which this chapter is primarily concerned.
We shall find that the use of matrices permits a unified treatment
of the different methods, including simpler proofs and some extensions of the classical results. We begin by reviewing briefly in
Section 1.1 some fundamental notation and definitions needed for
the remainder of the chapter. In Section 1.2 the crucial idea of
a companion matrix associated with a polynomial is introduced. In
particular, the g.c.d. can be determined in an elegant way by constructing a polynomial in a companion matrix. This procedure is
described in Section 1.3, including an extension to *direct* determination of the g.c.d. of more than two polynomials. In Section
1.4, appropriate properties of Sylvester's matrix are developed
from the companion matrix results, and both these approaches are
related to the bezoutian matrix in Section 1.5. A useful recur-
sive form of Euclid's algorithm is an array named after Routh, set
out in Section 1.6.1. A different tabular scheme is outlined in
Section 1.6.2, and a modified form of Routh's array which reduces
coefficient growth is given in Section 1.6.3. Finally, the solution of an important polynomial equation is considered in Section

1.6.4. The g.c.d. procedures will be encountered in a somewhat different guise in Chapter 3, relating to the location of roots of polynomials in the complex plane, and some extensions for polynomial matrices will be described in Chapter 4.

1.1 REVIEW OF BASIC CONCEPTS

A *polynomial* in the variable λ is a function of the form

$$a(\lambda) = a_0\lambda^n + a_1\lambda^{n-1} + \cdots + a_{n-1}\lambda + a_n \qquad (1.1)$$

and if $a_0 \neq 0$, has *degree* n (a nonnegative integer) denoted by δa. The coefficients a_0, a_1, ..., a_n belong to some number field F, which in this book will be either the field of real or of complex numbers. The *zero* polynomial has $a_i = 0$, for all i, and when $a_0 = 1$, $a(\lambda)$ is termed *monic*. Let $b(\lambda)$ be a second polynomial of degree m:

$$b(\lambda) = b_0\lambda^m + b_1\lambda^{m-1} + \cdots + b_m, \qquad b_0 \neq 0 \qquad (1.2)$$

If $m \leqslant n$, then it is possible to divide $a(\lambda)$ by $b(\lambda)$, to produce

$$a(\lambda) \equiv b(\lambda)q(\lambda) + r(\lambda) \qquad (1.3)$$

where the quotient polynomial $q(\lambda)$ and remainder polynomial $r(\lambda)$ are unique and $\delta r < m$. If $r(\lambda) \equiv 0$, $a(\lambda)$ is divisible by $b(\lambda)$ (i.e., without remainder) and the notation $b(\lambda)|a(\lambda)$ is often used. This denotes that $b(\lambda)$ is a *factor* of $a(\lambda)$. In particular, when $b(\lambda)$ is a linear expression $\lambda - c$, where $c \in F$, then (1.3) becomes

$$a(\lambda) \equiv (\lambda - c)q(\lambda) + r \qquad (1.4)$$

where $\delta r < 1$, so that r is a scalar. In fact, setting $\lambda = c$ in the identity (1.4) reveals that $r = a(c)$ (this result, on division by a linear term, is usually called the *Remainder Theorem*).

A simple procedure for obtaining $q(\lambda)$ and r in (1.4) is provided by *Horner's scheme*. This is illustrated in Table 1.1. The coefficients of $a(\lambda)$ in (1.1) are written out in the first row, and the second row is constructed from left to right using the

Table 1.1   Horner's array

$$\begin{array}{c|ccccc} & a_0 & a_1 & a_2 & a_3 & \cdots & a_n \\ \hline c & q_0 & q_1 & q_2 & q_3 & \cdots & q_n \end{array}$$

relationships

$$q_0 = a_0, \quad q_1 = cq_0 + a_1, \quad q_2 = cq_1 + a_2, \quad \cdots, \quad q_n = cq_{n-1} + a_n \tag{1.5}$$

It is left as an easy exercise for the reader to show that in (1.4)

$$q(\lambda) = a_0 \lambda^{n-1} + q_1 \lambda^{n-2} + \cdots + q_{n-1}, \quad r = q_n$$

Example 1   The polynomial

$$a(\lambda) = \lambda^3 - 3\lambda^2 + \lambda + 5 \tag{1.6}$$

is to be divided by $\lambda + 2$. Horner's scheme gives

$$\begin{array}{c|cccc} & 1 & -3 & 1 & 5 \\ \hline -2 & 1 & -5 & 11 & -17 \end{array}$$

Hence in (1.4) we have

$$q(\lambda) = \lambda^2 - 5\lambda + 11, \quad r = -17$$

When c is such that $r = a(c) = 0$, then c is called a *zero* (or *root*) of the polynomial $a(\lambda)$, and $a(\lambda)$ is thus divisible by $\lambda - c$. If $a(\lambda)$ is divisible by $(\lambda - c)^k$ $(k \geqslant 1)$, but not divisible by $(\lambda - c)^{k+1}$, then c is called a root of *multiplicity* k (or, said to be *repeated* k times). When $k = 1$, the root is *simple*. The "Fundamental Theorem of Algebra" asserts that every polynomial of degree greater than zero has at least one root, which in general will be complex. It then follows by induction that $a(\lambda)$ can be expressed as a product of linear factors:

$$a(\lambda) = a_0 (\lambda - \lambda_1)(\lambda - \lambda_2) \cdots (\lambda - \lambda_n) \tag{1.7}$$

where the complex numbers $\lambda_1, \lambda_2, \ldots, \lambda_n$ are the n roots of $a(\lambda)$. Of course, some of these roots may be purely real, and they may not

be *distinct* (i.e., all different from one another). Suppose that
just p ($\leqslant$ n) of these roots are distinct, and that

$$a(\lambda) = a_0(\lambda - \lambda_1)^{s_1}(\lambda - \lambda_2)^{s_2} \cdots (\lambda - \lambda_p)^{s_p} \qquad (1.8)$$

Then in (1.8), $\lambda_1, \ldots, \lambda_p$ are the roots of $a(\lambda)$ with multiplicities $s_1, \ldots, s_p$, respectively, and $s_1 + s_2 + \cdots + s_p = n$.

The *derivative* polynomial $a'(\lambda) \overset{\triangle}{=} da(\lambda)/d\lambda$ is obtained from
(1.1) as

$$a'(\lambda) = na_0\lambda^{n-1} + (n - 1)a_1\lambda^{n-2} + \cdots + a_{n-1} \qquad (1.9)$$

It is easily seen from the expression in (1.8) that $a'(\lambda)$ has roots
$\lambda_1, \lambda_2, \ldots, \lambda_p$ with multiplicities $s_1 - 1, s_2 - 1, \ldots, s_p - 1$,
respectively [if $s_i = 1$, then λ_i is *not* a root of $a'(\lambda)$]. It
therefore follows that

$$\left[\frac{d^j a(\lambda)}{d\lambda^j}\right]_{\lambda=\lambda_i} \begin{cases} = 0, & j = 1, 2, \ldots, s_i - 1 \\ \neq 0, & j = s_i \end{cases}$$

for i = 1, 2, ..., p.

Example 2   (a) Consider again the polynomial in (1.6). Since

$$a(-1) = -1 - 3 - 1 + 5 = 0$$

it follows from the Remainder Theorem that $\lambda + 1$ is a factor of
$a(\lambda)$. Division of $a(\lambda)$ by $\lambda + 1$ using Horner's method gives

	1	-3	1	5
-1	1	-4	5	0

. showing that

$$a(\lambda) = (\lambda + 1)(\lambda^2 - 4\lambda + 5)$$

which can be simplified further as

$$a(\lambda) = (\lambda + 1)(\lambda - 2 + i)(\lambda - 2 - i) \qquad (1.10)$$

Thus the roots of $a(\lambda)$ are $\lambda_1 = -1$, $\lambda_2 = 2 - i$, $\lambda_3 = 2 + i$.
Notice that although the coefficients of $a(\lambda)$ in (1.6) are all

real, the factorization (1.7) must in general be carried out over the field of complex numbers.

(b) Consider

$$b(\lambda) = \lambda^5 - 4\lambda^4 + \lambda^3 + 10\lambda^2 - 4\lambda - 8 \qquad (1.11)$$

Again, since $b(-1) = 0$, we deduce that $\lambda_1 = -1$ is a root of $b(\lambda)$. Also, the derivative of (1.11) satisfies $b'(-1) = 0$, so this root has multiplicity at least 2. However, a second differentiation shows that $b''(-1) \neq 0$, so $\lambda_1 = 1$ has multiplicity exactly 2. Similarly, the reader can check that $b(2) = 0$, $b'(2) = 0$, $b''(2) = 0$, $b'''(2) \neq 0$, showing that the second root $\lambda_2 = 2$ of $b(\lambda)$ has multiplicity 3. Since $b(\lambda)$ has degree 5, there are no other factors, so we can write

$$b(\lambda) = (\lambda + 1)^2(\lambda - 2)^3 \qquad (1.12)$$

Alternatively, Horner's method can be used to determine the multiplicity of a root. For example, repeated division of (1.11) by $\lambda - 2$ produces the following array:

	1	-4	1	10	-4	-8
2	1	-2	-3	4	4	0
	1	0	-3	-2	0	
	1	2	1	0		
	1	4	9			

After each division the subsequent quotient is again divided by $\lambda - 2$. The three zero remainders show that $(\lambda - 2)^3$ is a factor of $b(\lambda)$, and from the penultimate row we can write down

$$b(\lambda) = (\lambda - 2)^3(\lambda^2 + 2\lambda + 1)$$

If each of the polynomials $a(\lambda)$ and $b(\lambda)$ in (1.1) and (1.2), respectively, is divisible by a polynomial $\phi(\lambda)$, then $\phi(\lambda)$ is called a *common divisor* of $a(\lambda)$ and $b(\lambda)$. A *greatest common divisor* (*g.c.d.*) $d(\lambda)$ of $a(\lambda)$ and $b(\lambda)$ is a common divisor which is divisible by all other common divisors. A g.c.d. always exists, and is unique up to multiplication by a constant factor. The

notation $d(\lambda) = \big(a(\lambda), b(\lambda)\big)$ is sometimes used. It is often
convenient to specify that $d(\lambda)$ is monic, so that it then makes
sense to refer to *the* g.c.d. If the factorization of $a(\lambda)$ as the
product (1.8) is known, together with the corresponding expression
for $b(\lambda)$,

$$b(\lambda) = b_o(\lambda - \mu_1)^{t_1}(\lambda - \mu_2)^{t_2} \cdots (\lambda - \mu_q)^{t_q} \qquad (1.13)$$

where $t_1 + t_2 + \cdots + t_q = m$, then the g.c.d. can be determined by
inspection. Simply pick out those values (if any) of the λ's which
coincide with the μ's, say λ_α, λ_β, ..., and then

$$d(\lambda) = (\lambda - \lambda_\alpha)^{u_\alpha}(\lambda - \lambda_\beta)^{u_\beta} \cdots$$

where $u_\alpha = \min(s_\alpha, t_\alpha)$, and so on.

If there are more than two polynomials, the definition of
g.c.d. is extended in an obvious fashion. To find this g.c.d. we
calculate the sequence

$$d_1(\lambda) = \big(a_1(\lambda), a_2(\lambda)\big), \quad d_2(\lambda) = \big(d_1(\lambda), a_3(\lambda)\big),$$
$$d_3(\lambda) = \big(d_2(\lambda), a_4(\lambda)\big), \quad \ldots, \quad d_{r-1}(\lambda) = \big(d_{r-2}(\lambda), a_r(\lambda)\big)$$

and $d_{r-1}(\lambda)$ is the g.c.d. of the set of polynomials $a_1(\lambda)$, $a_2(\lambda)$,
..., $a_r(\lambda)$.

If the g.c.d. of $a(\lambda)$ and $b(\lambda)$ is 1, then they are said to be
relatively prime (or *coprime*). From the preceding discussion it
follows that this is equivalent to their having no root in common.

Example 3   (a) Consider $a(\lambda)$ and $b(\lambda)$ in (1.6) and (1.11),
respectively. From their decompositions in (1.10) and (1.12) it is
clear that their g.c.d. is $d(\lambda) = \lambda + 1$.

(b)   If instead we take

$$b(\lambda) = 2\lambda^2 - 3\lambda + 1$$
$$= 2(\lambda - 1)(\lambda - \tfrac{1}{2}) \qquad (1.14)$$

then in this case $a(\lambda)$ in (1.6) and $b(\lambda)$ are relatively prime.

(c)   If

$$a(\lambda) = (\lambda + 1)^3(\lambda - 2)(\lambda + 3) \tag{1.15}$$

then the g.c.d. of $a(\lambda)$ in (1.15) and $b(\lambda)$ in (1.12) is
$d(\lambda) = (\lambda + 1)^2(\lambda - 2)$.   If a third polynomial is

$$\alpha(\lambda) = (\lambda + 1)(\lambda - 2)(\lambda + 3) \tag{1.16}$$

then the g.c.d. of (1.12), (1.15), and (1.16) is

$$d(\lambda) = (\lambda + 1)(\lambda - 2)$$

A *resultant* $R(a, b)$ of $a(\lambda)$ and $b(\lambda)$ in (1.1) and (1.2) is a
scalar which is nonzero if and only if $a(\lambda)$ and $b(\lambda)$ are relatively
prime; in other words, $R(a, b) = 0$ if and only if $a(\lambda)$ and $b(\lambda)$
have at least one common root.   A convenient definition is

$$R(a, b) \triangleq a_o^m b_o^n \prod_{i=1}^{n} \prod_{j=1}^{m} (\lambda_i - \mu_j) \tag{1.17}$$

where neither $a(\lambda)$ nor $b(\lambda)$ need necessarily have distinct roots,
so $a(\lambda)$ is taken to have the form (1.7), and similarly

$$b(\lambda) = b_o(\lambda - \mu_1)(\lambda - \mu_2) \cdots (\lambda - \mu_m) \tag{1.18}$$

Use of (1.7) or (1.18) then shows immediately that

$$R(a, b) = a_o^m b(\lambda_1) \, b(\lambda_2) \cdots b(\lambda_n)$$

$$= (-1)^{mn} b_o^n a(\mu_1) \, a(\mu_2) \cdots a(\mu_m) \tag{1.19}$$

In particular, if $b(\lambda)$ is set equal to the derivative $a'(\lambda)$, then
$R(a, a') = 0$ if and only if $a(\lambda)$ has at least one repeated root.
The term *discriminant* is used for the resultant in this case,
together with the notation

$$D(a) = (1/a_o)R(a, a') \tag{1.20}$$

It follows from (1.17) and (1.20) that

$$D(a) = \varepsilon(n) \, a_o^{2n-2} \prod_{1 \leqslant i < j \leqslant n} (\lambda_i - \lambda_j)^2 \tag{1.21}$$

where

$$\varepsilon(n) = (-1)^{n(n-1)/2} \qquad\qquad (1.22)$$

Example 4 For the polynomials $a(\lambda)$ and $b(\lambda)$ in (1.6) and (1.14), respectively, we have $n = 3$, $m = 2$, $b_o = 2$, $\mu_1 = 1$, $\mu_2 = \frac{1}{2}$, so (1.19) gives

$$R(a, b) = (-1)^6 2^3 a(1) a(\tfrac{1}{2}) = 156$$

Problems

1.1 Prove that if all the coefficients of $a(\lambda)$ in (1.1) are real numbers, then if c is a root of $a(\lambda)$, so is $\bar{c}$, the complex conjugate of c.

1.2 Use Horner's scheme to show that

$$a(\lambda) = \lambda^4 - 7\lambda^3 + 9\lambda^2 + 8\lambda + 16$$

has a root 4 of multiplicity 2, and hence express $a(\lambda)$ in the form (1.8).

1.3 Factorize

$$a(\lambda) = \lambda^5 + \lambda^4 - 5\lambda^3 - \lambda^2 + 8\lambda - 4$$

$$b(\lambda) = \lambda^6 + 14\lambda^5 + 65\lambda^4 + 120\lambda^3 + 40\lambda^2 - 128\lambda - 112$$

as in (1.8), (1.13). Hence determine their g.c.d.

1.4 Compute the resultant (1.17) for the polynomials

$$a(\lambda) = \lambda^3 + 4\lambda^2 - 2\lambda - 3, \qquad b(\lambda) = 4\lambda^2 + 5\lambda + 1$$

1.5 The n × n *Vandermonde* matrix is defined by

$$V_n \triangleq \begin{bmatrix} 1 & 1 & \cdots & 1 \\ \lambda_1 & \lambda_2 & \cdots & \lambda_n \\ \lambda_1^2 & \lambda_2^2 & \cdots & \lambda_n^2 \\ \vdots & \vdots & & \vdots \\ \lambda_1^{n-1} & \lambda_2^{n-1} & \cdots & \lambda_n^{n-1} \end{bmatrix} \qquad\qquad (1.23)$$

Show that det $V_2 = (\lambda_2 - \lambda_1)$, and by reducing the last column of det V_3 to 1, 0, 0, show that

$$\det V_3 = (\lambda_3 - \lambda_2)(\lambda_3 - \lambda_1) \det V_2$$

Similarly, prove that

$$\det V_n = (\lambda_n - \lambda_{n-1})(\lambda_n - \lambda_{n-2}) \cdots (\lambda_n - \lambda_1) \det V_{n-1}$$

and hence deduce that

$$\det V_n = \prod_{1 \leqslant i < j \leqslant n} (\lambda_j - \lambda_i) \qquad (1.24)$$

If the λ_i in (1.23) are the roots of $a(\lambda)$ in (1.1), use (1.21) and (1.24) to show that

$$D(a) = \epsilon(n) a_0^{2n-2} \det U_n \qquad (1.25)$$

where $U_n = V_n V_n^T$ has its i, j element u_{ij} given by

$$u_{ij} = \sum_{k=1}^{n} \lambda_k^{i+j-2}, \qquad i, j = 1, 2, \ldots, n$$

1.6 A problem of *interpolation* is to determine a polynomial

$$y(x) = a_0 x^{n-1} + a_1 x^{n-2} + \cdots + a_{n-2} x + a_{n-1}$$

so that it passes through n given points with cartesian coordinates (λ_1, y_1), (λ_2, y_2), ..., (λ_n, y_n). Show that the coefficients are determined by the equation $aV_n = y$, where $a = [a_{n-1}, a_{n-2}, \ldots, a_0]$, $y = [y_1, y_2, \ldots, y_n]$, and V_n is the Vandermonde matrix (1.23). Use (1.24) to deduce that if all the λ_i's are distinct, and the y_i's are not all zero, then the solution for $y(x)$ is unique and nontrivial.

1.7 Horner's scheme can also be used to express $a(\lambda)$ in (1.1) as a polynomial in powers of $\lambda - c$, i.e.,

$$a(\lambda) = \alpha_0 (\lambda - c)^n + \alpha_1 (\lambda - c)^{n-1} + \cdots + \alpha_{n-1} (\lambda - c) + \alpha_n$$

This is achieved by noticing that in (1.4) the remainder r is equal to α_n. Since $\alpha_n = a(c)$, this also provides a way of evaluating $a(\lambda)$ at $\lambda = c$. Similarly, if $q(\lambda)$ in (1.4) is divided by $\lambda - c$, the remainder is α_{n-1}, and so on. For example, Horner's array in Example 1 with $c = -2$ can be continued to give

	1	-3	1	5
-2	1	-5	11	-17
	1	-7	25	
	1	-9		
	1			

showing that the expansion for $a(\lambda)$ in (1.6) is in this case

$$a(\lambda) = (\lambda + 2)^3 - 9(\lambda + 2)^2 + 25(\lambda + 2) - 17$$

[notice also that $a(-2) = -17$, $a'(-2) = 25$, $a''(-2) = -9.2!$, $a'''(-2) = 1.3!$]. Apply this procedure to express $a(\lambda) = \lambda^4 - 5\lambda^3 - 3\lambda^2 + 9$ as a polynomial in powers of $\lambda - 3$.

1.2 COMPANION MATRIX AND PROPERTIES

It is now convenient to assume that $a(\lambda)$ in (1.1) is monic; if it is not, it can be made so by simply dividing each coefficient a_i by a_0. A *companion matrix* associated with $a(\lambda)$ is the $n \times n$ matrix

$$C_n \triangleq \begin{bmatrix} 0 & 1 & 0 & & \cdots & 0 \\ 0 & 0 & 1 & & \cdots & \vdots \\ \vdots & \vdots & \vdots & \ddots & 0 & 1 \\ -a_n & -a_{n-1} & -a_{n-2} & \cdots & -a_2 & -a_1 \end{bmatrix} \qquad (1.26)$$

$$= \begin{bmatrix} 0 & \vdots & I_{n-1} \\ \text{---} & \text{+} \text{---------------} \\ -a_n & \vdots -a_{n-1} & \cdots & -a_1 \end{bmatrix} \qquad (1.27)$$

where I_n denotes the $n \times n$ *unit* (or *identity*) *matrix*. The matrix C has the crucial property that its characteristic polynomial is $a(\lambda)$ in (1.1):

Theorem 1.1

$$\det(\lambda I_n - C_n) \equiv \lambda^n + a_1\lambda^{n-1} + \cdots + a_{n-1}\lambda + a_n \tag{1.28}$$

Proof. We use induction. The result is trivially established for n = 2. Expand $\det(\lambda I_n - C_n)$ by the first column using cofactors, to obtain

$$\det(\lambda I_n - C_n) = \lambda \det(\lambda I_{n-1} - C_{n-1}) + (-1)^{n+1}a_n \begin{vmatrix} -1 & 0 & \cdots & 0 \\ \lambda & -1 & \cdots & 0 \\ & & \ddots & \\ 0 & & \lambda & -1 \end{vmatrix}$$

$$= \lambda \det(\lambda I_{n-1} - C_{n-1}) + a_n$$

$$= \lambda(\lambda^{n-1} + a_1\lambda^{n-2} + \cdots + a_{n-1}) + a_n$$

using the induction hypothesis. ∎

Remarks on Theorem 1.1   (1) The identity (1.28) shows that the roots of $a(\lambda)$ are the eigenvalues of C_n, so in particular

$$\det C_n = \lambda_1\lambda_2 \cdots \lambda_n \tag{1.29}$$

showing that C_n is singular if and only if $a(\lambda)$ has a zero root.

(2)   We shall drop the suffix n on C_n when it is unnecessary to specify its dimensions (this practice will also be followed for I_n).   Also, the matrix C was deliberately not described as *the* companion matrix, since there are three other forms having the same properties.   These are:

(i)   The transpose of C:

$$C_I \overset{\Delta}{=} C^T = \begin{bmatrix} 0 & \vline & -a_n \\ \hline & \vline & -a_{n-1} \\ I_{n-1} & \vline & \vdots \\ & \vline & -a_1 \end{bmatrix} \tag{1.30}$$

(ii) The matrix obtained by reversing the orders of the rows and columns of C:

$$
C_{II} \triangleq
\left[
\begin{array}{ccccc:c}
-a_1 & -a_2 & \cdots & & -a_{n-1} & -a_n \\
\hdashline
& & I_{n-1} & & & 0
\end{array}
\right]
\tag{1.31}
$$

It is useful to notice that (1.31) can be written in the form $J_n C J_n$, where J_n is the *reverse unit matrix*, having 1's on the secondary diagonal and zeros elsewhere: for example,

$$
J_3 =
\begin{bmatrix}
0 & 0 & 1 \\
0 & 1 & 0 \\
1 & 0 & 0
\end{bmatrix}
$$

Premultiplying an n-rowed matrix by J_n reverses the order of its rows, whereas postmultiplying an n-columned matrix by J_n reverses the order of its columns. It is easy to verify that $J_n^2 = I_n$, which implies that $J_n^{-1} = J_n$; and to show that

$$
\det J_n = \varepsilon(n) \tag{1.32}
$$

where $\varepsilon(n)$ was defined in (1.22). In view of these properties, it follows that $C_{II} = J_n C J_n^{-1}$, showing that C_{II} is similar to C, and hence has the same characteristic polynomial $a(\lambda)$.

(iii) The fourth type of companion matrix is obtained from C^T in an analogous fashion, i.e.,

$$
C_{III} \triangleq J_n C^T J_n =
\left[
\begin{array}{c:c}
\begin{matrix} -a_1 \\ -a_2 \\ \vdots \\ \vdots \\ -a_{n-1} \end{matrix} & I_{n-1} \\
\hdashline
-a_n & 0
\end{array}
\right]
\tag{1.33}
$$

The forms displayed in (1.26) and (1.30) will be the ones used

most frequently in this book, but the other forms are also useful
in some circumstances.

(3)   A similarity transformation between C and C^T is

$$C^T = TCT^{-1} \tag{1.34}$$

where

$$T = \begin{bmatrix} a_{n-1} & a_{n-2} & \cdots & a_1 & 1 \\ a_{n-2} & a_{n-3} & \cdots & 1 & 0 \\ a_{n-3} & a_{n-4} & & & \\ \vdots & \vdots & & & \\ a_1 & 1 & & \mathbf{0} & \\ 1 & 0 & & & \end{bmatrix} \tag{1.35}$$

This result is easily proved by comparing both sides of the
identity

$$C^T T \equiv TC \tag{1.36}$$

Incidentally, notice that T has *striped* form — the elements along
the secondary diagonal, and along each line parallel to it, are
equal; matrices having such a pattern arise in many applications
(see also Problem 1.16). We shall encounter T in a different role
in Section 1.5.

Another similarity transformation between C and C^T is given
in (1.57).

(4)   In order to present two other important properties of
companion matrices, we must first introduce some definitions, which
actually apply to an arbitrary n × n matrix A: Form the characteristic matrix λI_n - A, and let $d_j(\lambda)$ denote the g.c.d. of all minors
of order j of λI - A, j = 1, 2, ..., n. These polynomials are
called the *determinantal divisors* of λI - A, and it follows that
the quotients $i_j(\lambda) = d_j(\lambda)/d_{j-1}(\lambda)$, j = 1, 2, ..., n $(d_o \overset{\Delta}{=} 1)$ are
also polynomials, called the *similarity invariants* of A.   A matrix
A is *nonderogatory* if and only if its first n - 1 similarity
invariants are unity.   Using this definition, it is easy to prove

that companion matrices are nonderogatory, which is the first fact
we wish to record.

Second, a standard theorem states that two $n \times n$ matrices
are similar if and only if they have identical similarity
invariants. It therefore follows that A is similar to a companion
matrix associated with its characteristic polynomial $k(\lambda) =$
$\det(\lambda I - A)$ if and only if A is nonderogatory.

An important, alternative, necessary, and sufficient condition
for A to be nonderogatory is that its *minimum polynomial* is equal
to its characteristic polynomial $k(\lambda)$; i.e., there is no polynomial
$m(\lambda)$ of degree less than n such that $m(A) \equiv 0$ [the Cayley-Hamilton
theorem states that $k(A) \equiv 0$].

A key feature for our development of the g.c.d. problem, and
other applications, is the study of polynomials in the matrix C.
That is, if $b(\lambda)$ is the polynomial in (1.2), we construct

$$b(C) \triangleq b_o C^m + b_1 C^{m-1} + \cdots + b_m I_n \tag{1.37}$$

Let e_i denote the ith row of I_n. It is easy to see that the *first*
rows of C, C^2, C^3, ..., C^{n-1} are e_2, e_3, e_4, ..., e_n. Hence it
follows that provided $\delta b = m < n$, then the first row (r_1, say) of
$b(C)$ is

$$r_1 = b_m e_1 + b_{m-1} e_2 + \cdots + b_o e_{m+1}$$
$$= [b_m, b_{m-1}, \ldots, b_o, 0, \ldots, 0] \tag{1.38}$$

Furthermore, inspection of (1.26) shows that the $(i - 1)$th row of
C is e_i, for i = 2, 3, ..., n; this property can be conveniently
written as

$$e_{i-1} C = e_i, \quad i = 2, 3, \ldots, n \tag{1.39}$$

[premultiplication of any matrix by e_{i-1} simply picks out its
$(i - 1)$th row]. Let the remaining rows of $b(C)$ be denoted by r_2,
r_3, ..., r_n. Then by definition

$$r_i = e_i b(C) \tag{1.40}$$

and substitution of (1.39) into (1.40) gives

$$r_i = e_{i-1} Cb(C)$$
$$= e_{i-1} b(C) C \qquad (1.41)$$

using the fact that C commutes with $b(C)$. Combining (1.40) and
(1.41) shows that the latter can be written as

$$r_i = r_{i-1} C, \quad i = 2, 3, \ldots, n \qquad (1.42)$$

Since r_1 is given by (1.38), the recurrence relation (1.42)
provides a very simple method of constructing $b(C)$.

   Another way of writing $b(C)$ is obtained by noting that (1.42)
implies $r_2 = r_1 C$, $r_3 = r_2 C = r_1 C^2$, and so on, so that we have
established:

Theorem 1.2   For $b(\lambda)$ in (1.2) with $\lambda b < n$, and C in (1.26), the
matrix $b(C)$ has rows

$$r_1, r_1 C, r_1 C^2, \ldots, r_1 C^{n-1} \qquad (1.43)$$

where r_1 is given by (1.38).

   We shall see in Chapter 2 that the expression (1.43) provides
some interesting links with the theory of linear control systems.

   We can now express the resultant $R(a, b)$ in terms of this
companion matrix formulation.   First, note that a standard result
states that if $\lambda_1, \lambda_2, \ldots, \lambda_n$ are the eigenvalues of C, then those
of $b(C)$ are $b(\lambda_1), b(\lambda_2), \ldots, b(\lambda_n)$.   Second, the determinant of a
matrix is equal to the product of its eigenvalues.   It therefore
follows that

$$\det b(C) = b(\lambda_1) b(\lambda_2) \cdots b(\lambda_n) \qquad (1.44)$$

Assuming, as usual, that $a_0 = 1$, comparison of (1.19) and (1.44)
reveals:

Theorem 1.3   The resultant of $a(\lambda)$ in (1.1) and $b(\lambda)$ in (1.2) is
given by

$$R(a, b) = \det b(C) \qquad (1.45)$$

where C is the matrix in (1.26).   In other words, $b(C)$ is nonsingular if and only if these polynomials are relatively prime.

Remarks on Theorems 1.2 and 1.3   (1) The derivation of (1.45) does *not* require the condition $\delta b < \delta a$.

(2) The result (1.45) clearly holds for *any* matrix C whose characteristic polynomial is $a(\lambda)$. A generalization for this case is given later, in Theorem 1.17 (Section 1.6.4). However, the advantage of taking C in companion form is that $b(C)$ can be constructed from Theorem 1.2.

(3) We shall refer to $b(C)$ as a companion matrix form of *resultant matrix*.

Example 5   Return to the polynomials used in Example 4, namely $a(\lambda)$ in (1.6) and $b(\lambda)$ in (1.14). The companion matrix (1.26) of $a(\lambda)$ is

$$C = \begin{bmatrix} 0 & 1 & 0 \\ 0 & 0 & 1 \\ -5 & -1 & 3 \end{bmatrix} \tag{1.46}$$

and $b(C) = 2C^2 - 3C + I$ has rows r_1, r_2, r_3. From (1.38) we have $r_1 = [1, -3, 2]$, and from (1.42)

$$r_2 = r_1 C = [-10, -1, 3], \qquad r_3 = r_2 C = [-15, -13, 8]$$

whence

$$b(C) = \begin{bmatrix} 1 & -3 & 2 \\ -10 & -1 & 3 \\ -15 & -13 & 8 \end{bmatrix} \tag{1.47}$$

It is easy to compute det $b(C) = 156$, which agrees with the value of $R(a, b)$ found in Example 4.

(4) Since the first row of $b(C)$ is given by (1.38), it follows that $b(C) \equiv 0$ if and only if $r_1 \equiv 0$, which is equivalent to having $b(\lambda) \equiv 0$. Thus there is *no* nontrivial polynomial of degree less than n such that $b(C) \equiv 0$. This is a simple derivation of the fact, stated earlier, that C is nonderogatory.

(5) Theorem 1.1 and the Cayley-Hamilton theorem together imply

$$a(C) \equiv C^n + a_1 C^{n-1} + \cdots + a_{n-1}C + a_n I_n \equiv 0 \tag{1.48}$$

In Theorem 1.2 it was assumed that $\delta b < \delta a$, but (1.48) gives us a
way of removing this restriction. Suppose that $\delta b = m \geqslant n$; then as
in (1.3) but with the roles of $a(\lambda)$ and $b(\lambda)$ interchanged, we can
divide $b(\lambda)$ by $a(\lambda)$ to give

$$b(\lambda) \equiv a(\lambda)q_1(\lambda) + r(\lambda) \tag{1.49}$$

where $\delta r < n$. Since (1.49) is an *identity*, we can replace λ by C
to give

$$\begin{aligned} b(C) &\equiv a(C)q_1(C) + r(C) \\ &\equiv r(C) \end{aligned} \tag{1.50}$$

by virtue of (1.48). Hence to compute $b(C)$ when $\delta b \geqslant \delta a$, simply
divide $b(\lambda)$ by $a(\lambda)$, and the desired expression is obtained by
applying Theorem 1.2 to the remainder polynomial $r(C)$. In
particular, when $\delta b = \delta a$ then

$$r(\lambda) = (b_1 - b_0 a_1)\lambda^{n-1} + (b_2 - b_0 a_2)\lambda^{n-2} + \cdots + (b_n - b_0 a_n)$$

so $b(C)$ has first row

$$r_1 = [b_n - b_0 a_n, \ldots, b_2 - b_0 a_2, b_1 - b_0 a_1] \tag{1.51}$$

In addition, (1.50) shows that if $a(\lambda)|b(\lambda)$, then $b(\lambda) \equiv 0$.

Example 6   Let $a(\lambda)$ again be the polynomial in (1.6) with companion
matrix C in (1.46). If

$$b(\lambda) = \lambda^4 + 5\lambda^3 - 20\lambda^2 + 6\lambda + 44 \tag{1.52}$$

then division by $a(\lambda)$ gives

$$b(\lambda) = a(\lambda)(\lambda + 8) + 3\lambda^2 - 7\lambda + 4$$

so by (1.50)

$$b(C) = 3C^2 - 7C + 4I_3$$

Theorem 1.2 then shows that $b(C)$ has rows

$$r_1 = [4, -7, 3], \qquad r_2 = r_1 C = [-15, 1, 2],$$
$$r_3 = r_2 C = [-10, -17, 7]$$

By Theorem 1.3, the resultant of $a(\lambda)$ and $b(\lambda)$ is $R(a, b) =$ det $b(C) = 364$.

An alternative way of obtaining the resultant when $\delta b > \delta a$ is to use the companion matrix of $b(\lambda)$ (B, say) and form $a(B)$ according to Theorem 1.2. Then corresponding to (1.44) we have

$$\det a(B) = a(\mu_1)a(\mu_2) \cdots a(\mu_m)$$

where $\mu_1, \ldots, \mu_m$ are the roots of $b(\lambda)$. Provided that $b_o = 1$, it then follows from (1.19) that $R(a, b) = (-1)^{mn}\det a(B)$ (the reader is invited to work out the details when $b_o \neq 1$).

Example 6 (continued) The companion matrix B of $b(\lambda)$ in (1.52) has last row $[-44, -6, 20, -5]$ and with $a(\lambda)$ given by (1.6), the rows of $a(B)$ are, respectively,

$$r_1 = [5, 1, -3, 1]$$

$$r_2 = r_1 B = [-44, -1, 21, -8]$$

$$r_3 = r_2 B = [352, 4, -161, 61]$$

$$r_4 = r_3 B = [-2684, -14, 1224, -466]$$

It is then routine to evaluate $\det a(B)$, and to confirm that the same value for $R(a, b)$ is obtained as before.

Instead of the procedure described above, another way of evaluating C_n^m, when $m \geqslant n$, is provided by writing

$$C^m = C^{n-1}C^{m-n+1} \tag{1.53}$$

By (1.43) with $r_1 = e_n$ (the nth row of I_n), the rows of C^{n-1} are $e_n, e_n C, \ldots, e_n C^{n-1}$, so (1.53) becomes

$$C^m = \begin{bmatrix} e_n C^{m-n+1} \\ e_n C^{m-n+2} \\ \vdots \\ e_n C^m \end{bmatrix} \tag{1.54}$$

Clearly, the rows s_1, ..., s_n of C^m in (1.54) satisfy the relationship $s_i = s_{i-1}C$, $i = 2$, ..., n. Thus to compute C^m for $m \geqslant n$,
simply apply the recurrence formula $r_i = r_{i-1}C$, with i running from
2 to m + 1, and $r_1 = e_n$. The vectors r_{m-n+2}, r_{m-n+3}, ..., r_{m+1} in
the sequence thereby generated are the rows of C^m.

Example 7   Using the 3 × 3 companion matrix C in (1.46) we have
$r_1 = [0, 0, 1]$,

$$r_2 = r_1C = [-5, -1, 3]$$

$$r_3 = r_2C = [-15, -8, 8]$$

$$r_4 = r_3C = [-40, -23, 16]$$

$$r_5 = r_4C = [-80, -56, 25]$$

and so on.   For example,

$$C^3 = \begin{bmatrix} r_2 \\ r_3 \\ r_4 \end{bmatrix}, \quad C^4 = \begin{bmatrix} r_3 \\ r_4 \\ r_5 \end{bmatrix}$$

A way of extending this procedure to determine b(C) when
$\delta b \geqslant n$ is best explained through a specific example.

Example 8   We repeat the problem of Example 6, which was to obtain
b(C), where C is given in (1.46) and b(λ) in (1.52). We can write

$$b(C) = C^4 + 5C^3 - 20C^2 + 6C + 44I$$

$$= [(C^2 + 5C + 0I)C^2] + (-20C^2 + 6C + 44I) \qquad (1.55)$$

The second group of terms in (1.55) has degree less than 3, so its
rows are obtained from Theorem 1.2 as

$$r_1' = [44, 6, -20], \quad r_1'C, \quad r_1'C^2$$

Let the first group of terms in (1.55) have rows r_1'', r_2'', r_3''. Then
by again appealing to Theorem 1.2 we see that $r_1'' = s_1C^2$, $r_2'' = s_1C^3 = r_1''C$, $r_3'' = s_1C^4 = r_1''C^2$, where $s_1 = [0, 5, 1]$. Thus b(C) has rows

$r_1 = r_1' + r_1''$, r_1C, r_1C^2, showing that we can construct $b(C)$ from the recurrence formula (1.42) with first row $r_1 = r_1' + s_1C^2 =$ [4, -7, 3]. This agrees with Example 6.

It is interesting to note that as a by-product, we have produced a method for determining the remainder $r(\lambda)$ in (1.49) on division of $b(\lambda)$ by $a(\lambda)$, without actually having to carry out any polynomial manipulations. This is because (1.50) shows that the *first row* of $b(C)$ is equal to the first row of $r(C)$, which since $\delta r < n$, is equal to the coefficients of $r(\lambda)$ in reverse order [compare with (1.38)]. Thus in Example 8, the coefficients of the remainder on division of $b(\lambda)$ in (1.52) by $a(\lambda)$ in (1.6) are given by the elements of r_1, i.e., $r(\lambda) = 3\lambda^2 - 7\lambda + 4$.

Problems

1.8 Using the polynomials in Problem 1.4, determine $b(C)$, where C is the companion matrix of $a(\lambda)$, and hence recalculate the resultant.

1.9 Determine $b(C)$, where C is the companion matrix of $a(\lambda)$ in Problem 1.4, and $b(\lambda)$ is given in Problem 1.3, using the method of Example 6.

1.10 Let C be the companion matrix of $a(\lambda)$ in Problem 1.4.
     (i) Using the method of Example 7, determine C^6.
     (ii) Using the method of Example 8, determine the remainder on division of $b(\lambda)$ in Problem 1.3 by $a(\lambda)$.

1.11 Verify that if λ_i is an eigenvalue of C_n in (1.26), then an associated eigenvector is the ith column of the Vandermonde matrix V_n defined in (1.23). Hence show that if all the roots of $a(\lambda)$ are distinct, then

$$V_n^{-1}C_nV_n = \text{diag}[\lambda_1, \lambda_2, \ldots, \lambda_n] \qquad (1.56)$$

Deduce that in this case

$$(V_nV_n^T)^{-1}C_n(V_nV_n^T) = C_n^T \qquad (1.57)$$

1.12 Show that $a(\lambda)$ in (1.1) has a repeated root if and only if
     $a'(C)$ is singular, where C is defined in (1.26) [thus
     det $a'(C)$ is the discriminant of $a(\lambda)$, defined in (1.20),
     except possibly for a difference in sign]. Hence show that
     the discriminants of $a\lambda^2 + b\lambda + c$ and $\lambda^3 + 3b\lambda + c$ are
     $b^2 - 4ac$ and $27(c^2 + 4b^3)$, respectively.

1.13 Suppose that C_n defined in (1.26) is nonsingular. Show that
     C_n^{-1} has the form of C_{II} in (1.31), with first row

     $[-a_{n-1}/a_n, \ -a_{n-2}/a_n, \ \ldots, \ -a_1/a_n, \ -1/a_n]$

     Deduce the characteristic equation of C_n^{-1} without evaluating
     $\det(\lambda I_n - C_n^{-1})$.

1.14 Show that C_{III} in (1.33) and C in (1.26) are related via the
     similarity transformation $C_{III} = T_1 C T_1^{-1}$, where $T_1 = JT$ and T
     is defined in (1.35).

1.15 A well-known result in matrix theory, called *Gershgorin's
     theorem*, states that if $A = [a_{ij}]$ is any $n \times n$ real or complex
     matrix, then its eigenvalues lie in the region of the complex
     z-plane consisting of all the n disks

     $|z - a_{ii}| \leqslant \rho_i, \quad i = 1, 2, \ldots, n$

     where

     $$\rho_i = \sum_{\substack{j=1 \\ j \neq i}}^{n} |a_{ij}|$$

     Apply this result to C_n in (1.26) and C_I in (1.30) to show
     that the roots of $a(\lambda)$ in (1.1) lie in the union of the disks

     $|z| \leqslant 1, \quad |z| \leqslant |a_n|, \quad |z| \leqslant 1 + |a_i|, \quad i = 2, 3, \ldots, n-1$
     $|z + a_1| \leqslant 1, \quad |z + a_1| \leqslant \sum_{j=2}^{n} |a_j|$

     This provides a simple way of obtaining some bounds on the
     roots of $a(\lambda)$.

1.16 Show that any $n \times n$ matrix X satisfying the equation $CX = XC^T$,
     where C is defined in (1.26), has the same striped pattern as

T in (1.35), including lines *below* the secondary diagonal.
Such a matrix X is often called *orthosymmetric* (or *Hankel*).
   If the stripes run parallel to the principal diagonal,
the matrix is called *Toeplitz*. Show that both JX and XJ in
this case are Toeplitz.

1.17 Let $a(\lambda)$ and C be as defined in (1.28). Let $b(\lambda)$ be a given
polynomial having the property that $a(\lambda)$ and $b(\lambda)x(\lambda) + 1$ are
relatively prime for all possible polynomials $x(\lambda)$.

   (i) Prove that $\det(I + b(C)x(C)) \neq 0$, for all $x(\lambda)$.

   (ii) By considering eigenvalues of $I + b(C)x(C)$, deduce that
        every root of $a(\lambda)$ is also a root of $b(\lambda)$.

1.18 Consider an $n \times n$ matrix $B = [b_{ij}]$ having $b_{ni} = y_i$, $b_{in} = x_i$,
$i = 1, 2, \ldots, n - 1$ and all other $b_{ij} = 0$, $i \neq j$. By
expanding $b(\lambda) = \det(\lambda I - B)$ by its last row, show that

$$b(\lambda) = \prod_{i=1}^{n} (\lambda - b_{ii}) - \sum_{i=1}^{n-1} x_i y_i p_i(\lambda)$$

where

$$p_i(\lambda) = \prod_{\substack{j=1 \\ j \neq i}}^{n-1} (\lambda - b_{jj})$$

Hence show that if $b_{11}, b_{22}, \ldots, b_{n-1,n-1}$ are $n - 1$ distinct
numbers, and in addition

$$b_{nn} = -a_1 - \sum_{i=1}^{n-1} b_{ii}, \quad x_i y_i = -a(b_{ii})/p_i(b_{ii}), \quad i = 1, 2, \ldots,$$
$$n - 1$$

where $a(\lambda)$ is the polynomial in (1.1) with $a_o = 1$, then
$b(\lambda) \equiv a(\lambda)$. [Hint: Show that $b(b_{kk}) - a(b_{kk}) = 0$, $k = 1$,
$\ldots, n - 1$.]

This provides an alternative way of constructing a matrix
whose characteristic polynomial is $a(\lambda)$, and can be used in
conjunction with Gershgorin's theorem (Problem 1.15).

1.19 Consider the rn × rn block partitioned matrix

$$C(r) = \begin{bmatrix} C & K & 0 & 0 & . & . \\ 0 & C & K & 0 & . & . \\ & & \cdot & \ddots & & \\ & & & \cdot & C \cdot & K \\ & 0 & & & 0 & C \end{bmatrix}$$

where C is defined in (1.26), K is an n × n matrix having its
n, 1 element equal to unity and all other elements zero, and
there are r block rows and columns. Prove:

(i) C(r) has characteristic polynomial $(a(\lambda))^r$, where $a(\lambda)$
is that of C.

(ii) C(r) is nonderogatory.

1.20 If θ is a root of $a(\lambda)$ in (1.1) (assumed monic) and C_I is the
companion matrix in (1.30), show that

$$[\theta^j, \theta^{j+1}, \ldots, \theta^{j+n-1}] = [1, \theta, \theta^2, \ldots, \theta^{n-1}] C_I^j,$$
$$j = 1, 2, 3, \ldots$$

Hence, by using the Cayley-Hamilton theorem, prove that if A
is an n × n matrix whose characteristic polynomial is $a(\lambda)$,
then

$$[A^j, A^{j+1}, \ldots, A^{j+n-1}] = [I_n, A, A^2, \ldots, A^{n-1}] (C_I^j \otimes I_n)$$

where θ denotes Kronecker product.

1.3 G.C.D. USING COMPANION MATRIX

We have seen in Theorem 1.3 that det b(C) forms a resultant of $a(\lambda)$
and $b(\lambda)$ in (1.1) and (1.2), respectively. However it turns out,
perhaps surprisingly, that if $a(\lambda)$ and $b(\lambda)$ are not relatively
prime, then their g.c.d. itself can be obtained in a very simple
way from the matrix b(C). In this section we shall use the
companion form C_I in (1.30), in which the coefficients of $a(\lambda)$ form
the last column. Since $C_I = C^T$, it follows that

$$R_o \triangleq b(C_I) = b(C^T) = [b(C)]^T$$

so the *columns* $\ell_1, \ell_2, \ldots, \ell_n$ of R_o are equal to the transpose of
the *rows* of b(C). If $\delta b = m < n = \delta a$, then (1.38) gives

$$\ell_1 = [b_m, b_{m-1}, \ldots, b_o, 0, \ldots, 0]^T \qquad (1.58)$$

and (1.42) and (1.43) become, respectively,

$$\ell_i = C^T \ell_{i-1}, \quad i = 2, 3, \ldots, n \qquad (1.59)$$

and Theorem 1.2 implies that

$$R_o \triangleq b(C^T) = [\ell_1, C^T\ell_1, (C^T)^2\ell_1, \ldots, (C^T)^{n-1}\ell_1] \qquad (1.60)$$

The sole reason for using C^T instead of C is that it is more convenient to display (1.60) on the printed page, rather than its transpose. Our first result is on the degree of the g.c.d. of $a(\lambda)$ and $b(\lambda)$, and we assume without loss of generality that $\delta a \geqslant \delta b$.

Theorem 1.4 The degree δd of the g.c.d. $d(\lambda)$ of $a(\lambda)$ and $b(\lambda)$ in (1.1) and (1.2), respectively, with $\delta a \geqslant \delta b$, is equal to $n - \text{rank } R_o$.

Proof. Let the Jordan form of C^T be denoted by J. It is a standard result of matrix algebra that since C^T is nonderogatory, there is only *one* Jordan block in J associated with each distinct root $\lambda_1, \ldots, \lambda_p$ of $a(\lambda)$ in its factorized form (1.8).
Thus we can write

$$J = \text{diag}[J_1, J_2, \ldots, J_p] \qquad (1.61)$$

where J_i is an $s_i \times s_i$ Jordan block associated with λ_i. Also, from the factorized form of $b(\lambda)$ in (1.13), we have

$$\text{rank } R_o = \text{rank}(J - \mu_1 I)^{t_1}(J - \mu_2 I)^{t_2} \cdots (J - \mu_q I)^{t_q} \qquad (1.62)$$

(assuming that $b_o \neq 0$). Clearly, from (1.61),

$$(J - \mu_j I_n)^{t_j} = (\text{diag}[J_1 - \mu_j I_{s_1}, J_2 - \mu_j I_{s_2}, \ldots,$$
$$J_p - \mu_j I_{s_p}])^{t_j}$$
$$= \text{diag}[(J_1 - \mu_j I)^{t_j}, (J_2 - \mu_j I)^{t_j}, \ldots,$$
$$(J_p - \mu_j I)^{t_j}]$$

so on multiplying these factors together, (1.62) becomes

$$\text{rank } R_o = \text{rank}\left\{\text{diag}\left[\prod_{j=1}^{q} (J_1 - \mu_j I)^{t_j}, \prod_j (J_2 - \mu_j I)^{t_j}, \ldots, \right.\right.$$

$$\left.\left. \prod_j (J_p - \mu_j I)^{t_j}\right]\right\}$$

$$= \text{rank } \prod_j (J_1 - \mu_j I)^{t_j} + \text{rank } \prod_j (J_2 - \mu_j I)^{t_j} + \cdots$$

$$+ \text{rank } \prod_j (J_p - \mu_j I)^{t_j} \qquad (1.63)$$

the last step following from the fact that the matrix within braces
is block diagonal. Now each matrix $(J_i - \mu_j I)$ is singular if and
only if $\mu_j \equiv \lambda_i$, in which case it is straightforward to compute
that

$$\left.\begin{array}{ll} \text{rank}(J_i - \mu_j I)^{t_j} = s_i - t_j, & t_j < s_i \\ \qquad\qquad\quad = 0, & t_j \geqslant s_i \end{array}\right\} \qquad (1.64)$$

Only at most one such matrix can be singular in each term
$\prod_j (J_i - \mu_j I)^{t_j}$ in (1.63), so (1.64) implies for (1.63) that

$$\text{rank } R_o = \sum (s_i - t_j), \quad \text{for } \mu_j = \lambda_i, \quad s_i > t_j \qquad (1.65)$$

However, by inspection, the degree of the g.c.d. of $a(\lambda)$ in (1.8)
and $b(\lambda)$ in (1.13) is

$$\delta d = \sum \min(s_i, t_j) \quad \text{for } \mu_j = \lambda_i$$

$$= \sum_{s_i > t_j} t_j + \sum_{s_i \leqslant t_j} s_i \quad \text{for } \mu_j = \lambda_i \qquad (1.66)$$

Adding together (1.65) and (1.66) thus gives

$$\text{Rank } R_o + \delta d = \sum_{s_i > t_j} s_i + \sum_{s_i \leqslant t_j} s_i$$

$$= s_1 + s_2 + \cdots + s_n = n \quad \blacksquare$$

Example 9 Consider again $a(\lambda)$ in (1.6), whose companion matrix in the form (1.30) is

$$
C^T = \begin{bmatrix} 0 & 0 & -5 \\ 1 & 0 & -1 \\ 0 & 1 & 3 \end{bmatrix}
\tag{1.67}
$$

Let

$$
b(\lambda) = \lambda^2 - 4\lambda + 5
$$

so that from (1.58) and (1.59) we obtain

$$
R_o = b(C^T) = \begin{bmatrix} 5 & -5 & 5 \\ -4 & 4 & -4 \\ 1 & -1 & 1 \end{bmatrix}
\tag{1.68}
$$

It is clear from (1.68) that rank R_o = 1, so the degree of the g.c.d. $d(\lambda)$ is 3 - 1 = 2. In fact, it is obvious from the factorized form of $a(\lambda)$ in Example 2 that $d(\lambda) \equiv b(\lambda) = \lambda^2 - 4\lambda + 5$. We also notice that in (1.68) the *rows* r_1, r_2, r_3 of R_o satisfy $r_2 = -4r_3$, $r_1 = 5r_3$, these coefficients being the same as those in $d(\lambda)$. This is no coincidence, as the following result shows:

Theorem 1.5 Let

$$
d(\lambda) = \lambda^k + d_1\lambda^{k-1} + \cdots + d_k, \qquad k < n
\tag{1.69}
$$

be the monic g.c.d. of $a(\lambda)$ and $b(\lambda)$ in (1.1) and (1.2), respectively, with $\delta a \geqslant \delta b$. Then the rows r_1, ..., r_n of $R_o = b(C^T)$ are such that r_{k+1}, r_{k+2}, ..., r_n are linearly independent, and furthermore

$$
r_i = d_{k+1-i}r_{k+1} + \sum_{j=k+2}^{n} x_{ij}r_j, \qquad i = 1, 2, \ldots, k
\tag{1.70}
$$

for some x_{ij}.

Proof. Let $b(\lambda) = d(\lambda)\,b_1(\lambda)$, and consider

$$d(C^T) = (C^T)^k + d_1(C^T)^{k-1} + \cdots + d_k I$$

The g.c.d. of $a(\lambda)$ and $d(\lambda)$ is of course $d(\lambda)$, so by Theorem 1.4
the rank of $d(C^T)$ is $n - k$. Moreover, from (1.58) and (1.59) it is
easy to show that the first $n - k$ columns of $d(C^T)$ are

$$\begin{bmatrix} d_k & 0 & . & 0 \\ d_{k-1} & d_k & . & 0 \\ \vdots & \vdots & \vdots & \vdots \\ d_2 & d_3 & . & . \\ d_1 & d_2 & . & . \\ 1 & d_1 & . & . \\ 0 & 1 & . & . \\ \vdots & 0 & . & . \\ . & . & . & d_1 \\ 0 & 0 & . & 1 \end{bmatrix} \tag{1.71}$$

If the rows of $d(C^T)$ are denoted by $s_1, s_2, \ldots, s_n$, then it is
obvious from (1.71) that $s_{k+1}, \ldots, s_n$ are linearly independent,
and that

$$s_i = d_{k+1-i} s_{k+1} + \sum_{j=k+2}^{n} y_{ij} s_j, \quad i = 1, 2, \ldots, k \tag{1.72}$$

for some y_{ij}. Also, $b(C^T) = d(C^T) b_1(C^T)$, which implies that
$r_i = s_i b_1(C^T)$, for all i. Hence, postmultiplication of (1.72) by
$b_1(C^T)$ [which is nonsingular by Theorem 1.3, since $a(\lambda)$ and $b_1(\lambda)$
are relatively prime] produces the desired result (1.70) (and
incidentally shows that $x_{ij} = y_{ij}$, for all i, j). ∎

Example 10   (a) In Example 9, $n = 3$, $k = 2$, and $r_1 = 5r_3$, $r_2 = -4r_3$
by inspection of (1.68), so from (1.70) we conclude that $d_1 = -4$,
$d_2 = 5$, which is in fact what we noticed earlier.
    (b) Again let $a(\lambda)$ be the polynomial in (1.6) with companion
matrix (1.67), and let

$$b(\lambda) = \lambda^2 + 3\lambda + 2 \quad [\equiv (\lambda + 1)(\lambda + 2)] \tag{1.73}$$

so from (1.58) and (1.59)

$$R_0 = b(C^T) = \begin{bmatrix} 2 & -5 & -30 \\ 3 & 1 & -11 \\ 1 & 6 & 19 \end{bmatrix} \begin{matrix} r_1 \\ r_2 \\ r_3 \end{matrix}$$

Here rank $R_0 = 2$, and $r_1 = r_2 - r_3$, so from (1.70) the g.c.d. of $a(\lambda)$ and $b(\lambda)$ is $d(\lambda) = \lambda + 1$ [this agrees with (1.73) and the factorized form of $a(\lambda)$ in (1.10)].

(c) If

$$a(\lambda) = \lambda^5 + \lambda^4 - 9\lambda^3 - 5\lambda^2 + 16\lambda + 12 \tag{1.74}$$

$$[\equiv (\lambda + 1)^2(\lambda + 3)(\lambda - 2)^2]$$

$$b(\lambda) = \lambda^4 - 3\lambda^3 + \lambda^2 + 3\lambda - 2 \tag{1.75}$$

$$[\equiv (\lambda + 1)(\lambda - 1)^2(\lambda - 2)]$$

then using (1.58) and (1.59) we obtain

$$R_0 = b(C^T) = \begin{bmatrix} -2 & -12 & 48 & -168 & 504 \\ 3 & -18 & 52 & -176 & 504 \\ 1 & 8 & -38 & 122 & -386 \\ -3 & 10 & -28 & 88 & -256 \\ 1 & -4 & 14 & -42 & 130 \end{bmatrix} \begin{matrix} r_1 \\ r_2 \\ r_3 \\ r_4 \\ r_5 \end{matrix} \tag{1.76}$$

where C^T is the companion matrix for $a(\lambda)$. It is routine to calculate that for the matrix in (1.76), rank $R_0 = 3$, and

$$r_1 = -2r_3 - 2r_4 - 6r_5, \quad r_2 = -r_3 - 3r_4 - 5r_5$$

so by (1.70) the g.c.d. is

$$d(\lambda) = \lambda^2 - \lambda - 2 \quad [\equiv (\lambda + 1)(\lambda - 2]$$

This agrees with inspection of the factorized forms of (1.74) and (1.75).

Remarks on Theorem 1.5   (1) If $b(C)$ is used instead of $b(C^T)$, the rows r_i in the statement of the theorem become the *columns* $c_1, \ldots, c_n$ of $b(C)$. In particular, (1.70) is replaced by

$$c_i = d_{k+1-i} c_{k+1} + \sum_{j=k+2}^{n} x_{ij} c_j \qquad (1.77)$$

(2) The g.c.d. of two polynomials can be found by computing rank R_o, and solving the set of linear algebraic equations represented by (1.70). Both these problems can be solved using standard techniques of numerical linear algebra, for example gaussian elimination. The procedure is attractive since (i) it removes the need to carry out polynomial manipulations, and (ii) it can be extended to deal directly with more than two polynomials. This is in contrast with the classical approach indicated in Section 1.1, which requires the polynomials to be taken only two at a time.

Let Π denote a set of $h + 1$ polynomials whose maximum degree is n, so we can assume that the set consists of $a(\lambda)$ in (1.1) and

$$b_i(\lambda) = b_{io} \lambda^n + b_{i1} \lambda^{n-1} + \cdots + b_{in}, \qquad i = 1, 2, \ldots, h \quad (1.78)$$

Let $d(\lambda)$ in (1.69) *now* denote the monic g.c.d. of this set Π, and construct an $n \times nh$ matrix

$$R' \triangleq [b_1(C^T), b_2(C^T), \ldots, b_h(C^T)] \qquad (1.79)$$

where C^T is the companion matrix (1.30) of $a(\lambda)$; again, we note the convenience of using C^T instead of C for display purposes. By Theorem 1.2 the columns of $b_i(C^T)$ are

$$\ell_{i1}, \; C^T \ell_{i1}, \; (C^T)^2 \ell_{i1}, \; \ldots, \; (C^T)^{n-1} \ell_{i1}, \qquad i = 1, 2, \ldots, h$$

where ℓ_{i1} is given by (1.58) or (1.51) according as b_{io} is zero or nonzero. A straightforward extension of the proof when $h = 1$ leads to the result that the g.c.d. is obtained from the *same* expression as in Theorem 1.5, but with the rows $r_1, \ldots, r_n$ being those of (1.79). The relationships between these rows are unaltered by a permutation of the columns in (1.79). In particular, take together all the first columns $\ell_{11}, \ell_{21}, \ldots, \ell_{n1}$ in each block of R', then all the second columns of each block, and so on, to obtain the $n \times nh$ matrix

$$R \overset{\triangle}{=} [B, \ C^TB, \ (C^T)^2B, \ \ldots, \ (C^T)^{n-1}B] \qquad (1.80)$$

where

$$B = [\ell_{11}, \ \ell_{21}, \ \ldots, \ \ell_{h1}] \qquad (1.81)$$

Theorem 1.6   The degree δd of the g.c.d. of $a(\lambda)$ in (1.1) and $b_1(\lambda), \ \ldots, \ b_h(\lambda)$ in (1.78) is equal to n - rank R, and the coefficients of $d(\lambda)$ are given by the *same* relationship (1.70) as for the case h = 1, where the r_i are now the rows of R in (1.80).

Example 11   Let Π consist of the polynomials $a(\lambda)$ in (1.74), $b(\lambda)$ in (1.75) and

$$b_2(\lambda) = \lambda^2 + 5\lambda - 14 \quad [\equiv (\lambda - 2)(\lambda + 7)]$$

From (1.58), (1.80), and (1.81) we obtain

$$R = \begin{bmatrix} -2 & -14 & -12 & 0 & 48 & 0 & -168 & -12 & 504 & -48 \\ 3 & 5 & -18 & -14 & 52 & 0 & -176 & -16 & 504 & -76 \\ 1 & 1 & 8 & 5 & -38 & -14 & 122 & 5 & -368 & 4 \\ -3 & 0 & 10 & 1 & -28 & 5 & 88 & -5 & -256 & 41 \\ 1 & 0 & -4 & 0 & 14 & 1 & -42 & 4 & 130 & -9 \end{bmatrix} \begin{matrix} r_1 \\ r_2 \\ r_3 \\ r_4 \\ r_5 \end{matrix}$$
$$\quad\quad B \qquad C^TB \qquad (C^T)^2B \qquad (C^T)^3B \qquad (C^T)^4B$$

Notice that each 5 × 2 block is obtained by premultiplying the preceding block by C^T. The rank of R is found to be 4, and

$$r_1 = -2r_2 - 4r_3 - 8r_4 - 16r_5$$

and so by Theorem 1.6 the g.c.d. is $\lambda - 2$, which can be confirmed immediately from the factorized forms of the three polynomials.

Problems

1.21 Use Theorems 1.4 and 1.5 to find the g.c.d. of each of the following pairs of polynomials.

(i) $\lambda^4 + \lambda^3 - 3\lambda^2 - 4\lambda - 1, \ \lambda^3 + \lambda^2 - \lambda - 1$

(ii) $\lambda^5 + \lambda^4 - \lambda^3 - 2\lambda - 1, \ 3\lambda^4 + 2\lambda^3 + \lambda^2 + 2\lambda - 2$

(iii) $\lambda^5 - 2\lambda^4 + \lambda^3 + 7\lambda^2 - 12\lambda + 10$, $3\lambda^4 - 6\lambda^3 + 5\lambda^2 + 2\lambda - 2$

1.22 Use Theorem 1.6 to find the g.c.d. of each of the following
sets of polynomials.

(i) $\lambda^3 + 3\lambda^2 - \lambda - 3$, $\lambda^4 + 2\lambda^3 - 2\lambda^2 + 2\lambda - 3$,

$\lambda^4 + 2\lambda^3 - 5\lambda^2 - 4\lambda + 6$, $2\lambda^4 + 5\lambda^3 - 2\lambda^2 + \lambda - 6$

(ii) $\lambda^5 + \lambda^4 - 5\lambda^3 - \lambda^2 + 8\lambda - 4$, $\lambda^3 + 4\lambda^2 + \lambda - 6$,

$\lambda^5 + 6\lambda^4 + 11\lambda^3 + 2\lambda^2 - 12\lambda - 8$

[Remember to choose as $a(\lambda)$ a polynomial having highest
degree.]

1.23 Suppose that $a(\lambda)$ in (1.1) when factorized as in (1.8) has h_i
distinct factors with multiplicity i, for i = 1, 2, ..., ℓ, so
that $n = h_1 + 2h_2 + 3h_3 + \cdots + \ell h_\ell$. By considering the
g.c.d. of $a(\lambda)$ and $a'(\lambda)$, use Theorem 1.4 to prove that the
number of *distinct* factors of $a(\lambda)$ (i.e., $h_1 + h_2 + \cdots + h_\ell$)
is equal to the rank of $a'(C^T)$. Use this result to show that
$\lambda^4 + 4\lambda^3 + 8\lambda^2 + 8\lambda + 3$ has three distinct roots.

1.4   SYLVESTER'S RESULTANT MATRIX

Consider again the polynomials $a(\lambda)$ in (1.1) and $b(\lambda)$ in (1.2), and
suppose that $a_o = 1$, $b_o \neq 0$ and $\delta b \leqslant \delta a$. If the polynomials have a
common factor, this means that there exists a value of λ for which
the equations $a(\lambda) = 0$, $b(\lambda) = 0$ are simultaneously satisfied.
Regard the "unknowns" in these equations as the powers of λ, and
multiply the first equation by λ^{m-1}, λ^{m-2}, ..., λ^1, λ^o, respectively, to obtain the m equations

$$
\left.
\begin{aligned}
\lambda^{m+n-1} + a_1\lambda^{m+n-2} + a_2\lambda^{m+n-3} + \cdots + a_n\lambda^{m-1} &= 0 \\
\lambda^{m+n-2} + a_1\lambda^{m+n-3} + \cdots + a_{n-1}\lambda^{m-1} + a_n\lambda^{m-2} &= 0 \\
\vdots \\
\lambda^n + a_1\lambda^{n-1} + \cdots + a_n &= 0
\end{aligned}
\right\}
$$

(1.82)

Similarly, multiply $b(\lambda) = 0$ by λ^{n-1}, λ^{n-2}, ..., λ^1, λ^o,
respectively, to obtain the n equations

$$
\left.\begin{array}{l}
b_0\lambda^{m+n-1} + b_1\lambda^{m+n-2} + \cdots + b_m\lambda^{n-1} = 0 \\
\quad b_0\lambda^{m+n-2} + \cdots + b_{m-1}\lambda^{n-1} + b_m\lambda^{n-2} = 0 \\
\qquad\qquad\qquad \vdots \\
\qquad b_0\lambda^m + b_1\lambda^{m-1} + \cdots + b_m = 0
\end{array}\right\} \qquad (1.83)
$$

In order that the $(m + n)$ equations (1.82) and (1.83) in the $(m + n)$ unknowns λ^0, λ^1, ..., λ^{m+n-1} have a nontrivial solution, it follows by a basic result in the theory of linear equations that the matrix of coefficients, namely

$$
S = \begin{bmatrix}
1 & a_1 & a_2 & \cdots & a_n & 0 & . & . & 0 & 0 \\
0 & 1 & a_1 & \cdots & a_{n-1} & a_n & . & . & 0 & 0 \\
. & . & . & \cdots & & & & & & . \\
0 & 0 & 0 & 1 & \cdots & & . & . & a_{n-1} & a_n \\
b_0 & b_1 & b_2 & \cdots & b_m & 0 & 0 & . & . & 0 & 0 \\
0 & b_0 & b_1 & \cdots & b_{m-1} & b_m & 0 & . & . & 0 & 0 \\
. & . & . & \cdots & & & & & & . \\
0 & 0 & 0 & b_0 & \cdots & & . & . & b_{m-1} & b_m
\end{bmatrix}
\begin{array}{l} \Big\} m \text{ rows} \\ \\ \Big\} n \text{ rows} \end{array}
\qquad (1.84)
$$

must be singular. The $(m + n) \times (m + n)$ matrix S in (1.84) is *Sylvester's resultant matrix* associated with the polynomials $a(\lambda)$ and $b(\lambda)$, and we have thus shown that a necessary condition for the polynomials to have a common factor is that det $S = 0$. In fact, this condition is also sufficient, and we shall prove this by demonstrating the relationship between S and the matrix $b(C)$ in Theorem 1.3. Partition (1.84) as follows:

$$
S = \begin{array}{c} \\ \begin{bmatrix} T_1 & T_2 \\ T_3 & T_4 \end{bmatrix} \end{array}
\begin{array}{l} m \\ n \end{array}
\qquad (1.85)
$$

Theorem 1.7   The Sylvester matrix (1.85) satisfies the identity

$$
\begin{array}{c}
m \\ n
\end{array}
\begin{bmatrix}
I_m & 0 \\
K & I_n
\end{bmatrix}
S =
\begin{bmatrix}
T_1 & T_2 \\
0 & J_n b(C) J_n
\end{bmatrix}
\begin{array}{l} m \\ n \end{array}
\qquad (1.86)
$$

where $b(C)$ is defined by Theorems 1.2 and 1.3, and K is defined below. Hence S is nonsingular if and only if $a(\lambda)$ in (1.1) and $b(\lambda)$ in (1.2) are relatively prime.

Proof. Suppose first that $m = n$. On substituting (1.85) into (1.86), the left side becomes

$$\begin{bmatrix} T_1 & T_2 \\ (KT_1 + T_3) & (KT_2 + T_4) \end{bmatrix}$$

so it is first necessary to show that

$$KT_1 + T_3 = 0 \tag{1.87}$$

This is achieved by taking K to be an upper triangular Toeplitz matrix with first row $[-b_o, -z_1, -z_2, \ldots, -z_{n-1}]$, where

$$z_k = b_k - a_k b_o - a_{k-1} z_1 - \cdots - a_1 z_{k-1}, \quad k = 1, \ldots,$$
$$n - 1 \quad (z_o = b_o)$$

as the reader can readily check. It remains to show that

$$KT_2 + T_4 \equiv Jb(C)J \tag{1.88}$$

and this is done by comparing rows on both sides. The result for $m < n$ is easily deduced by suppressing appropriate rows and columns in (1.87).

For the second statement in the theorem, take the determinant of both sides of (1.86) to obtain

$$\det S = \det T_1 \det Jb(C)J$$
$$= \det T_1 (\det J)^2 \det b(C) = \det b(C)$$

on using (1.32) and the fact that $\det T_1 = 1$. Hence, by Theorem 1.3, $R(a, b) = \det S$. ∎

Remarks on Theorem 1.7   (1) Recall that $J_n^{-1} = J_n$, so it follows that

$$Jb(C)J = b(JCJ) = b(C_{II})$$

where C_{II} is the matrix defined in (1.31). Thus if C_{II} is used instead of C, the factors J disappear.

(2) The fine details of the proof of Theorem 1.7 have been omitted, since they are of little concern here. What is more interesting, although not in our main line of development, is to note that in (1.84) the $m \times m$ leading principal submatrix T_1 is nonsingular. Thus (1.87) gives

$$K = -T_3 T_1^{-1} \tag{1.89}$$

and substitution of (1.89) into (1.88) produces

$$Jb(C)J = T_4 - T_3 T_1^{-1} T_2 \overset{\triangle}{=} (S/T_1) \tag{1.90}$$

The matrix (S/T_1) defined in (1.90) is known as the *Schur complement* of T_1 in the matrix S in (1.85). This has interesting properties, some of which are explored in Problems 1.26 to 1.28 [note that the definition applies for *any* matrix partitioned in the form (1.85), with T_1 nonsingular].

(3) In some applications it is more convenient to use a slightly different form of S in which the last n rows in (1.84) are reversed in order. In this case, construct centrally situated submatrices by successively deleting a row and column all the way round. For example, if n = 3, m = 2 we obtain

$$
S =
\begin{bmatrix}
1 & a_1 & a_2 & a_3 & 0 \\
0 & 1 & a_1 & a_2 & a_3 \\
0 & 0 & b_0 & b_1 & b_2 \\
0 & b_0 & b_1 & b_2 & 0 \\
b_0 & b_1 & b_2 & 0 & 0
\end{bmatrix}
$$

The matrices within the dashed lines are called the *inners* of S, and their determinants are *subresultants*. In general, the inners S_1, S_2, ... have dimensions $m + n - 2$, $m + n - 4$, ..., and the g.c.d. of $a(\lambda)$ and $b(\lambda)$ has degree k (> 0) if and only if det S = 0, det $S_1 = 0$, det $S_2 = 0$, ..., det $S_{k-1} = 0$, det $S_k \neq 0$. We shall encounter an important application of inners in Theorem 3.11.

(4) It is clear from (1.86) that

rank S = rank T_1 + rank $Jb(C)J$

= m + rank $b(C)$

so by Theorem 1.4 we can immediately deduce:

Theorem 1.8   The degree δd of the g.c.d. of $a(\lambda)$ in (1.1) and $b(\lambda)$ in (1.2) is equal to $m + n$ - rank S, where S is defined in (1.84).

Example 12   Consider again the polynomials in Example 9, namely

$$a(\lambda) = \lambda^3 - 3\lambda^2 + \lambda + 5, \quad b(\lambda) = \lambda^2 - 4\lambda + 5$$

The associated Sylvester matrix (1.84) is

$$S = \begin{bmatrix} 1 & -3 & 1 & 5 & 0 \\ 0 & 1 & -3 & 1 & 5 \\ 1 & -4 & 5 & 0 & 0 \\ 0 & 1 & -4 & 5 & 0 \\ 0 & 0 & 1 & -4 & 5 \end{bmatrix} \qquad (1.91)$$

To determine the rank of (1.91), apply the sequence of row operations: (row 3) - (row 1); (row 3) + (row 2), (row 4) - (row 2); (row 4) + (row 3), (row 5) - (row 3), to end up with

$$\begin{bmatrix} 1 & -3 & 1 & 5 & 0 \\ 0 & 1 & -3 & 1 & 5 \\ 0 & 0 & 1 & -4 & 5 \\ 0 & 0 & 0 & 0 & 0 \\ 0 & 0 & 0 & 0 & 0 \end{bmatrix} \qquad (1.92)$$

showing that rank $S = 3$, so by Theorem 1.8, $\delta d = 3 + 2 - 3 = 2$, agreeing with what we found earlier in Example 9. However, on looking back at that example the reader will notice that the coefficients of the last nonvanishing row in (1.92) are precisely those of the g.c.d., namely $d(\lambda) = \lambda^2 - 4\lambda + 5$. Again this is not a coincidence, as the following result makes clear.

Theorem 1.9   Let the Sylvester matrix S in (1.84) be put into row echelon form, using only elementary row transformations.   Then the last nonvanishing row gives the coefficients of a g.c.d. of $a(\lambda)$ in (1.1) and $b(\lambda)$ in (1.2).

Proof.   Premultiply both sides of (1.86) as follows:

$$\begin{bmatrix} I_m & 0 \\ 0 & UJ_n \end{bmatrix} \begin{bmatrix} I_m & 0 \\ K & I_n \end{bmatrix} S = \begin{bmatrix} T_1 & T_2 \\ 0 & UR_o^T J_n \end{bmatrix} \tag{1.93}$$

where U is an $n \times n$ matrix to be determined below, and $R_o = b(C^T)$. The left-hand side of (1.93) represents a set of row transformations applied to S.   On the right-hand side T_1 is upper triangular.   Thus, to complete the desired reduction of S, it remains only to consider the choice of U so as to reduce the lower right block $UR_o^T J_n$ to row echelon form.

To study this, go back to (1.70), and recall that the r_i are the rows of R_o, so the set of equations represented by (1.70) can be written in the matrix form

$$R_o = PQ \tag{1.94}$$

where

$$P = \begin{bmatrix} d_k & x_{1,k+2} & \cdots & x_{1n} \\ d_{k-1} & x_{2,k+2} & \cdots & x_{2n} \\ \vdots & \vdots & \cdots & \vdots \\ d_1 & x_{k,k+2} & \cdots & x_{kn} \\ \hline & & I_{n-k} & \end{bmatrix}, \quad Q = \begin{bmatrix} r_{k+1} \\ r_{k+2} \\ \vdots \\ r_n \end{bmatrix} \tag{1.95}$$

Since by Theorem 1.5, the $n \times (n - k)$ matrix Q^T has rank $(n - k)$, a standard theorem in matrix algebra states that there exists a non-singular matrix U such that

$$UQ^T = \begin{bmatrix} D \\ 0 \end{bmatrix} \begin{matrix} n-k \\ k \end{matrix} \tag{1.96}$$

where D is a nonsingular diagonal matrix of order n - k. Combining
(1.94) and (1.96) produces

$$
UR_o^T = \begin{bmatrix} D \\ 0 \end{bmatrix} P^T \tag{1.97}
$$

Hence, on using the form of P in (1.95), equation (1.97) becomes

$$
UR_o^T J_n = \begin{bmatrix} D \\ 0 \end{bmatrix} P^T J_n
$$

$$
= \begin{bmatrix}
0 & . & . & 0 & 0 & \theta & (\theta d_1) & (\theta d_2) & \cdots & (\theta d_k) \\
0 & . & . & 0 & x & x & . & . & \cdots & x \\
0 & . & 0 & x & x & x & . & . & \cdots & x \\
. & . & . & . & . & . & . & & \cdots & . \\
0 & 0 & 0 & . & . & . & . & . & \cdots & 0
\end{bmatrix} \tag{1.98}
$$

where θ denotes the (1, 1) element of D, and the x's indicate
elements whose values are not of importance. An obvious final set
of row transformations then puts (1.98) into row echelon form, and
the last nonvanishing row is $\theta(1, d_1, \ldots, d_k)$, which gives the
coefficients of a g.c.d. (note that it need not necessarily turn
out monic). ∎

Example 13 We repeat Example 10b, using the method of Theorem 1.9.
For $a(\lambda)$ in (1.6) and $b(\lambda)$ in (1.73), the Sylvester matrix (1.84)
is

$$
S = \begin{bmatrix}
1 & -3 & 1 & 5 & 0 \\
0 & 1 & -3 & 1 & 5 \\
1 & 3 & 2 & 0 & 0 \\
0 & 1 & 3 & 2 & 0 \\
0 & 0 & 1 & 3 & 2
\end{bmatrix}
$$

The reader can check that the following sequence of row operations:
(row 3) - (row 1); (row 3) - 6(row 2), (row 4) - (row 2); inter-
change (row 3) and (row 5); (row 4) - 6(row 3), (row 5) - 19(row 3);
(row 5) - 4(row 4) reduces S to

$$\begin{bmatrix} 1 & -3 & 1 & 5 & 0 \\ 0 & 1 & -3 & 1 & 5 \\ 0 & 0 & 1 & 3 & 2 \\ 0 & 0 & 0 & -17 & -17 \\ 0 & 0 & 0 & 0 & 0 \end{bmatrix}$$

showing that a g.c.d. is $-17\lambda - 17$, agreeing with the monic form $\lambda + 1$ found earlier. Of course, the row operations and the consequent echelon form, are not unique.

An advantage of using Theorem 1.9 instead of Theorem 1.5 is that S in (1.84) can be written down immediately, compared with the construction of R_o by Theorem 1.2. However, a disadvantage is that S has dimensions $m + n$, compared with only $\max(m, n)$ for R_o. The latter becomes more apparent when we turn to the extension for more than two polynomials. As before, let Π denote the set of $h + 1$ polynomials defined in (1.1) and (1.78). Suppose that the maximum degree among $b_1(\lambda)$, ..., $b_h(\lambda)$ in (1.78) is $p \leqslant n$, i.e., $b_{k,n-p} \neq 0$ for at least one value of k, but $b_{kj} = 0$ for all $j < n - p$ and all k. Define a $p \times (n + p)$ matrix associated with $a(\lambda)$:

$$S_o = \begin{bmatrix} 1 & a_1 & a_2 & \cdots & a_n & & \cdot & \cdot & 0 \\ 0 & 1 & a_1 & \cdots & a_{n-1} & a_n & & \cdot & 0 \\ \cdot & \cdot & \cdot & & \cdots & & & & \cdot \\ 0 & 0 & 0 & 1 & \cdots & & \cdot & a_{n-1} & a_n \end{bmatrix} \qquad (1.99)$$

and an $n \times (n + p)$ matrix associated with $b_i(\lambda)$ in (1.78):

$$S_i = \begin{bmatrix} b_{i,n-p} & b_{i,n-p+1} & \cdots & b_{in} & 0 & \cdot\cdot & 0 & 0 \\ 0 & b_{i,n-p} & \cdots & b_{i,n-1} & b_{in} & \cdot\cdot & 0 & 0 \\ \cdot & \cdot & \cdots & & \cdot & \cdot\cdot & \cdot & \cdot \\ 0 & 0 & b_{i,n-p} & \cdot & & \cdot\cdot & b_{i,n-1} & b_{in} \end{bmatrix}$$

$$(1.100)$$

for each i = 1, 2, ..., h. An *extended Sylvester matrix* for the set Π is

$$S_e \triangleq \begin{bmatrix} S_o \\ S_1 \\ S_2 \\ \vdots \\ S_h \end{bmatrix} \tag{1.101}$$

and has dimensions $(nh + p) \times (n + p)$, compared with $n \times nh$ for R in (1.80). When there are only two polynomials, then $h = 1$ and $p = m$, and S_e in (1.101) coincides with S in (1.84). It is interesting that the properties of S carry over to S_e without alteration:

Theorem 1.10   The degree of the g.c.d. of the set Π is equal to $n + p - \text{rank } S_e$. Furthermore, if S_e is put into row echelon form using only row transformations, then the last nonvanishing row gives the coefficients of a g.c.d.

Proof.   Using (1.86), with $K = K_i$ corresponding to $b_i(C^T)$, for $i = 1, 2, \ldots, h$, it follows that the relationship between S_e and the matrix R' in (1.79) is

$$\begin{bmatrix} I_p & 0 & 0 & . & 0 \\ K_1 & I_n & 0 & . & . \\ K_2 & 0 & I_n & . & . \\ . & . & . & . & . \\ K_h & 0 & 0 & . & I_n \end{bmatrix} S_e = \begin{bmatrix} T_1 & T_2 \\ 0 & Jb_1(C^T)J \\ 0 & Jb_2(C^T)J \\ \vdots & \vdots \\ 0 & Jb_h(C^T)J \end{bmatrix}$$

$$= \begin{bmatrix} T_1 & T_2 \\ 0 & \begin{pmatrix} J & 0 \\ & \ddots & \\ 0 & & J \end{pmatrix}(R')^T \end{bmatrix} \tag{1.102}$$

Since rank $T_1 = p$, (1.102) implies that

$$\text{rank } S_e = p + \text{rank} \begin{bmatrix} J & 0 \\ & \ddots & \\ 0 & & J \end{bmatrix}(R')^T$$

$$= p + \text{rank } R'$$

By Theorem 1.6, the degree k of the g.c.d. is equal to $n - \text{rank } R'$,

so $k = n + p - \text{rank } S_e$. The proof of the second part closely
follows that for Theorem 1.9. ∎

Example 14 We apply the preceding theorem to the polynomials

$$\lambda^3 + 6\lambda^2 + 11\lambda + 6, \quad \lambda^2 + 2\lambda - 3, \quad \lambda^2 + \lambda + 6 \qquad (1.103)$$

The highest-degree polynomial is chosen as $a(\lambda)$, so $n = 3$ and
$p = 2$. From (1.99)

$$S_o = \begin{bmatrix} 1 & 6 & 11 & 6 & 0 \\ 0 & 1 & 6 & 11 & 6 \end{bmatrix}$$

and from (1.100)

$$S_1 = \begin{bmatrix} 1 & 2 & -3 & 0 & 0 \\ 0 & 1 & 2 & -3 & 0 \\ 0 & 0 & 1 & 2 & -3 \end{bmatrix}, \quad S_2 = \begin{bmatrix} 1 & 1 & 6 & 0 & 0 \\ 0 & 1 & 1 & 6 & 0 \\ 0 & 0 & 1 & 1 & 6 \end{bmatrix}$$

so that in (1.101) we obtain

$$S_e = \begin{bmatrix} S_o \\ S_1 \\ S_2 \end{bmatrix}$$

It is straightforward, albeit tedious, to reduce S to the row
echelon form (not unique):

$$\begin{bmatrix} 1 & 6 & 11 & 6 & 0 \\ 0 & 1 & 6 & 11 & 6 \\ 0 & 0 & 1 & 2 & -3 \\ 0 & 0 & 0 & -6 & -18 \\ & & 0 & & \end{bmatrix} \updownarrow 4$$

so by Theorem 1.10 a g.c.d. is

$$-6\lambda - 18 = -6(\lambda + 3)$$

This can easily be confirmed by factorizing the polynomials in
(1.103).

   To end this section we demonstrate a different application of
S. Suppose that $\delta b \leqslant \delta a$, and the equation

$$a(\lambda)x(\lambda) + b(\lambda)y(\lambda) = 1 \qquad\qquad (1.104)$$

is to be solved for the polynomials $x(\lambda)$ and $y(\lambda)$, whose degrees are assumed to satisfy the conditions $\delta x < \delta b$, $\delta y < \delta a$. Equations of the form (1.104) are sometimes called *diophantine* equations, after the classical problem when a, b, x, y are integers. Let us write

$$x(\lambda) = x_o \lambda^{m-1} + x_1 \lambda^{m-2} + \cdots + x_{m-1}$$

$$y(\lambda) = y_o \lambda^{n-1} + y_1 \lambda^{n-2} + \cdots + y_{n-1}$$

and substitute these expressions into (1.104). Equating coefficients of powers of λ gives, on using (1.1) and (1.2):

$$
\left.
\begin{aligned}
\lambda^{m+n-1}: &\quad x_o + b_o y_o = 0 \\[4pt]
\lambda^{m+n-2}: &\quad a_1 x_o + x_1 + b_1 y_o + b_o y_1 = 0 \\[4pt]
&\quad \cdot \qquad \cdot \qquad \cdot \qquad \cdot \qquad \cdot \qquad \cdot \\[4pt]
\lambda^1: &\quad a_n x_{m-2} + a_{n-1} x_{m-1} + b_m y_{n-2} + b_{m-1} y_n = 0 \\[4pt]
\lambda^o: &\quad a_n x_{m-1} + b_n y_{n-1} = 1
\end{aligned}
\right\} \qquad (1.105)
$$

These $m + n$ equations in the $m + n$ unknown coefficients of $x(\lambda)$ and $y(\lambda)$ can be written in the form

$$[x_o, \; x_1, \; x_2, \; \ldots, \; x_{m-1}, \; y_o, \; y_1, \; \ldots, \; y_{n-1}] S = [0, \; 0, \; \ldots, \; 0, \; 1]$$
$$(1.106)$$

where S is the Sylvester matrix in (1.84). If $a(\lambda)$ and $b(\lambda)$ are relatively prime, then by Theorem 1.7 S is nonsingular, so (1.106) has the unique solution

$$[x_o, \; \ldots, \; x_{m-1}, \; y_o, \; \ldots, \; y_{n-1}] = [0, \; \ldots, \; 0, \; 1] S^{-1}$$
$$= \text{last row of } S^{-1}$$

The last row of S^{-1} can be obtained, for example, by standard gaussian-elimination-type methods. A different approach to solving (1.104) will be discussed in Section 1.6.4. However, it should be noted that the converse can also be proved. The complete result is:

<u>Theorem 1.11</u> The polynomials $a(\lambda)$ in (1.1) and $b(\lambda)$ in (1.2) are relatively prime if and only if there exist polynomials $x(\lambda)$, $y(\lambda)$ satisfying (1.104). Moreover, when $a(\lambda)$ and $b(\lambda)$ are relatively prime, there is a unique pair of polynomials in (1.104) satisfying $\delta x < \delta b$, $\delta y < \delta a$.

<u>Remark</u>. We shall encounter *two* different generalizations of Theorem 1.11 in Chapter 4 (see Theorems 4.18 and 4.25).

Problems
1.24 Repeat Problem 1.21 using Theorems 1.8 and 1.9.
1.25 Write a computer program to determine the rank and row echelon form of the extended Sylvester matrix defined in (1.101), for h, p, n $\leqslant$ 5, as required in Theorem 1.10. Test your program by using it to resolve Problem 1.22.
1.26 Premultiply S in (1.85) by

$$\begin{bmatrix} T_1^{-1} & 0 \\ -T_3 T_1^{-1} & I_n \end{bmatrix}$$

and hence obtain *Schur's determinantal formula*:

$$\det S = \det T_1 \det(S/T_1)$$

Use Theorem 1.8 to deduce that $a(\lambda)$ and $b(\lambda)$ are relatively prime if and only if the Schur complement (S/T_1) is nonsingular.
1.27 Let an n × n hermitian matrix A, and an n × n matrix L be partitioned as

$$A = \begin{bmatrix} A_1 & A_2 \\ A_2^* & A_3 \end{bmatrix} \begin{matrix} k \\ n-k \end{matrix}, \quad L = \begin{bmatrix} I_k & -A_1^{-1}A_2 \\ 0 & I_{n-k} \end{bmatrix}$$

where it is assumed that A_1 is nonsingular. Verify that

$$L^*AL = \begin{bmatrix} A_1 & 0 \\ 0 & (A/A_1) \end{bmatrix}$$

1.28 Consider any matrix S partitioned in the form (1.85) with T_1 nonsingular, and suppose that it is factorized as

$$\begin{bmatrix} T_1 & T_2 \\ T_3 & T_4 \end{bmatrix} = \begin{bmatrix} L_1 & 0 \\ L_2 & L_3 \end{bmatrix} \begin{bmatrix} U_1 & U_2 \\ 0 & U_3 \end{bmatrix}$$

Prove that $(S/T_1) = L_3 U_3$.

1.29 (i) Show that if $x_0(\lambda)$, $y_0(\lambda)$ is a particular solution of (1.104), then the pair

$$x(\lambda) = x_0(\lambda) + t(\lambda)b(\lambda), \qquad y(\lambda) = y_0(\lambda) - t(\lambda)a(\lambda)$$

is also a solution, for an arbitrary polynomial $t(\lambda)$.

(ii) If $d(\lambda)$ is a g.c.d. of $a(\lambda)$ and $b(\lambda)$, deduce from Theorem 1.11 that there exists a unique pair of polynomials $x(\lambda)$, $y(\lambda)$ such that

$$\delta x < \delta b - \delta d, \qquad \delta y < \delta a - \delta d, \qquad ax + by = d$$

1.30 Let $a(\lambda)$, $b(\lambda)$ be the polynomials in (1.1) and (1.2), respectively, and let p be any positive integer.

(i) Prove from Theorem 1.3 that

$$R[a(\lambda^p), b(\lambda^p)] = \{R[a(\lambda), b(\lambda)]\}^p$$

where $R[a(\lambda), b(\lambda)]$ is the resultant for $a(\lambda)$ and $b(\lambda)$.

(ii) If S is the Sylvester matrix in (1.84), and S(p) denotes the corresponding matrix for $a(\lambda^p)$ and $b(\lambda^p)$, prove that $S(p) = S \otimes I_p$, and hence deduce that det $S(p) = (\det S)^p$.

1.5 BEZOUTIAN MATRIX

Consider again the nth-degree polynomial $a(\lambda)$ in (1.1), but now assume that

$$b(\lambda) = b_0\lambda^n + b_1\lambda^{n-1} + \cdots + b_{n-1}\lambda + b_n \qquad (1.107)$$

Notice that if $\delta b = m < n$, we set

$$b_j = 0, \quad j < n - m \qquad (1.108)$$

in (1.107). Now form the polynomial

$$a(\lambda)b(\mu) - a(\mu)b(\lambda) \qquad (1.109)$$

It is obvious that when $\lambda \equiv \mu$ this vanishes, so by the Remainder Theorem (Section 1.1) the expression (1.109) must have $\lambda - \mu$ as a factor. In other words, the quotient

$$F[a, b] = \frac{a(\lambda)b(\mu) - a(\mu)b(\lambda)}{\lambda - \mu} \qquad (1.110)$$

is itself a polynomial in λ of degree $n - 1$, and each coefficient of the powers of λ will be a polynomial in μ of degree $n - 1$. Thus we can write (1.110) in the form

$$F[a, b] = \sum_{i=1}^{n} \sum_{j=1}^{n} z_{ij}\lambda^{i-1}\mu^{j-1} \qquad (1.111)$$

and the $n \times n$ matrix $Z = [z_{ij}]$ defined by (1.111) is the *bezoutian matrix* (named after Bézout) associated with $a(\lambda)$ and $b(\lambda)$. From (1.110) it can be seen that $F[a, b]$ is symmetric in λ and μ, which implies that Z is a symmetric matrix.

Example 15 Let $n = 3$. Direct expansion of (1.110) and comparison of terms in (1.111) reveals that

$$Z = \begin{bmatrix} |a_2b_3| & |a_1b_3| & |a_0b_3| \\ |a_1b_3| & |a_0b_3| + |a_1b_2| & |a_0b_2| \\ |a_0b_3| & |a_0b_2| & |a_0b_1| \end{bmatrix} \qquad (1.112)$$

where

$$|a_rb_s| \overset{\triangle}{=} \begin{vmatrix} a_r & b_r \\ a_s & b_s \end{vmatrix} = a_rb_s - a_sb_r \qquad (1.113)$$

If in this example $\delta b = 2$, say, (1.108) would imply that $b_0 = 0$, so for example $|a_0b_3| = a_0b_3 - a_3b_0 = a_0b_3$.

By carrying out the same procedure in general, it is found that the elements of Z are given by

$$z_{ij} = |a_{n-i-j+1}b_n| + |a_{n-i-j+2}b_{n-1}| + \cdots + |a_{n-i-j+k+1}b_{n-k}|$$

$$(1.114)$$

where

$$k = \min(i - 1, j - 1) \qquad (1.115)$$

In particular, when $i = n$, then (1.115) gives $k = j - 1$, so from (1.114) the elements of the last row (and, by symmetry, of the last column) of Z are

$$z_{nj} = z_{jn} = |a_0b_{n-j+1}|, \qquad j = 1, 2, \ldots, n \qquad (1.116)$$

Note that if $\delta b < n$, the formula (1.116) still applies, subject to (1.108).

An alternative way of constructing Z is obtained using the companion matrix C of $a(\lambda)$ in (1.26) (note that from now on we assume that $a_0 = 1$). In fact, the rows $z_1, z_2, \ldots, z_n$ of Z can be constructed from the recurrence formula

$$z_i = z_{i+1}C + a_{n-i}z_n, \qquad i = n - 1, n - 2, \ldots, 2, 1 \qquad (1.117)$$

where from (1.116)

$$z_n = \begin{bmatrix} |a_0b_n|, & |a_0b_{n-1}|, & \ldots, & |a_0b_1| \end{bmatrix} \qquad (1.118)$$

To prove (1.117), consider the jth elements on both sides of that equation:

$$z_{ij} - (z_{i+1}C)_j - a_{n-i}z_{nj} = z_{ij} - z_{i+1,j-1} + a_{n-j+1}z_{i+1,n}$$

$$- a_{n-i}z_{nj} \quad [\text{using } (1.26)]$$

$$= z_{ij} - z_{i+1,j-1} + a_{n-j+1} \left| a_o b_{n-i} \right|$$

$$- a_{n-i} \left| a_o b_{n-j+1} \right|$$

$$= z_{ij} - z_{i+1,j-1} + \left| a_{n-j+1} b_{n-i} \right|$$

$$(1.119)$$

using (1.113). On substituting (1.114) into (1.119) there are three separate cases to consider according to whether i is less than, equal to, or greater than j - 1, but in each it is routine to verify that the expression (1.119) is identically zero, thus verifying (1.117).

The recurrence formula (1.117) is important because it enables us to obtain a simple relationship between the bezoutian matrix and the companion form of the matrix.

Theorem 1.12 Let Z be the bezoutian matrix associated with the nth-degree polynomials $a(\lambda)$ in (1.1) and $b(\lambda)$ in (1.107). Then

$$Z = Tb(C) \qquad\qquad (1.120)$$

where T is the triangular matrix in (1.35), and C is the companion matrix (1.26). Furthermore, the degree of the g.c.d. of $a(\lambda)$ and $b(\lambda)$ is equal to n - rank Z.

Proof. As in Section 1.2, denote the rows of b(C) defined in (1.37) by $r_1, \ldots, r_n$. Since $\delta b = \delta a$, r_1 is given by (1.51), which is seen to be identical to (1.118), since we have taken $a_o = 1$. If we denote the rows of Tb(C) by $\rho_1, \ldots, \rho_n$, then using the form of T in (1.35) shows that

$$\rho_n = r_1 \quad (\equiv z_n) \qquad\qquad (1.121)$$

$$\rho_i = a_{n-i}r_1 + a_{n-i-1}r_2 + \cdots + r_{n-i+1}, \qquad i = 1, 2, \ldots, n-1$$
$$(1.122)$$

Because of the recurrence formula (1.42) between the rows of $b(C)$, we can write (1.122) as

$$\rho_i = a_{n-i}r_1 + (a_{n-i-1}r_1 C + a_{n-i-2}r_2 C + \cdots + r_{n-i}C) \qquad (1.123)$$

and by replacing i by $i + 1$ in (1.122), the term within parentheses in (1.123) is seen to be $\rho_{i+1}C$. Thus (1.123) can be written as

$$\rho_i = a_{n-i}\rho_n + \rho_{i+1}C \qquad (1.124)$$

where we have substituted for r_1 from (1.121). Equation (1.124) shows that the rows of $Tb(C)$ satisfy the same relationship as do those of Z in (1.117), and since $\rho_n \equiv z_n$, this establishes (1.120).

The second part of the theorem follows immediately from Theorem 1.4 and the fact that T is nonsingular, so from (1.120) rank Z = rank $b(C)$. ∎

Remarks on Theorem 1.12   (1) In particular, it follows that $a(\lambda)$ and $b(\lambda)$ are relatively prime if and only if Z is nonsingular. Indeed, since det $T = \varepsilon(n)$ [defined in (1.22)], equations (1.45) and (1.120) show that

$$\det Z = \varepsilon(n) R(a, b) \qquad (1.125)$$

(2) We now have three different ways of obtaining the resultant of two polynomials, or the degree of their g.c.d.: (i) the companion matrix expression of Theorems 1.3 and 1.4, (ii) the Sylvester matrix of Theorems 1.7 and 1.8, (iii) the bezoutian matrix. An advantage of methods (i) and (iii) is that only $n \times n$ matrices are involved, as compared with $(m + n) \times (m + n)$ for method (ii), although this is partially offset by the fact that the Sylvester matrix can be written down directly.

(3) Equation (1.120) can be interpreted as showing that T is a *left symmetrizer* of $b(C)$, since premultiplication of $b(C)$ by T produces a symmetric matrix.

Example 16   Return to the problem considered in Example 5, where
$a(\lambda)$ is given in (1.6) and $b(\lambda)$ in (1.14). Substitution of the
values of the coefficients of $a(\lambda)$ and $b(\lambda)$ into (1.112), with
$b_0 = 0$, gives

$$Z = \begin{bmatrix} 16 & -13 & 1 \\ -13 & 8 & -3 \\ 1 & -3 & 2 \end{bmatrix} \qquad (1.126)$$

and det $Z = -156$. Since from (1.22) $\varepsilon(n) = -1$, (1.125) agrees with
the value of $R(a, b)$ found before in Example 5. Also, (1.35) is
here

$$T = \begin{bmatrix} 1 & -3 & 1 \\ -3 & 1 & 0 \\ 1 & 0 & 0 \end{bmatrix}$$

and premultiplication of $b(C)$ in (1.47) by T produces (1.126),
confirming (1.120) for this example.

(4) We have shown how the Sylvester and bezoutian matrix
results can be deduced from the companion matrix approach, using
Theorems 1.7 and 1.12, respectively. The interrelationships are
completed by the following:

Theorem 1.13   The bezoutian matrix Z defined by (1.111), and the
Sylvester matrix S defined in (1.85), with $m = n$, are related by

$$\begin{bmatrix} I_n & 0 \\ -T_3 & T_1 \end{bmatrix} S = \begin{matrix} & n & n \\ & \begin{bmatrix} T_1 & T_2 \\ 0 & ZJ \end{bmatrix} & \begin{matrix} n \\ n \end{matrix} \end{matrix} \qquad (1.127)$$

Proof. Consider (1.86), where C is the companion matrix (1.26),
and K is obtained from (1.89) as $K = -T_3 T_1^{-1}$. Substitute $b(C) =$
$T^{-1}Z$ from (1.120) into (1.86), and premultiply both sides of the
latter as follows:

$$
\begin{bmatrix} I & 0 \\ 0 & TJ \end{bmatrix} \begin{bmatrix} I & 0 \\ K & I \end{bmatrix} S = \begin{bmatrix} I & 0 \\ 0 & TJ \end{bmatrix} \begin{bmatrix} T_1 & T_2 \\ 0 & JT^{-1}ZJ \end{bmatrix}
$$

This reduces to

$$
\begin{bmatrix} I & 0 \\ TJK & TJ \end{bmatrix} S = \begin{bmatrix} T_1 & T_2 \\ 0 & ZJ \end{bmatrix} \tag{1.128}
$$

on using $J^2 = I$. In order to show that (1.128) is the same as the desired result (1.127), we need to prove that $TJ = T_1$, and $TJK = -T_3$. Since T_1 is the $n \times n$ leading principal submatrix of S in (1.84) (with $m = n$), comparison with T in (1.35) reveals that T_1 is just T with its column reversed in order, i.e., $T_1 = TJ$, as required. In addition, together with (1.89), this shows that

$$
TJK = -T_1 T_3 T_1^{-1} \tag{1.129}
$$

Now, T_3 consists of the last n rows and first n columns of S in (1.84) (with $m = n$). Therefore, both T_1 and T_3 have upper triangular Toeplitz form (defined in Problem 1.16), and it is easy to verify (Problem 1.32) that $T_1 T_3 = T_3 T_1$. This reduces (1.129) to $TJK = -T_3$. ∎

Remarks on Theorem 1.13   (1) Notice that ZJ, rather than Z, occurs in (1.127). This matrix is Z with the order of its columns reversed, and indeed some authors use this as the definition of bezoutian matrix, although it is then no longer symmetric (see Problem 1.29).

    (2) Expanding (1.127) shows that

$$
Z = T_1(T_4 J) - T_3(T_2 J) \tag{1.130}
$$

which gives an alternative way of constructing Z.

Example 17   When n = 3, (1.127) is

$$
\begin{bmatrix}
1 & 0 & 0 & 0 & 0 & 0 \\
0 & 1 & 0 & 0 & 0 & 0 \\
0 & 0 & 1 & 0 & 0 & 0 \\
-b_0 & -b_1 & -b_2 & 1 & a_1 & a_2 \\
0 & -b_0 & -b_1 & 0 & 1 & a_1 \\
0 & 0 & -b_0 & 0 & 0 & 1
\end{bmatrix}
\begin{bmatrix}
1 & a_1 & a_2 & a_3 & 0 & 0 \\
0 & 1 & a_1 & a_2 & a_3 & 0 \\
0 & 0 & 1 & a_1 & a_2 & a_3 \\
b_0 & b_1 & b_2 & b_3 & 0 & 0 \\
0 & b_0 & b_1 & b_2 & b_3 & 0 \\
0 & 0 & b_0 & b_1 & b_2 & b_3
\end{bmatrix}
$$

$$
=
\left[
\begin{array}{ccc:ccc}
1 & a_1 & a_2 & a_3 & 0 & 0 \\
0 & 1 & a_1 & a_2 & a_3 & 0 \\
0 & 0 & 1 & a_1 & a_2 & a_3 \\
\hdashline
 & 0 & & & ZJ &
\end{array}
\right]
$$

where Z is the matrix in (1.112).   In addition (1.130) gives

$$
Z =
\begin{bmatrix}
a_0 & a_1 & a_2 \\
0 & a_0 & a_1 \\
0 & 0 & a_0
\end{bmatrix}
\begin{bmatrix}
0 & 0 & b_3 \\
0 & b_3 & b_2 \\
b_3 & b_2 & b_1
\end{bmatrix}
-
\begin{bmatrix}
b_0 & b_1 & b_2 \\
0 & b_0 & b_1 \\
0 & 0 & b_0
\end{bmatrix}
\begin{bmatrix}
0 & 0 & a_3 \\
0 & a_3 & a_2 \\
a_3 & a_2 & a_1
\end{bmatrix}
$$

$$(1.131)$$

A further interesting consequence of Theorem 1.12 can now be derived:

Theorem 1.14   When $a(\lambda)$ and $b(\lambda)$ are relatively prime, so that their associated bezoutian matrix Z defined by (1.111) is non-singular, then Z^{-1} is *orthosymmetric* (i.e., Z^{-1} has equal elements along each line parallel to the secondary diagonal).

Proof.   From (1.120)

$$
\begin{aligned}
C^T Z &= C^T Tb(C) \\
&= TCb(C), \quad \text{on using (1.36)} \\
&= Tb(C)C \\
&= ZC
\end{aligned}
$$

$$(1.132)$$

the penultimate step following from the fact that C commutes with
b(C). When Z is nonsingular, pre- and postmultiplication of
(1.132) by Z^{-1} gives $CZ^{-1} = Z^{-1}C^T$, and it then follows from the
first part of Problem 1.16 that Z^{-1} is orthosymmetric. ∎

Problems

1.31 Prove that if $Z_1 = ZJ$, where Z is the bezoutian matrix, then
$JZ_1J = Z_1^T$ (any matrix X satisfying the condition $J_n XJ_n = X^T$ is
termed *persymmetric*). Prove also that when Z_1 is nonsingular
its inverse is persymmetric. Use Theorem 1.14 and Problem
1.16 to show that Z_1^{-1} is a Toeplitz matrix.

1.32 Let $A = [a_{ij}]$, $B = [b_{ij}]$ be two n × n upper triangular
matrices defined by

$$a_{ij} = a_{j-i}, \quad b_{ij} = b_{j-i}, \quad i, j = 1, 2, \ldots, n$$

with $a_k = 0$, $b_k = 0$ for k < 0. (Notice that A and B have
Toeplitz form.) Prove that AB = BA.

1.33 Use (1.90) and (1.120) to show that the bezoutian matrix Z and
Schur complement (S/T_1) are related by $Z = T_1(S/T_1)J$.

1.6 RECURSIVE ALGORITHMS

1.6.1 The Routh Array

The oldest method for determining the g.c.d. of two polynomials is
Euclid's algorithm: for $a(\lambda)$ and $b(\lambda)$ defined in (1.1) and (1.2),
with m ≤ n, perform the sequence of divisions

$$f_i(\lambda) = f_{i+1}(\lambda)q_{i+1}(\lambda) + f_{i+2}(\lambda), \quad i = 0, 1, 2, \ldots \quad (1.133)$$

where $f_0(\lambda) \equiv a(\lambda)$ and $f_1(\lambda) \equiv b(\lambda)$ [the case i = 0 is set out in
(1.3)]. The polynomials q_i, f_{i+2} are unique, and $\delta(f_{i+2}) < \delta(f_{i+1})$.
The set of polynomials $c_2 f_2$, $c_3 f_3$, $c_4 f_4$, $\ldots$, where the c_i are
arbitrary nonzero constants, is called a *polynomial remainder
sequence (p.r.s.)*. If k is the smallest integer for which
$f_k(\lambda) \equiv 0$, then $f_{k-1}(\lambda)$ is a g.c.d. of $a(\lambda)$ and $b(\lambda)$. The scheme
is very well known, and further details can be found in many stan-

dard textbooks. It is interesting, however, to consider a
relationship with Sylvester's matrix S in (1.84), in the case when
$a(\lambda)$ and $b(\lambda)$ are relatively prime. Suppose that S is transformed
using elementary row operations to give an upper triangular matrix
$P = [r_{i,j-i+1}]$, i, j = 1, 2, 3, ..., with $r_{ik} = 0$, k $\leqslant$ 0. Then it
can be shown that the rows of P are given by

$$r_{ij} = - \begin{vmatrix} r_{i-2,1} & r_{i-2,j+1} \\ r_{i-1,1} & r_{i-1,j+1} \end{vmatrix} \Bigg/ r_{i-1,1}, \quad i = 2, 3, \ldots \quad (1.134)$$

where

$$r_{oj} \overset{\Delta}{=} a_{j-1}, \quad r_{1j} \overset{\Delta}{=} b_{j-1}, \quad j = 1, 2, 3, \ldots \quad (1.135)$$

It is assumed that P is *regular*, i.e., all r_{i1} are nonzero.
Furthermore, alternate rows in P provide a p.r.s., namely

$$g_i(\lambda) = r_{2i-1,1}\lambda^{n-i} + r_{2i-1,2}\lambda^{n-i-1} + \cdots + r_{2i-n,n-i+1},$$

$$i = 2, 3, \ldots \quad (1.136)$$

and the euclidean remainders in (1.133) are given by

$$f_2(\lambda) = \frac{r_{21}}{r_{11}} g_2(\lambda)$$

$$f_i(\lambda) = \frac{r_{2i-5,1}r_{2i-2,1}}{r_{2i-4,1}r_{2i-3,1}} g_i(\lambda), \quad i > 2 \quad (1.137)$$

apart from possible differences in sign. The scheme defined by
(1.134) for computing the r_{ij} can be set out in the form of a
tabular array shown in Table 1.2. Each row is constructed from
the *preceding two rows*, by forming all 2 × 2 determinants
involving the first column, shown within dashed lines in Table 1.2.

Table 1.2   Routh's array

$\{r_{0j}\}$	a_0		a_1	a_2		.	.		.
$\{r_{1j}\}$	b_0		b_1	b_2		.	.		.
.		.		.	.		.		.
$\{r_{i-2,j}\}$	$\boxed{r_{i-2,1}}$		.	.		.	.		$\boxed{r_{i-2,j+1}}$
$\{r_{i-1,j}\}$	$\boxed{r_{i-1,1}}$		.	.		.	.		$\boxed{r_{i-1,j+1}}$
$\{r_{ij}\}$	r_{i1}		.	.	.	r_{ij}			.
	↑					↑			↑
	column 1					column j			column j + 1

Example 18   Consider the polynomials

$$a(\lambda) = \lambda^4 - 3\lambda^3 + 5\lambda^2 + 7\lambda + 2$$

$$b(\lambda) = 2\lambda^3 + 5\lambda^2 + \lambda + 3$$

It is easy to perform (1.133) to obtain the euclidean remainders

$$f_2(\lambda) = \frac{73}{4}\lambda^2 + \frac{33}{4}\lambda + \frac{41}{4}, \quad f_3(\lambda) = -\frac{10,524}{5329}\lambda + \frac{3728}{5329}$$

The first six rows of the array defined by (1.134) and (1.135) are
given below. The arithmetical details have been set out in full
below only for the third and fourth rows, after which the procedure
should be clear. From (1.136) we have

$$g_2(\lambda) = \frac{73}{11}\lambda^2 + 3\lambda + \frac{41}{11}, \quad g_3(\lambda) = -\frac{2631}{510}\lambda + \frac{932}{510}$$

and it is then easily seen that $f_2 = (11/4)g_2$, $f_3 = (2040/5329)g_2$,
agreeing with (1.137).

$$r_{oj} \qquad 1 \qquad\qquad -3 \qquad\qquad 5 \qquad\qquad 7 \qquad\qquad 2$$

$$r_{1j} \qquad 2 \qquad\qquad 5 \qquad\qquad 1 \qquad\qquad 3$$

$$r_{2j} \; -\begin{vmatrix} 1 & -3 \\ 2 & 5 \end{vmatrix} \Bigg/ 2, \quad -\begin{vmatrix} 1 & 5 \\ 2 & 1 \end{vmatrix} \Bigg/ 2, \quad -\begin{vmatrix} 1 & 7 \\ 2 & 3 \end{vmatrix} \Bigg/ 2, \quad -\begin{vmatrix} 1 & 2 \\ 2 & 0 \end{vmatrix} \Bigg/ 2$$

$$= \frac{-11}{2} \qquad\qquad = \frac{9}{2} \qquad\qquad = \frac{11}{2} \qquad\qquad = 2$$

$$r_{3j} \; -\begin{vmatrix} 2 & 5 \\ \frac{-11}{2} & 9 \end{vmatrix} \Bigg/ \left(\frac{-11}{2}\right), \quad -\begin{vmatrix} 2 & 1 \\ \frac{-11}{2} & \frac{11}{2} \end{vmatrix} \Bigg/ \left(\frac{-11}{2}\right), \quad -\begin{vmatrix} 2 & 3 \\ \frac{-11}{2} & 2 \end{vmatrix} \Bigg/ \left(\frac{-11}{2}\right)$$

$$= \frac{73}{11} \qquad\qquad\qquad = 3 \qquad\qquad\qquad = \frac{41}{11}$$

$$r_{4j} \quad \frac{510}{73} \qquad \frac{627}{73} \qquad 2$$

$$r_{5j} \quad -\frac{2631}{510} \qquad \frac{932}{510}$$

We saw in Theorem 1.9 that transforming S into row echelon form gives a g.c.d. of $a(\lambda)$ and $b(\lambda)$. Having now observed that the elements in this echelon form can be generated via (1.134), we can therefore state:

Theorem 1.15 Construct the *Routh array* defined by (1.134) and (1.135), associated with the polynomials $a(\lambda)$ in (1.1) and $b(\lambda)$ in (1.2).

(i) If at some stage a complete row of zeros is obtained, then the preceding row gives the coefficients of a g.c.d. of $a(\lambda)$ and $b(\lambda)$.

(ii) If the first, but not all the elements in a row are zero, the row is shifted to the left until a nonzero element appears in the first position. A g.c.d. is then obtained by continuing as in (i).

Example 19   Consider the two polynomials in Example 10c, given in (1.74) and (1.75).   The array (1.134), (1.135) is

r_{oj} 1   1   -9   -5   16   12   [from (1.74)]

r_{1j} 1   -3   1   3   -2   0   [from (1.75)]

r_{2j} 4   -10 -8   18   12

r_{3j} $-\dfrac{1}{2}$ 3 $-\dfrac{3}{2}$ -5

r_{4j} 14   -20 -22 12

r_{5j} $\dfrac{16}{7}$ $\dfrac{-16}{7}$ $\dfrac{-32}{7}$

r_{6j} -6   6   12

r_{7j} 0   0

From part (i) of the theorem we conclude that a g.c.d. is

$$d(\lambda) = -6\lambda^2 + 6\lambda + 12 = -6(\lambda^2 - \lambda - 2) \tag{1.138}$$

agreeing with the result in Example 10c, using the companion matrix method.

Remarks on Theorem 1.15   (1) The attribution to Routh is because of his application to stability of polynomials (see Theorem 3.3).

(2) It follows that if the array is regular, and thus terminates with a single nonzero element, then $a(\lambda)$ and $b(\lambda)$ are relatively prime.   In particular, if $m = n$ or $n - 1$, then in view of the equivalence of the matrices S and P it is easy to show that the resultant $R(a, b)$ is equal to det $P = r_{11}r_{21} \cdots r_{m+n,1}$ (apart from a possible difference in sign).

(3) It should be clear that any row of the array can be multiplied by an arbitrary nonzero scaling factor before the next row is calculated, without doing more than introduce an extra constant factor into the form of the resulting g.c.d.   In particular, it is not necessary for the purposes of Theorem 1.15 to either carry out the divisions by $r_{i-1,1}$ in (1.134), or to use the factor -1.   These have been retained, however, for uniformity

with the stability result, mentioned above.

Example 20   Using (1.134), the third row of the Routh array for

$$a(\lambda) = \lambda^4 + 2\lambda^3 + \lambda^2 + 14\lambda + 6, \quad b(\lambda) = 2\lambda^3 + 4\lambda^2 + \lambda + 21$$

is $\{r_{2j}\} = \left\{0, \frac{1}{2}, \frac{7}{2}, 6\right\}$. Following part (ii) of the theorem, we discard the initial zero, and for convenience we also remove the factor $\left(\frac{1}{2}\right)$, to continue the array as follows:

r_{1j}	2	4	1	21
r_{2j}	1	7	12	
r_{3j}	-10	-23	21	
r_{4j}	47	141		(removing a factor 1/10)
r_{5j}	$\frac{329}{47}$	$\frac{987}{47}$		
r_{6j}	0			

From part (i) of the theorem we conclude that a g.c.d. of $a(\lambda)$ and $b(\lambda)$ is

$$\frac{329}{47}\lambda + \frac{987}{47} = \frac{329}{47}(\lambda + 3)$$

(4) Suppose that $a(\lambda)$ and $b(\lambda)$ have degrees n, n - 1, respectively, and that their g.c.d. is $d(\lambda)$. The quotient polynomials $\hat{a}(\lambda) = a/d$, $\hat{b}(\lambda) = b/d$ can also be computed from the array (assumed regular), by defining

$$h_i = \frac{r_{i-1,1}}{r_{i1}}, \quad i = 1, 2, 3, \ldots, 2k \tag{1.139}$$

$$p_{2k+1,1} = 1, \quad p_{i1} = h_i p_{i+1,1}, \quad i = 2k, 2k-1, \ldots, 1 \tag{1.140}$$

$$p_{ij} = p_{i+2,j-1} + h_i p_{i+1,j}, \quad i = 2k-1, \ldots, 1 \tag{1.141}$$

where the number of nonzero h's is 2k (i.e., the last nonvanishing row in the array is $\{r_{2k,j}\}$). Then

$$\hat{a}(\lambda) = p_{11}\lambda^k + p_{12}\lambda^{k-1} + \cdots + p_{1,k+1}$$

$$\hat{b}(\lambda) = p_{21}\lambda^{k-1} + \cdots \ p_{2k} \qquad\qquad (1.142)$$

Example 21 Return to Example 19, where we see from the array that k = 3. From (1.139) we obtain

$$h_1 = 1, \quad h_2 = 1/4, \quad h_3 = -8, \quad h_4 = -1/28, \quad h_5 = 49/8,$$

$$h_6 = -8/21$$

whence from (1.140)

$$p_{71} = 1, \quad p_{61} = -\frac{8}{21}, \quad p_{51} = -\frac{7}{3}, \quad p_{41} = \frac{1}{12}, \quad p_{31} = -\frac{2}{3},$$

$$p_{21} = -\frac{1}{6}, \quad p_{11} - = \frac{1}{6}$$

We then use (1.141) to compute

$$p_{52} = 1, \quad p_{42} = p_{61} + h_4 p_{52} = -\frac{5}{12}, \quad p_{32} = p_{51} + h_3 p_{42} = 1$$

and so on, to finally obtain from (1.142)

$$\hat{a}(\lambda) = -\frac{1}{6}(\lambda^3 + 2\lambda^2 - 5\lambda - 6), \quad \hat{b}(\lambda) = -\frac{1}{6}(\lambda^2 - 2\lambda + 1)$$

The reader can verify that these are the correct quotients corresponding to $d(\lambda)$ in (1.138).

1.6.2 An Alternative Array

We now turn to an alternative form of iterative procedure, which is best introduced by considering a specific case, say m = n = 3 when

$$a(\lambda) = \sum_{i=0}^{3} a_i \lambda^{n-i}, \quad b(\lambda) = \sum_{i=0}^{3} b_i \lambda^{n-i}$$

Construct two second-degree polynomials

$$c_2(\lambda) = -b_o a(\lambda) + a_o b(\lambda)$$

$$= c_{21}\lambda^2 + c_{22}\lambda + c_{23}$$

$$d_2(\lambda) = [b_3 a(\lambda) - a_3 b(\lambda)]/\lambda$$

$$= d_{21}\lambda^2 + d_{22}\lambda + d_{23}$$

It is easy to verify that

$$c_{2i} = \begin{vmatrix} a_o & a_i \\ b_o & b_i \end{vmatrix}, \quad d_{2i} = \begin{vmatrix} a_{i-1} & a_3 \\ b_{i-1} & b_3 \end{vmatrix} \qquad (1.143)$$

The process could then be repeated to construct two first-degree polynomials. In general, for two nth-degree polynomials, we can set out the calculations in the form of an array having rows alternately labeled $\{c_{ij}\}$, $\{d_{ij}\}$, where $c_{1i} \triangleq a_{i-1}$, $d_{1i} \triangleq b_{i-1}$, and where each pair of rows is obtained by forming all the minors involving the first and last columns of the *preceding pair*, i.e.,

$$c_{ji} = \begin{vmatrix} c_{j-1,1} & c_{j-1,i+1} \\ d_{j-1,1} & d_{j-1,i+1} \end{vmatrix}, \quad d_{ji} = \begin{vmatrix} c_{j-1,i} & c_{j-1,n-j+3} \\ d_{j-1,i} & d_{j-1,n-j+3} \end{vmatrix}$$

$$(1.144)$$

for $i = 1, 2, \ldots, n - j + 2$; $j = 2, 3, \ldots, n + 1$. For $n = 3$, $j = 2$, the expressions (1.144) reduce to (1.143). When $a(\lambda)$ and $b(\lambda)$ are relatively prime the last element in the array $c_{n+1,1}$ is equal to

$$(d_{21})^{n-2}(d_{31})^{n-3} \cdots (d_{n-2,1})^2 d_{n-1,1} R(a, b)$$

Otherwise, an identically zero row is obtained, and as for the Routh array, the preceding nonzero row gives a g.c.d.

Example 22   We repeat Example 10b, where $a(\lambda)$ is the third-degree polynomial in (1.6), and $b(\lambda)$ is given in (1.73). Since the latter has degree two, it is first necessary to multiply it by λ so that $d_{11} = b_o \neq 0$; this does not affect the g.c.d. of $a(\lambda)$ and $b(\lambda)$. The array defined by (1.143) is

$$c_{1i} \quad 1 \quad -3 \quad 1 \quad 5 \Big\} \quad \text{[from (1.6)]}$$

$$d_{1i} \quad 1 \quad 3 \quad 2 \quad 0 \Big\} \quad \text{[from (1.73)]}$$

$$c_{2i} \quad 6 \quad 1 \quad -5 \Big\}$$

$$d_{2i} \quad -5 \quad -15 \quad -10 \Big\}$$

$$c_{3i} \quad -85 \quad -85 \Big\}$$

$$d_{3i} \quad -85 \quad -85 \Big\}$$

$$c_{4i} \quad 0$$

A g.c.d. is thus $-85\lambda - 85 = -85(\lambda + 1)$, agreeing with what we found before.

Any *pair* of zero first column elements is dealt with as for the Routh array by shifting both rows to the left. The reader would soon find on tackling problems by each of the two tabular methods that an irregular Routh array may not produce an irregularity for the second type of array; and conversely. This second type of array is not generally used to compute a g.c.d., partly because the intermediate polynomials generated do not form a p.r.s. However, as we shall see in Theorem 3.12, it is important because of its application to stability problems.

1.6.3 Coefficient Growth

We now consider polynomials defined over a unique factorization domain D. This includes cases of practical importance when the coefficients a_i, b_i in (1.1) and (1.2) are integers, or polynomials in one or more variables. An obvious approach would be to define a Routh array $\{s_{ij}\}$ *without divisions* which has the same two rows as before, namely

$$s_{0j} = a_{j-1}, \quad s_{1j} = b_{j-1}, \quad j = 1, 2, 3, \ldots \qquad (1.145)$$

and subsequent rows defined by

$$s_{ij} = - \begin{vmatrix} s_{i-2,1} & s_{i-2,j+1} \\ s_{i-1,1} & s_{i-1,j+1} \end{vmatrix}, \quad j = 1, 2, 3, \ldots \qquad (1.146)$$

Comparison of (1.146) with (1.134) shows that the only difference is the absence of division by the first column element. All elements generated by (1.146) remain within D, so the array can be termed *fraction free*.

<u>Example 23</u> We construct the array defined by (1.145) and (1.146) for the polynomials

$$a(\lambda) = \lambda^4 + 5\lambda^3 + 2\lambda^2 + 4\lambda + 3, \quad b(\lambda) = 4\lambda^3 + 3\lambda^2 + \lambda + 2$$

$$(1.147)$$

s_{0j}	1		5	2	4	3
s_{1j}	4		3	1	2	
s_{2j}	17		7	14	12	
s_{3j}	23		-39	-14		
s_{4j}	824		560	276		
s_{5j}	$-45,016$		$-17,884$			
s_{6j}	$-104,772,544$	$-12,424,416$				
s_{7j}	$-372,006,533,760$					

Example 23 illustrates dramatically the phenomenon of rapid coefficient growth, which presents severe computational difficulties. One suggestion for reducing the growth would be to divide the elements of each row, after generation, by their g.c.d., but this involves considerable extra effort. We can be more precise about the nature of the growth: Regarding the s_{ij} as polynomials in the coefficients a_k and b_k, with total degree ν_i, then from (1.146) we have

$$\nu_0 = 1, \quad \nu_1 = 1, \quad \nu_i = \nu_{i-1} + \nu_{i-2}, \quad i = 2, 3, \ldots$$

Thus $\{\nu_i\}$ is the *Fibonacci sequence*, which is known to grow exponentially. We state without proof a recently derived result, which is the best possible.

Theorem 1.16 Minimal coefficient growth v_i = i is achieved by the
optimal fraction-free Routh array, for which $t_{oj} = a_{j-1}$, $t_{1j} = b_{j-1}$,
j = 1, 2, 3, ..., and

$$t_{ij} = -\frac{1}{d_i} \begin{vmatrix} t_{i-2,1} & t_{i-2,j+1} \\ t_{i-1,1} & t_{i-1,j+1} \end{vmatrix} \qquad (1.148)$$

where d_2 = 1, d_3 = 1, and

$$d_i = t_{i-3,1}, \qquad i \geq 4 \qquad (1.149)$$

Example 24 We construct the array defined by (1.148) and (1.149)
for the polynomials (1.147) in Example 23. The first four rows are
unaltered, and the remaining rows are

t_{4j}	206	140	69
t_{5j}	-662	-263	
t_{6j}	-1674	-1986	
t_{7j}	-4245		

showing a remarkable reduction in the magnitudes of the elements,
as compared with the array in Example 23.

Remarks on Theorem 1.16 (1) The elements t_{ij} can be shown equal to
the values of certain minors of the Sylvester matrix. This property will be explored in more detail in Remark 3 on Theorem 3.4.

(2) An alternative array with larger, but still *linear*, growth
is given in Problem 1.39.

The scheme of Section 1.6.2 produces slightly smaller
coefficient growth than the standard Routh array, but still of
exponential nature. The reader may care to verify that (1.144)
applied to the polynomials in Example 23 leads to a final entry
c_{41} = 950,880. The question of optimality has not been investigated for this case.

1.6.4   The Equation ax + by = 1

We can now return to the diophantine equation encountered in
Section 1.4, namely

$$a(\lambda)x(\lambda) + b(\lambda)y(\lambda) = 1 \qquad\qquad (1.150)$$

and see how Euclid's algorithm leads to a solution $x(\lambda)$, $y(\lambda)$, when
$a(\lambda)$ and $b(\lambda)$ are relatively prime (assume that $\delta b < \delta a$). The
first step $i = 0$ in (1.133) is $a = bq_1 + f_2$, which we can write as

$$f_2 = aw_1 + bv_1, \quad w_1 = 1, \quad v_1 = -q_1 \qquad\qquad (1.151)$$

Similarly, setting $i = 1$ in (1.133), and using (1.151), produces

$$f_3 = aw_2 + bv_2, \quad w_2 = -q_2 w_1, \quad v_2 = 1 - q_2 v_1$$

Continuing this procedure shows that in general

$$f_{i+1} = aw_i + bv_i, \quad i = 1, 2, 3, \ldots \qquad\qquad (1.152)$$

where

$$w_i = \begin{vmatrix} w_{i-2} & w_{i-1} \\ q_i & 1 \end{vmatrix}, \quad v_i = \begin{vmatrix} v_{i-2} & v_{i-1} \\ q_i & 1 \end{vmatrix}, \quad i = 2, 3, \ldots$$

$$(1.153)$$

subject to

$$w_0 = 0, \quad w_1 = 1, \quad v_0 = 1, \quad v_1 = -q_1 \qquad\qquad (1.154)$$

The algorithm terminates when $i = p$, say, with f_{p+1} a nonzero
constant. Division by f_{p+1} of both sides of (1.152), with $i = p$,
shows that a solution of (1.150) is

$$x(\lambda) = \frac{w_p(\lambda)}{f_{p+1}}, \quad y(\lambda) = \frac{v_p(\lambda)}{f_{p+1}} \qquad\qquad (1.155)$$

Example 25   We solve (1.150) with $a(\lambda)$, $b(\lambda)$ the polynomials in
(1.6) and (1.14), respectively. These were found to be relatively
prime in Example 3b. Performing the divisions in (1.133), we
obtain

$$q_1 = \frac{1}{2}\lambda - \frac{3}{4}, \quad q_2 = -\frac{8}{7}\lambda - \frac{100}{49}, \quad f_3 = \frac{624}{9}$$

Hence from (1.155), with $p = 2$, the desired solution of (1.150) is

$$x(\lambda) = \frac{w_2}{f_3} = \frac{-q_2(\lambda)}{f_3}, \quad \text{by (1.153) and (1.154)}$$

$$= (14\lambda + 25)/156$$

$$y(\lambda) = \frac{v_2}{f_3} = \frac{1 + q_1 q_2}{f_3}, \quad \text{by (1.153) and (1.154)}$$

$$= -(7\lambda^2 + 2\lambda - 31)/156$$

The reader can check that $ax + by = 1$, as required.

Recall the fundamental property of Euclid's algorithm, that if a zero remainder is obtained, then the preceding nonzero remainder $f_{k-1}(\lambda)$, say, is a g.c.d. $d(\lambda)$ of $a(\lambda)$ and $b(\lambda)$. Setting $i = k - 2$ in (1.152), this is equivalent to stating that there exist poly- nomials $w_{k-2}(\lambda)$, $v_{k-2}(\lambda)$ such that a g.c.d. $d(\lambda)$ of $a(\lambda)$ and $b(\lambda)$ can be expressed as

$$d(\lambda) = a(\lambda)w_{k-2}(\lambda) + b(\lambda)v_{k-2}(\lambda) \tag{1.156}$$

This fact enables us to obtain the following generalization of Theorem 1.3, thereby nicely rounding off this chapter.

Theorem 1.17 Let A be an $n \times n$ matrix with minimum polynomial $m(\lambda)$. Then $m(\lambda)$ and $b(\lambda)$ in (1.2) are relatively prime if and only if $b(A)$ is nonsingular.

Proof. Suppose first that $m(\lambda)$ and $b(\lambda)$ have a g.c.d. $\alpha(\lambda)$ with $\delta\alpha > 0$, so that

$$m(\lambda) = m_1(\lambda)\alpha(\lambda), \quad b(\lambda) = b_1(\lambda)\alpha(\lambda) \tag{1.157}$$

By definition $m(A) \equiv 0$, so replacing λ by A in (1.157) gives

$$m_1(A)\alpha(A) = 0, \quad b(A) = b_1(A)\alpha(A) \tag{1.158}$$

It follows from the first expression in (1.158) that $\alpha(A)$ is singular. For if not, postmultiplication of (1.158) by $[\alpha(A)]^{-1}$ would give $m_1(A) \equiv 0$, which is invalid, because $\delta m_1 < \delta m$. Hence, taking determinants in the second expression in (1.158) shows that det $b(A) = 0$.

Conversely, suppose that $b(A)$ is singular. We wish to show that if $\beta(\lambda)$ is a g.c.d. of $m(\lambda)$ and $b(\lambda)$, then $\delta\beta > 0$ [i.e., $m(\lambda)$ and $b(\lambda)$ are not relatively prime]. Now, $\beta(\lambda)$ can be expressed as in (1.156); i.e., there exist polynomials $x(\lambda)$, $y(\lambda)$ such that

$$\beta(\lambda) = m(\lambda)x(\lambda) + b(\lambda)y(\lambda) \tag{1.159}$$

Replacing λ by A in the identity (1.159), and using the fact that $m(A) \equiv 0$, produces

$$\beta(A) = b(A)y(A) \tag{1.160}$$

Taking determinants in (1.160) shows that $\beta(A)$ is singular, from which it follows that $\beta(\lambda)$ cannot be a constant. ∎

Remarks on Theorem 1.17   (1) When A is a companion matrix C, then $m(\lambda)$ is precisely the characteristic polynomial of C, and the result reduces to Theorem 1.3.

(2) The proof given holds for polynomials over *any* field, not just that of complex numbers.

Problems

1.34 Repeat Problem 1.21 using Theorem 1.15.

1.35 Show that $a(\lambda) = \lambda^4 + 4\lambda^3 + 4\lambda^2 + 1$ has no repeated roots by applying Routh's array to $a(\lambda)$ and $a'(\lambda)$.

1.36 Repeat Problem 1.21 using the tabular array defined in (1.144).

1.37 Repeat Problem 1.21(ii) using the Routh array without divisions in (1.146), and the optimal array in (1.148).

1.38 Prove by induction that the elements r_{ij} in the standard Routh array (1.134) are related to the optimal array (1.148) by $r_{ij} = t_{ij}/t_{i-1,1}$, $i \geq 2$.

1.39 Another fraction-free array can be defined by $u_{0j} = a_{j-1}$,
$u_{1j} = b_{j-1}$,

$$u_{ij} = -\frac{u_{i-1,1}}{e_i} \begin{vmatrix} u_{i-2,1} & u_{i-2,j+1} \\ u_{i-1,1} & u_{i-1,j+1} \end{vmatrix}, \qquad j = 1, 2, 3, \ldots$$

where $e_2 = 1$, $e_i = (u_{i-2,1})^2$, $i \geq 3$. Show that the degree of u_{ij} in the a's and b's is equal to $2i - 1$.

1.40 Solve the equation (1.150) when:

(i) $a(\lambda) = 3\lambda^3 - 2\lambda^2 + \lambda + 2$, $b(\lambda) = \lambda^2 - \lambda + 1$

(ii) $a(\lambda) = 2\lambda^4 + 3\lambda^3 - 3\lambda^2 - 5\lambda + 2$

  $b(\lambda) = 2\lambda^3 + \lambda^2 - \lambda - 1$

1.41 If the values are given of all the elements r_{01}, r_{11}, $\ldots$, r_{N1} in the first column of a regular Routh array defined by (1.134), the remainder of the array can be calculated by the following scheme. For $k = 1, 2, \ldots, N$, construct an $N \times N$ matrix H_k with elements $H_k(i, j)$ defined by

$$H_k(i, i) = \begin{cases} h_{k+1-i}, & \text{for } i = 1, 2, \ldots, k \\ 1, & \text{for } i = k + 1, \ldots, N, \text{ for } k < N \end{cases}$$

$$H_k(i, i + 1) = 1, \quad \text{for } i = 1, 2, \ldots, k - 1$$

all other elements being zero, where the h's are defined in (1.139). Compute the (upper triangular) matrix $H = H_N H_{N-1} \cdots H_1$. Then the row $\{r_{ij}\}$ of the Routh array is given by the $(i + 1)$th row of $r_{N1}H$, starting from the element on the principal diagonal, with h_j replaced by h_{j+i}, for $i = 0, 1, 2, \ldots, N - 1$, $j = 1, 2, \ldots, N - 1$.
Determine H when $N = 5$, and hence compute the array whose first column elements are 1, 4, 2, -2, 1, -4.

BIBLIOGRAPHICAL NOTES

Section 1.0. It is not the intention in this book to consider
numerical aspects of analysis of polynomials, including the vast
subject of computation of roots; see, for example, the texts by
Dahlquist and Björck (1974), Henrici (1974), Householder (1970a),
and Knuth (1969), where many further references can be found. All
the prerequisite mathematical material can be found in Archbold
(1970), and relevant matrix algebra in Barnett (1979) and Lancaster
(1969).

Section 1.1. A standard old-style text on the "theory of
equations" is by Burnside and Panton (1912, reprinted 1960); a
contemporary treatment has been provided by Dobbs and Hanks (1980).
Also useful as sources of classical material are the books by
Bôcher (1907, reprinted 1964), Turnbull (1952), a translation from
Russian of Mishina and Proskuryakov (1965), and Chapter 14 of
Ledermann and Vajda (1980). Easily the most important reference
book on polynomials is by Marden (1966), who concentrates however
on the analytic, as opposed to algebraic, aspects. New proofs of
the fundamental theorem of algebra continue to appear, for example
Brenner and Lyndon (1981). Extensions of Horner's method are given
in Problems 5.20 to 5.23 at the end of Chapter 5.

Section 1.2. Theorem 1.2 was given by Kalman (1963), and Theorem
1.3 was first recorded by MacDuffee (1950a). Further material on
the companion form can be found in Brand (1964, 1968), Chapter 2 of
Barnett (1971a) and Barnett (1973c, 1976c). Balestrino (1979) has
given an explicit formula for $(\lambda I - C)^{-1}$. The procedure of Example
7 for determining C^m is a simplified version of Choudhury (1973);
for a different formulation, see Maroulas and Barnett (1979c). The
method for determining the remainder, illustrated in Example 8, was
suggested in a private communication by Kailath (Stanford). For
construction of polynomials in an arbitrary matrix, see Van Loan
(1979). The Vandermonde transformation in Problem 1.11, together
with the case of repeated eigenvalues, was set out by Power

(1967a). The result in Problem 1.16 is contained in Barnett and
Lancaster (1980a), and a survey on striped matrices has been given
by Roebuck and Barnett (1978). Problem 1.18 is based on László
(1981), and the matrix in Problem 1.19 can be developed into the
so-called *hypercompanion matrix* (Ayres, 1974). Problem 1.20 is
taken from Hartwig (1975).

Section 1.3. MacDuffee (1950a) contains Theorem 1.4 and the result
of Problem 1.22. Theorem 1.4 was rediscovered by Barnett (1970a)
using the somewhat different proof given here. Theorems 1.5 and
1.6 first appeared in Barnett (1970a) and Barnett (1971b),
respectively.

Section 1.4. Sylvester's resultant matrix can be found, for
example, in Bôcher (1964). Theorem 1.7 was given for the cases
$m = n - 1$ and $m = n$ in Barnett (1971b,c) respectively, and for the
general case in Barnett (1980). An appealing direct proof of the
second part of Theorem 1.7 was devised by Skala (1971). The Schur
complement was defined by Haynsworth (1968), where further development of the result of Problem 1.27 can be found, and some other
properties were given in Cottle (1974), Barnett and Jury (1978),
Ouellette (1981), and Huseyin and Jury (1981). An interesting
control theoretic interpretation of the Schur complement has been
made [Hung and MacFarlane (1981)]. The term inners was coined by
Jury (1971), who has developed many further applications (Jury,
1974). Theorem 1.9 was first published by Laidacker (1969) using
a different method of proof. The extended Sylvester matrix (1.101)
and Theorem 1.10 were given in Barnett (1980). A weaker version of
the first part of Theorem 1.10 on the degree of the g.c.d.,
appeared in Vardulakis and Stoyle (1978). As Sylvester's matrix
for two polynomials dates back to 1840, it is surprising that the
extension for several polynomials has been obtained only very
recently, although there is a little discussion of this case in
older textbooks, for example Scott (1880). Solution of the diophantine equation (1.104) has been applied by Kucera (1979) to the

design of control systems. Theorem 1.11 is covered in several of
the standard texts, for example Bôcher (1964). A nice exposition
of the involvement of Sylvester-type matrices in stability problems
(Chapter 3) was given by Householder (1968). His paper also
explained the concept of *bigradient* which provides another,
although rather clumsy way of obtaining a g.c.d. For references on
other applications of the Sylvester resultant, see Chapter 2 of
Barnett (1971a). Problem 1.28 follows Neumann (1981), who gave
further development of the result, and Problem 1.30 is based on
Wang (1980).

Section 1.5. The idea of Bézout's method of elimination dates back
to 1764, and an interesting review of subsequent developments,
again including stability applications, was given by Householder
(1970b). The formula (1.117) and Theorem 1.12 appeared in Barnett
(1972a). The relationship expressed in Theorem 1.13 is well known
(e.g., Turnbull, 1952), but the proof given here is new. Theorem
1.14 is well known (Gohberg and Feldman, 1974) but the proof given
here is simpler, and was presented by Barnett and Lancaster
(1980a). The result (1.132) was obtained independently by Anderson
(1972b).

Section 1.6.1. Euclid's algorithm for polynomials dates back to
the sixteenth century, but the original scheme for integers is
about 1900 years older. For material on computation of g.c.d.,
see Knuth (1969), appropriate references in Barnett (1974a), and
Brown (1978). The relationship between Sylvester's matrix and
Routh's array was given implicitly by Van Vleck (1899), and
explicitly by Fryer (1959), who also derived (1.137). The many
applications of Routh's array were surveyed on the hundredth
anniversary of the publication of Routh's original paper on
stability (Barnett and Siljak, 1977). The scheme in Remark 4 on
Theorem 1.15 is based on Shieh et al. (1978). Since the Routh
array relates to the Sylvester matrix, which is in turn related to
the bezoutian matrix (Theorem 1.13), it is not unexpected that the

Routh elements can themselves be derived from suitably defined bezoutian matrices (Barnett, 1977).

Section 1.6.2. A deeper investigation of the algorithm (1.144) has been made by Barnett (1974a), and its relationship to the Routh array was described in Barnett (1973b).

Section 1.6.3. The array (1.146) was suggested by Barnett (1973a), and Theorem 1.16 is due to Jeltsch (1979).

Section 1.6.4. The method for solving (1.150) can be found, for example, in Bôcher (1964). An interesting application of Euclid's algorithm to the solution of a different equation has been dis-covered by McEliece and Shearer (1978). Theorem 1.17 is due to Hensel, and our proof follows MacDuffee (1950b). The result in Problem 1.38 was obtained by Jeltsch (1979), who also pointed out that so-called p.r.s. algorithms (Brown, 1978) are equivalent to the scheme in Problem 1.39. The method in Problem 1.41 was obtained by Shieh et al. (1971).

2

Basic Properties of Control Systems

2.0 INTRODUCTION

In this chapter we study the second theme of this book: the theory
of linear control systems with constant coefficients. Our particular concern is to interpret this theory in terms of polynomials,
using matrix methods. A basic knowledge of the solution of linear
differential equations using the Laplace transform will be assumed,
and a similar acquaintance with the parallel method of the
z-transform for linear difference equations will be useful. As is
customary in the control literature, we shall often use s instead
of λ to denote the polynomial variable. We begin in Section 2.1
with a very brief introduction to the so-called "state space"
approach to control problems, whereby systems are described by
equations in matrix form. Application of the Laplace transform or
z-transform, relative to continuous or discrete time measurements,
respectively, then shows the way in which polynomials arise. The
fundamental concepts of controllability and observability (also,
reachability and reconstructibility) are defined in Section 2.2.
The appropriate criteria are derived, together with the important
"duality" property, which shows that controllability and observability are mathematically equivalent, although their physical
interpretations are quite different. In Section 2.3 we explore

relationships between controllability/observability and the
greatest common divisor of polynomials, the main link with Chapter
1 being provided by companion matrices.   In the remainder of the
chapter we restrict ourselves to linear control systems with a
single input and single output, results for the multivariable cases
being dealt with in Chapter 4.   We first consider canonical forms
for such systems in Section 2.4, and show how the bezoutian matrix
of Chapter 1 is unexpectedly involved; indeed, the bezoutian
matrix can be expressed in a variety of ways in terms of controllability and observability matrices.   This is followed in Section
2.5 by the problem of determining a set of equations which form a
"realization" of a given transfer function.   The condition for
such a set to contain a minimal number of state variables turns out
to be equivalent to the system being controllable and observable,
so that these various concepts are linked together in a rather nice
way.   Finally, we show in Section 2.6, using the appropriate
canonical form, that linear state feedback enables the characteristic polynomial of the system, obtained after the control has
been applied, to be selected at will.

  The results of Chapter 1, while undoubtedly interesting in
their own right, take on an extra dimension when the applications
and interactions in the control field are appreciated.

2.1 STATE SPACE CONCEPTS

In most areas of applied mathematics the general aim is to use a
mathematical model in order to *analyze* the situation under study.
To take a simple example, if an oven is being heated at a certain
rate, then applying standard assumptions about heat transfer,
radiation, and so on, the problem would be to calculate the
temperature of the oven interior after a certain time has elapsed.
A different, and intrinsically more interesting approach would be
to try to determine how to *control* the heat input of the oven so as
to achieve a certain desired temperature, perhaps in the shortest
possible time, or with minimum consumption of energy.   This is the

essence of control theory: given a mathematical description of a
system, to find a scheme for manipulating input variables in order
to obtain a satisfactory performance, according to some specified
criterion.   The term "system" can embrace a very wide variety of
circumstances, and control problems might be: the running of a
steelworks in an efficient fashion, the soft landing of an unmanned
spacecraft on the moon, the artificial aeration of a river so as to
reduce pollution, or the reduction of inflation in a country's
economy.   This wide diversity of applications is one of the
fascinations of control, and as the complexity of the contemporary
world increases, understanding such problems becomes ever more
important.   Very often, it is preferable for control to be applied
automatically, hence the term "automation".   Two familiar, simple,
examples are (i) the scheme whereby the water level in the tank of
a domestic water supply system is kept close to a preset level,
independently of water usage, or fluctuations in supply pressure;
and (ii) a heating/air-conditioning system, in which a room thermo-
stat controls the operation of the heating/cooling mechanism so as
to maintain a comfortable environment, irrespective of external
variations in temperature.   This book, however, is not the place in
which to describe the mathematical modeling of situations (see
Problems 2.1 to 2.3).   Instead, we assume that a mathematical

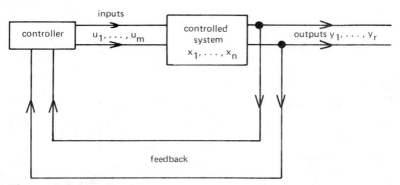

Figure 2.1   Feedback system.

description of a control system under study is available, posses-
sing the general features represented in Figure 2.1. The *state
variables* $x_1(t)$, $x_2(t)$, ..., $x_n(t)$ (where t represents time)
describe the condition or *state* of the system. Knowing the
mathematical equations describing the system, and the future values
of the *input* (or *control*) *variables* $u_1(t)$, $u_2(t)$, ..., $u_m(t)$, then
a knowledge of the state variables at some instant of time enables
us to determine completely the future behavior of the system. In
practice, it is often difficult or too expensive to measure the
state variables themselves, so only a knowledge of certain *output
variables* $y_1(t)$, ..., $y_r(t)$ is available, these being some
functions of the state variables. In this book we will not
consider the effect of any external disturbances of a statistical
nature, although these are often important in practice — for
example, gusty winds during the automatic landing of an aircraft.

If the controller, which is the device shown in Figure 2.1
manipulating the control variables, operates according to some
preset scheme, taking no account of the measured outputs, then the
system is called *open loop*, because the "loop" in Figure 2.1 is
incomplete. However, if there is *feedback* of measurements of the
outputs to the controller, which modifies the control action
accordingly, then the system is termed *closed loop*. Both the water
tank and room temperature system mentioned above are examples of
closed-loop systems, and both exhibit the important and characteristic property of being relatively unaffected by even quite large
external disturbances. Feedback loops are very common in nature:
for example, the way in which human body temperature or heartbeat
rate is regulated, or the balance control which enables a person to
ride a bicycle.

We shall consider only *linear time-invariant* (or *constant*)
systems, whose mathematical description takes one of two specific
forms. When time is a continuous variable, this will be a set of
linear, constant-coefficient *differential* equations

$$dx_1/dt \triangleq \dot{x}_1 = a_{11}x_1 + a_{12}x_2 + \cdots + a_{1n}x_n + b_{11}u_1 + \cdots + b_{1m}u_m$$

$$\dot{x}_2 = a_{21}x_1 + \cdots \qquad \cdots + a_{2n}x_n + b_{21}u_1 + \cdots + b_{2m}u_m$$

$$\cdots \qquad \cdots$$

$$\dot{x}_n = a_{n1}x_1 + \cdots \qquad \cdots + a_{nn}x_n + b_{n1}u_1 + \cdots + b_{nm}u_m$$

$$(2.1)$$

where the a_{ij}, b_{ij} are real constants. It is much more convenient to write (2.1) in matrix-vector form as

$$\dot{x}(t) = Ax(t) + Bu(t) \qquad (2.2)$$

where A = $[a_{ij}]$ is the n × n *state matrix*, B = $[b_{ij}]$ is n × m, and $x(t) = [x_1(t), x_2(t), \ldots, x_n(t)]^T$ is the *state vector*. The functions which comprise the *control vector* u = $[u_1(t), u_2(t), \ldots, u_m(t)]^T$ are assumed to be piecewise continuous. The n-dimensional space to which x belongs is called the *state space*, and this name has come to be used as a generic term for the mathematical study of control systems using matrix methods. Each output is assumed to be a *linear* combination of the states, so that

$$y(t) = Hx(t) \qquad (2.3)$$

where $y(t) = [y_1(t), \ldots, y_r(t)]^T$ is the *output vector* and H is a constant r × n matrix. It is usually the case in practice that m < n, r < n.

In many systems of practical importance the variables are known, or are of interest, only at discrete intervals of time, which we shall assume to be multiples of some basic unit T. Often this will arise from sampling an underlying continuous situation — for example, it may be feasible to only take a census of an animal population at intervals of a year or more (see Problem 2.3); a meteorologist will only measure the velocity of the wind at (say) hourly intervals; a savings bank will pay interest perhaps only once or twice a year.

We shall write $x(k)$, $y(k)$, $u(k)$ to denote the values of the
state, output, and control vectors at time kT, where $k = 0, \pm 1,$
$\pm 2, \ldots$. The differential equations (2.2) are replaced by the
linear *difference* equations

$$x(k + 1) = Ax(k) + Bu(k) \tag{2.4}$$

Equation (2.3) changes only by interpretation of the time variable,
to become $y(k) = Hx(k)$.

It will be useful to record here standard, easily derived,
expressions for the solution of (2.2) and (2.4), subject to given
initial conditions. A preliminary step is to represent the
solutions to the systems *without* control, namely

$$\dot{x}(t) = Ax(t), \quad x(t_0) = x_0 \tag{2.5}$$

$$x(k + 1) = Ax(k), \quad x(k_0) = x_0 \tag{2.6}$$

in the respective forms

$$x(t) = \Phi(t, t_0)x_0 \tag{2.7}$$

$$x(k) = \Phi(k, k_0)x_0 \tag{2.8}$$

where $\Phi(\cdot, \cdot)$ is the $n \times n$ *state transition matrix*, so called
because it provides the transition from the state at any time
$t_0(k_0)$ to any other time t (k). For (2.5) the transition matrix
is the *exponential matrix*

$$\Phi(t, t_0) = e^{A(t-t_0)}$$
$$\triangleq I + (t - t_0)A + (t - t_0)^2 A^2/2! + \cdots \tag{2.9}$$

and for (2.6)

$$\Phi(k, k_0) = A^{k-k_0} \tag{2.10}$$

where $k \geqslant k_0$ if $\det A = 0$. The two transition matrices have
analogous properties, summarized in Table 2.1 ($L\{\cdot\}$, $Z\{\cdot\}$ denote
Laplace and z-transforms, respectively). Notice that the matrix

Table 2.1 Properties of transition matrices

Continuous time	Discrete time
$\frac{d}{dt} \Phi(t, t_0) = A\Phi(t, t_0)$	$\Phi(k + 1, k_0) = A\Phi(k, k_0)$
$\Phi(t, t) = I$	$\Phi(k, k) = I$
$\Phi(t_0, t) = \Phi^{-1}(t, t_0)$	$\Phi(k_0, k) = \Phi^{-1}(k, k_0)$, provided that $\|A\| \neq 0$
$\Phi(t, t_0) = \Phi(t, t_1)\Phi(t_1, t_0)$	$\Phi(k, k_0) = \Phi(k, k_1)\Phi(k_1, k_0)$, $k \geq k_1 \geq k_0$
$L\{\Phi(t, 0)\} = (sI - A)^{-1} = L\{e^{At}\}$	$Z\{\Phi(k, 0)\} = z(zI - A)^{-1} = Z\{A^k\}$

in (2.9) is *always* nonsingular, but for discrete-time systems Φ^{-1} does not exist if A is singular, so it is not then possible to express $x(k_0)$ in terms of $x(k)$.

The general solutions of (2.2) and (2.4) can now be written, respectively, as

$$x(t) = \Phi(t, t_0)\left[x_0 + \int_{t_0}^{t} \Phi(t_0, \tau)Bu(\tau) \, d\tau\right] \tag{2.11}$$

and

$$x(k) = \Phi(k, k_0)\left[x_0 + \sum_{i=k_0}^{k-1} \Phi(k_0, i + 1)Bu(i)\right] \tag{2.12}$$

We can now show how polynomials are involved in properties of control systems by applying Laplace or z-transforms, respectively, to the differential and difference equations (2.2) and (2.4). First consider the former, and write $\bar{x}(s) \triangleq L\{x(t)\}$, where s is the Laplace transform variable, so that taking the transform of both sides of (2.2) gives

$$s\bar{x}(s) - x_0 = A\bar{x}(s) + B\bar{u}(s) \tag{2.13}$$

In particular, if $x_0 = 0$, then (2.13) can be rearranged in the form

$$(sI - A)\bar{x}(s) = B\bar{u}(s)$$

or

$$\bar{x}(s) = (sI - A)^{-1}B\bar{u}(s) \tag{2.14}$$

From (2.3) we have, after Laplace transformation, $\bar{y}(s) = H\bar{x}(s)$ so that by (2.14)

$$\bar{y}(s) = H(sI - A)^{-1}B\bar{u}(s)$$

$$= G(s)\bar{u}(s) \tag{2.15}$$

where

$$G(s) \overset{\Delta}{=} H(sI - A)^{-1}B \tag{2.16}$$

Consider carefully the implications of (2.16), which is a key expression. The matrix $G(s)$ has dimensions $r \times m$, and relates the Laplace transform of the output vector to that of the input vector. For this reason it is called the *transfer function matrix*. Furthermore, each element $g_{ij}(s)$ of $G(s)$ is a ratio of polynomials. To demonstrate this, write

$$(sI_n - A)^{-1} = \text{adj}(sI_n - A)/\det(sI_n - A) \tag{2.17}$$

Since each element of $\text{adj}(sI_n - A)$ is an $(n - 1) \times (n - 1)$ determinant having degree in s not more than $n - 1$, collecting together powers of s in (2.17) shows that

$$(sI_n - A)^{-1} = (s^{n-1}I_n + s^{n-2}B_1 + s^{n-3}B_2 + \cdots + B_{n-1})$$

$$/\det(sI_n - A) \tag{2.18}$$

where $B_1, \ldots, B_{n-1}$ are constant $n \times n$ matrices [the reader can easily check that only the diagonal elements of $\text{adj}(sI_n - A)$ have degree $n - 1$, which explains why the leading matrix coefficient in (2.18) is I_n; see Problem 2.5]. Substituting (2.18) into (2.16) gives

$$G(s) = (s^{n-1}G_0 + s^{n-2}G_1 + \cdots + G_{n-1})/k(s) \qquad (2.19)$$

where $G_0 = HB$, $G_i = HB_iB$, $i = 1, 2, \ldots, n - 1$ and

$$k(s) = \det(sI_n - A) \overset{\Delta}{=} s^n + k_1 s^{n-1} + \cdots + k_n \qquad (2.20)$$

Equations (2.19) and (2.20) show that each element $g_{ij}(s)$ is a proper rational function. Also, the numerators in (2.18) and (2.19) are examples of polynomial matrices, being polynomial expressions with matrix coefficients. These will be considered further in Chapter 4.

When there is only a *single input* and a *single output* to the system (i.e., $r = m = 1$), the matrix $G(s)$ reduces to a scalar rational function $g(s)$, the *transfer function*, on which much of the earlier control engineering work is based. Equation (2.15) can then be rewritten as

$$g(s) = \frac{\bar{y}(s)}{\bar{u}(s)} \qquad (2.21)$$

so that the transfer function is the ratio of the Laplace transforms of output and input. With this terminology, for the general case when $r > 1$, $m > 1$, equation (2.15) implies that the ith component of $\bar{y}(s)$ is

$$\bar{y}_i(s) = g_{i1}\bar{u}_1(s) + g_{i2}\bar{u}_2(s) + \cdots + g_{im}\bar{u}_m(s)$$

so that $g_{ij}(s)$ is the transfer function between the ith output and jth input, all other inputs being set equal to zero.

Turning now to the discrete-time system described by (2.4), everything carries over without alteration, provided that the Laplace transform is replaced by the z-transform $\tilde{x}(z) \overset{\Delta}{=} Z\{x(k)\}$. Under the assumption $x(k_0) = 0$, equation (2.15) is replaced by $\tilde{y}(z) = G(z)\tilde{u}(z)$, where $G(\cdot)$ is defined precisely as in (2.16).

The representation of linear control systems using transforms is called in the literature the *frequency-domain* approach, because of the idea of regarding the transform variables s or z as complex frequencies.

Problems

2.1 Consider a controlled environment containing predator and prey
     populations (e.g., wolves and rabbits on a large, grassy
     island; big and small fishes in a large lake), the numbers of
     each at time t being $x_1(t)$ and $x_2(t)$, respectively. Suppose
     that without the presence of the predators the prey would
     increase exponentially, but that the rate of growth of the
     prey population is reduced by an amount proportional to the
     number of predators. Similarly, suppose that without prey to
     feed on the predator population would decrease exponentially,
     but that the rate of growth of the predator population is
     increased by an amount proportional to the number of prey.
     Show that the system equations can be written in the form
     (2.5), in which A is a 2 × 2 matrix with $a_{11} < 0$, $a_{12} > 0$,
     $a_{21} < 0$, $a_{22} > 0$.
          Under certain values of the parameters, both populations
     can increase without limit (assuming an inexhaustible supply
     of food for the prey population). It is decided to attempt to
     control the environment by introducing a disease which does
     not affect the predators, but which reduces the rate of growth
     of the prey population by an amount $u(t)$. Obtain the new form
     of the system equations.
          This problem will be further investigated in Problems 2.9
     and 4.5.

2.2 The temperature in the interior of an oven is to be controlled
     by varying the current through an electric heating coil in the
     jacket, as shown in cross section in Figure 2.2. Let the
     parameters of the system be as follows:

	Heat capacity	Surface area	Radiation coefficient	Temperature
Oven interior	c_1	a_1	r_1	$T_1(t)$
Jacket	c_2	a_2	r_2	$T_2(t)$

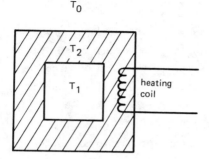

Figure 2.2   Oven problem.

Assume that there is uniform and instantaneous distribution of
temperature throughout, that the rate of loss of heat is
proportional to surface area, and that the temperature of the
surroundings is T_0. Define as state variables $x_1(t) = T_1 - T_0$,
$x_2(t) = T_2 - T_0$, and show that the system equations can be
written in the form (2.2) with

$$A = \begin{bmatrix} \dfrac{-a_1 r_1}{c_1} & \dfrac{a_1 r_1}{c_1} \\[2ex] \dfrac{a_1 r_1}{c_2} & \dfrac{-(a_1 r_1 + a_2 r_2)}{c_2} \end{bmatrix}, \quad B = \begin{bmatrix} 0 \\[2ex] \dfrac{1}{c_2} \end{bmatrix} \qquad (2.22)$$

This system is considered again in Problem 2.10.

2.3 Consider a simplified cattle ranching problem involving a
single breed of cattle. The discrete-time state variables are
defined by

$x_1(k)$ = number of one-year-old cattle (young)

$x_2(k)$ = number of two-year-old cattle (mature)

$x_3(k)$ = number of three-year-old and older cattle (old)

and $k = 0, 1, 2, 3, \ldots$ years. In the absence of
slaughtering, the assumptions concerning breeding and
mortality are:

(a)   Young cattle do not breed.

(b)   Each two-year-old produces 0.8 young cattle per year.

(c)   Old cattle each produce 0.4 young cattle per year.

(d)   Only old cattle die, at the rate of 30% per year.

Show that system can be described by the state equation (2.7) with

$$A = \begin{bmatrix} 0 & 0.8 & 0.4 \\ 1 & 0 & 0 \\ 0 & 1 & 0.7 \end{bmatrix}$$

For further development, see Problems 2.11 and 4.6.

2.4   The exponential matrix is defined by

$$e^A = I + A + \frac{1}{2!} A^2 + \frac{1}{3!} A^3 + \cdots$$

[see (2.9)]. Show that if D is a matrix which commutes with A, then $e^{A+D} = e^A e^D$. Hence show that $(e^A)^{-1} = e^{-A}$, and deduce that e^A is nonsingular for any n × n matrix A having finite elements.

2.5   By considering the identity

$$(sI - A) \, adj(sI - A) \equiv k(s)I$$

where $k(s)$ is defined in (2.20), obtain the following expressions for the matrices in (2.18):

$$B_1 = A + k_1 I, \quad B_i = AB_{i-1} + k_i I, \quad i = 2, 3, \ldots, n - 1$$

It can also be shown that

$$k_1 = -tr(A), \quad k_i = -\frac{1}{i} tr(AB_{i-1}), \quad i = 2, 3, \ldots, n$$

where

$$tr(A) \overset{\Delta}{=} a_{11} + a_{22} + \cdots + a_{nn}$$

The formulas are often named after *Leverrier* or *Faddeev*.

2.6   A continuous-time linear system is known to be described by
      (2.5) with n = 3, but the values of the elements of A are
      unknown. By appropriate measurements it is found that

$$x(0) = \begin{bmatrix} 1 \\ 0 \\ 1 \end{bmatrix}, \quad x(1) = \begin{bmatrix} 1 \\ 0 \\ 0 \end{bmatrix}, \quad x(2) = \begin{bmatrix} 0 \\ 1 \\ 0 \end{bmatrix}, \quad x(3) = \begin{bmatrix} 0 \\ 0 \\ 1 \end{bmatrix}$$

      Compute x(4).

2.7   A continuous-time linear system is known to be described by
      (2.5) with n = 2. It is possible to measure the state vector,
      but because of difficulties in setting up the equipment, this
      measurement can only be started after an unknown amount τ
      (> 2) of time has elapsed. It is then found that

$$x(\tau) = \begin{bmatrix} 1.0 \\ 1.0 \end{bmatrix}, \quad x(\tau + 1) = \begin{bmatrix} 1.5 \\ 1.6 \end{bmatrix}, \quad x(\tau + 2) = \begin{bmatrix} 1.8 \\ 2.1 \end{bmatrix}$$

      Compute $x(\tau - 2)$.

2.8   Show that if $\Phi(t, t_0)$ is defined by (2.9), then $[\Phi^{-1}(t, t_0)]^T$
      is the transition matrix for the system $\dot{x}(t) = -A^T x(t)$.

2.2   CONTROLLABILITY AND OBSERVABILITY

Before attempting to devise a suitable control scheme which will
achieve a desired objective, it is clearly sensible to determine
whether indeed such a control program actually exists. If not,
then either control will have to be applied through some different
configuration of inputs, or the objective itself will have to be
modified. To take a worldwide example, economists and politicians
make many claims as to methods for controlling the rate of economic
inflation, but there seems to have been little investigation as to
whether the variables they select, such as rates of interest,
taxes, money supply, and so on, can really be manipulated so as to
have the intended effect.

      There are many different definitions of "controllability" of a
system, according to the objective under study. Here we will

consider the widely used property of being able to transfer a
system from any given state to any other state by means of a
suitable choice of control functions. Specifically, consider first
the continuous-time system described by (2.2):

The system (2.2) is *completely controllable (c.c.)* if for any
initial state $x(0) = x_0$, and any given final state x_f, there exists
a finite time $t_f > 0$ and a control vector $u(t)$, defined on
$0 \leqslant t \leqslant t_f$, such that $x(t_f) = x_f$.

The qualification "completely" indicates that the property
holds for *all* x_0 and x_f. In fact, since we can always transfer the
origin of coordinates to the given final state, there is no loss of
generality in setting $x_f = 0$ in the definition. Note that in
general there will not be a *unique* control which performs the
required transfer of states.

Example 1   Consider a system described by

$$\dot{x}_1 = 2x_1 + 3x_2 + u(t)$$

$$\dot{x}_2 = -2x_2$$

Clearly, we have $x_2(t) = e^{-2t}x_2(0)$, for all $t \geqslant 0$, so $x_2(t)$ depends
only on $x_2(0)$. There is no way in which $x_2(t_f)$ can be made to take
a predetermined value by altering $u(t)$, so the system is not c.c.

The basic criterion for complete controllability is the
following "rank test":

Theorem 2.1   Define the n × nm *controllability matrix*

$$C(A, B) \triangleq [B, AB, A^2B, \ldots, A^{n-1}B] \qquad (2.23)$$

Then the system (2.2) is c.c. if and only if rank $C(A, B) = n$.

Proof:

(i) *Necessity.* We assume that the system is c.c., so that for
any x_0 there exists a control vector $u(\tau)$, $0 \leqslant \tau \leqslant t_f$ such that
$x(t_f) = 0$. After substituting this condition, together with
$t_0 = 0$, $t = t_f$, into (2.11) we obtain

$$0 = \Phi(t_f, 0)\left[x_0 + \int_0^{t_f} \Phi(0, \tau)Bu(\tau)\ d\tau\right]$$

$$= \Phi(t_f, 0)\left[x_0 + \int_0^{t_f} \exp(-A\tau)Bu(\tau)\ d\tau\right] \qquad (2.24)$$

where (2.24) results on using $\Phi(0, \tau) = \exp(-A\tau)$, from (2.9).
Notice that $\Phi(t_f, 0)$ is nonsingular, so this premultiplying factor
in (2.24) can be deleted. In addition [see (A5) in Appendix A], we
can write

$$\exp(-A\tau) = r_0 I + r_1 A + \cdots + r_{n-1} A^{n-1} \qquad (2.25)$$

where the r_i are functions of τ. Substitution of (2.25) into
(2.24) reduces it to

$$0 = x_0 + \int_0^{t_f} (r_0 B + r_1 AB + \cdots + r_{n-1} A^{n-1} B)u(\tau)\ d\tau \qquad (2.26)$$

We establish that rank $C(A, B) = n$ by showing that the assumption
rank $C < n$ produces a contradiction. For if the latter did hold,
there would exist a nonzero row n-vector q (say), such that $qC = 0$,
i.e.,

$$qB = 0, \quad qAB = 0, \quad \ldots, \quad qA^{n-1}B = 0 \qquad (2.27)$$

Premultiplying (2.26) by q and using (2.27) gives $qx_0 \equiv 0$.
However, since x_0 is arbitrary, it follows that $q \equiv 0$, the required
contradiction.

    (ii) *Sufficiency.* Suppose that rank $C(A, B) = n$. We wish to
show that for arbitrary x_0, there exists a function $u(\tau)$,
$0 \leqslant \tau \leqslant t_f$ such that (2.24) is satisfied. In (2.24), again omit
the nonsingular factor $\Phi(t_f, 0)$, and replace τ by $-\tau$ to obtain the
equivalent condition

$$\int_0^{t_f} \exp(A\tau)Bu(\tau)\ d\tau = x_0 \qquad (2.28)$$

which is to be satisfied by $u(\tau)$. Suppose, for simplicity of exposition, that A has n distinct eigenvalues $\lambda_1, \ldots, \lambda_n$. We need the fact that since $\exp(\lambda_1 t), \ldots, \exp(\lambda_n t)$ are linearly independent, the matrix M with i, j element

$$m_{ij} = \int_0^{t_f} e^{\lambda_i \tau} e^{\bar{\lambda}_j \tau} \, d\tau \tag{2.29}$$

(where the overbar denotes complex conjugate) is positive definite hermitian. Thus in particular M is nonsingular. It now follows that there exists a function $f_p(t)$ such that

$$\int_0^{t_f} e^{\lambda_i \tau} \bar{f}_p(\tau) \, d\tau = \lambda_i^p, \qquad i = 1, 2, \ldots, n \tag{2.30}$$

for any nonnegative integer p. For if we set

$$f_p(\tau) = d_1 e^{\lambda_1 \tau} + \cdots + d_n e^{\lambda_n \tau} \tag{2.31}$$

then by substituting (2.31) into (2.30), and using (2.29), we obtain

$$\bar{d}_1 m_{i1} + \bar{d}_2 m_{i2} + \cdots + \bar{d}_n m_{in} = \lambda_i^p, \qquad i = 1, 2, \ldots, n$$

or in matrix notation

$$[\bar{d}_1, \ldots, \bar{d}_n] M^T = [\lambda_1^p, \ldots, \lambda_n^p] \tag{2.32}$$

Since M^T is nonsingular, the d's [and hence also $f_p(\tau)$ in (2.31)] are uniquely determined.

To complete the proof, we also need to record that from the theory of functions of a matrix [see (A5) in Appendix A] we can write

$$\exp(A\tau) = \sum_{i=1}^{n} e^{\lambda_i \tau} Z_i, \qquad A^p = \sum_{i=1}^{n} \lambda_i^p Z_i \tag{2.33}$$

where the Z_i are constant n × n matrices determined entirely by A.

Armed with the above, consider (with e_j denoting the jth column of I_n)

$$\int_0^{t_f} \exp(A\tau)B\bar{f}_p(\tau)e_j \; d\tau = \int_0^{t_f} \sum_i Z_i Be_j e^{\lambda_i \tau} \bar{f}_p(\tau) \; d\tau, \qquad \text{by (2.33)}$$

$$= \sum_{i=1}^{n} Z_i Be_j \lambda_i^p, \qquad \text{by (2.30)}$$

$$= A^p Be_j, \qquad \text{by (2.33)} \qquad\qquad (2.34)$$

Since $A^p Be_j$ is a real vector, it follows by taking the real part of both sides of (2.34) that

$$\int_0^{t_f} \exp(A\tau)Bu_{pj}(\tau) \; d\tau = A^p Be_j$$

$$= \text{jth column of } A^p B \qquad\qquad (2.35)$$

where $u_{pj} = \text{Re}\{\bar{f}_p(\tau)e_j\}$. By assumption $[B, AB, \ldots, A^{n-1}B]$ contains at least one set of independent columns, so we can express any x_0 as a linear combination of this basis. The desired vector $u(\tau)$ satisfying (2.28) will then be the same combination of the corresponding vectors $u_{pj}(\tau)$ obtained from (2.35). ∎

Example 1 (continued)   We have

$$A = \begin{bmatrix} 2 & 3 \\ 0 & -2 \end{bmatrix}, \qquad B = \begin{bmatrix} 1 \\ 0 \end{bmatrix}, \qquad C(A, B) = \begin{bmatrix} 1 & 2 \\ 0 & 0 \end{bmatrix}$$

Since $C(A, B)$ has rank one, Theorem 2.1 confirms that the system is not c.c.

Example 2   Consider a system in the form (2.2), with

$$A = \begin{bmatrix} 0 & 1 & -1 \\ 4 & 0 & 2 \\ 3 & 2 & 1 \end{bmatrix}, \qquad B = \begin{bmatrix} 1 & 2 \\ -1 & 0 \\ 0 & 3 \end{bmatrix} \qquad\qquad (2.36)$$

It is easy to compute

$$C(A, B) = \begin{bmatrix} 1 & 2 & -1 & -3 & 3 & 5 \\ -1 & 0 & 4 & 14 & -2 & 6 \\ 0 & 3 & 1 & 9 & 6 & 28 \end{bmatrix} \tag{2.37}$$

It is readily checked that rank $C(A, B) = 3$, so the system
corresponding to (2.36) is c.c.

The following alternative necessary and sufficient condition
for controllability is sometimes useful, especially for theoretical
purposes.

<u>Theorem 2.2</u> The matrix $C(A, B)$ has rank n if and only if
$\text{rank}[\lambda_i I - A, B] = n$, for $i = 1, 2, \ldots, n$, where $\lambda_1, \ldots, \lambda_n$ are
the eigenvalues of A.

Proof. Suppose that rank $[\lambda_k I - A, B] < n$, for some value of k.
This implies that there exists a row n-vector $p \neq 0$ such that
$p[\lambda_k I - A, B] = 0$, i.e., $pA = \lambda_k p$, $pB = 0$, showing that p is a left
eigenvector corresponding to λ_k. Hence

$$pAB = \lambda_k pB = 0, \quad pA^2 B = \lambda_k pAB = 0$$

and so on, i.e.,

$$pA^j B = \lambda_k^j PB = 0, \quad j = 1, 2, \ldots, n - 1$$

Therefore,

$$p[B, AB, A^2 B, \ldots, A^{n-1}B] = 0$$

which implies, since $p \neq 0$, that the matrix within square brackets
[i.e., $C(A, B)$] has rank less than n.

For the converse, suppose that rank $C(A, B) < n$. This implies
that there exists a row n-vector $q \neq 0$ such that

$$qB = 0, qAB = 0, \ldots, qA^{n-1}B = 0 \tag{2.38}$$

Since $B \neq 0$, (2.38) can be interpreted to mean that the $n \times n$ matrix α, say, with rows q, qA, qA^2, ..., qA^{n-1} has rank less than n. We can therefore find another nonzero row n-vector $\phi = [\phi_1, \ldots, \phi_n]$ such that $\phi\alpha = 0$, i.e.,

$$\phi_1 q + \phi_2 qA + \cdots + \phi_n qA^{n-1} = 0$$

or

$$q(\phi_1 I + \phi_2 A + \cdots + \phi_n A^{n-1}) = 0$$

In other words, we can always determine a polynomial

$$\phi(\lambda) = \phi_1 + \phi_2 \lambda + \cdots + \phi_n \lambda^{n-1}$$

having the property that $\phi(A)$ is singular. Let $\phi(\lambda)$ be such a polynomial of *least* possible degree satisfying $q\phi(A) = 0$. Since the eigenvalues of $\phi(A)$ are $\phi(\lambda_1)$, ..., $\phi(\lambda_n)$, it follows that $\phi(\lambda_j) = 0$ for at least one j. Hence we can write

$$\phi(\lambda) \equiv \psi(\lambda)(\lambda_j - \lambda) \tag{2.39}$$

Replacing λ by A in the identity (2.39), and premultiplying by q shows that

$$q\phi(A) = 0 = q\psi(A)(\lambda_j I - A)$$

$$= \hat{q}(\lambda_j I - A), \quad \text{say}$$

Moreover,

$$\hat{q}B = q\psi(A)B$$

$$= q(\psi_0 B + \psi_1 AB + \psi_2 A^2 B + \cdots) = 0$$

on using (2.38). Thus we have proved that

$$\hat{q}[\lambda_j I - A, B] = 0 \tag{2.40}$$

By virtue of the definition of $\phi(\lambda)$ it follows that $\hat{q} = q\psi(A) \neq 0$, since in (2.39) $\delta\psi < \delta\phi$. Therefore, (2.40) establishes that $\text{rank}[\lambda_j I - A, B] < n$. ∎

Remarks on Theorems 2.1 and 2.2   (1) The proof given of the
sufficiency part of Theorem 2.1 is easily modified for repeated
eigenvalues, and shows that a suitable control vector can be found
from the class of piecewise continuous functions. The proofs
given in many textbooks require the use of delta functions.

(2) Actual determination of a control vector which achieves a
desired transfer of states lies outside the scope of our treatment.

(3) Note that in constructing $C(A, B)$, each block is obtained
from the previous one by premultiplication, i.e., $A^r B = A(A^{r-1}B)$,
$r = 1, 2, \ldots, n - 1$. In fact, if rank $B = p$, then it is only
necessary to go as far as $r = n - p$ (see Problem 2.23).
Alternatively, if the minimum polynomial of A has degree N, then
since A^N and all higher powers can be expressed in terms of I, A,
$A^2, \ldots, A^{N-1}$, it follows that we can stop with $r = N - 1$.

(4) It is often convenient to refer to the controllability of
the *pair* (A, B), it being appreciated that (A, B) is not regarded as
a single matrix.

(5) If a system is not c.c., it would be misleading to
describe it as "uncontrollable", since it can be shown that it is
still possible to transfer it from an initial state x_0 to a final
state x_f provided that both belong to the *column space* of $C(A, B)$
[the set of all linear combinations of the columns of $C(A, B)$,
often called the *range* of $C(A, B)$].

Example 3   If

$$A = \begin{bmatrix} 0 & 1 \\ 3 & -2 \end{bmatrix}, \quad B = \begin{bmatrix} 1 \\ -3 \end{bmatrix} \tag{2.41}$$

then

$$C(A, B) = \begin{bmatrix} 1 & -3 \\ -3 & 9 \end{bmatrix}$$

which has rank 1. Thus for transfer between x_0 and x_f to be
possible, both must have the form $\theta[1, -3]^T$, $\theta \neq 0$.

When we turn to the discrete-time system (2.4), the definition of complete controllability is the same, except that t_f = NT, and the control which achieves the transfer form x(0) to x(N) = 0 is a sequence u(0), u(1), ..., u(N - 1). However, it is not possible in general to obtain a discrete-time version of the proof of Theorem 2.1. This is because the step corresponding to the derivation of (2.26) from (2.24) breaks down when the transition matrix is singular, which can happen only for the discrete case. It is necessary to define a somewhat different concept:

The discrete system (2.4) is *completely reachable* (*c.r.*) if for any final state x_f, there exists a finite time t_f = NT and a control sequence u(0), u(1), ..., u(N - 1) such that if x(0) = 0, then x(N) = x_f. The same definition applies for the continuous case (2.2), with u(t) defined on $0 \leqslant t \leqslant t_f$.

Notice what this new definition says: We have to be able to reach any final state *from* the origin, whereas complete controllability requires transfer *to* the origin from any initial state. For the continuous system (2.2), since the transition matrix is always nonsingular, the two concepts are *identical*, and this is also true for the discrete-time equation (2.4) *except* when A is singular. For example, if A is such that A^M = 0 for some finite M, then with the zero input u(k) = 0, for all k, in (2.4) the solution is

$$x(M) = A^M x(0) = 0$$

for *any* x(0), so it is always possible to drive the system to the origin, even if C(A, B) has rank less than n; thus the rank condition is not a necessary one for complete controllability of (2.4).

However, the interesting and important fact for our development is that the criterion for complete *reachability* of (2.4) takes the same form as that for complete controllability/reachability of (2.2).

<u>Theorem 2.3</u>   The system (2.4) is c.r. if and only if the matrix
C(A, B) defined in (2.23) has rank n.

Proof. This time we begin with the solution (2.12) of (2.4) and
set k_0 = 0, k = N, x_0 = 0, to obtain

$$x_f = x(N) = \sum_{i=0}^{N-1} \Phi(N, i + 1)Bu(i)$$

$$= \sum_{i=0}^{N-1} A^{N-i-1}Bu(i), \quad \text{by (2.10)}$$

$$= [A^{N-1}B, A^{N-2}B, \ldots, AB, B] \begin{bmatrix} u(0) \\ \vdots \\ u(N - 1) \end{bmatrix} \quad (2.42)$$

If rank C = n, then for N $\geqslant$ n the first matrix on the right in
(2.42) will have rank n, so a set u(0), u(1), u(2), ... can always
be chosen to satisfy (2.42), for any x_f. Thus, by definition, the
system is c.r.

Conversely, if the system is c.r., then rank C(A, B) cannot be
less than n; for if it were, this would mean that the matrix
$[A^{N-1}B, \ldots, B]$ in (2.42) contains less than n linearly independent
columns. Hence there would be some vectors x_f for which (2.42)
could not be satisfied for any choice of u(0), ..., u(N - 1),
contradicting the definition of c.r. ■

<u>Remarks on Theorem 2.3</u>   (1) A control sequence which produces the
desired transfer from x(0) = 0 to x(N) = x_f can be computed by
solving (2.42).

(2) If controllability is considered, then x_f = 0, $x_0 \neq 0$,
and an equation like (2.42) is obtained with x_f replaced by $-A^N x_0$.
It therefore follows that a necessary and sufficient condition for
(2.4) to be c.c. is that the range of A^n must be contained within
the range of C(A, B). This can be shown equivalent to the
condition in Theorem 2.2 holding for all *nonzero* eigenvalues of A.

Turn now to another pair of concepts which have a completely different physical interpretation from controllability/reachability, and yet turn out to be closely linked in mathematical terms. As mentioned in Section 2.1, measurements are usually available only of the output vector $y(t)$ [or $y(k)$] and the problem is to compute the state vector; or rather, to first determine whether it is possible to compute the state, before setting up a scheme to actually do it. To continue with our economic example, if the values of current indicators, such as rate of inflation, volume of production, and balance of payments, are known, can the actual state of the economy (which is described by a very large number of variables in any realistic model) be determined? In general, it is necessary to know the control inputs over the time period concerned, and in our definitions there is no loss of generality in assuming that these are zero throughout the time interval.

The system $\dot{x}(t) = Ax(t)$, $y(t) = Hx(t)$ is (i) *completely observable* (*c.o.*) if for *any* initial state $x(0) = x_0$, there exists a finite time t_f such that a knowledge of $y(t)$, $0 \leqslant t \leqslant t_f$, is sufficient to uniquely determine x_0; (ii) *completely reconstructible* (*c.rec.*) if for *any* final state x_f, there exists a finite time t_f such that a knowledge of $y(t)$, $0 \leqslant t \leqslant t_f$, is sufficient to uniquely determine x_f.

The definitions apply also to the discrete system $x(k + 1) = Ax(k)$, $y(k) = Hx(k)$, with the modifications that $t_f = NT$, and the known output is $y(0)$, $y(1)$, ..., $y(N - 1)$.

Clearly, the idea of observability is that we can deduce the state at some given time from *future* measurements of output, whereas for reconstructibility this is done from *past* output data. Just as for controllability/reachability, when the transition matrix is nonsingular the two concepts are identical. Thus, for the continuous-time case, or the discrete-time case when A is nonsingular, complete observability and reconstructibility are equivalent.

Example 4   Consider the system in Example 1 obtained by setting
$u \equiv 0$, i.e.,

$$\dot{x}_1 = 2x_1 + 3x_2, \quad \dot{x}_2 = -2x_2 \tag{2.43}$$

and suppose that the output is $y(t) = x_2(t)$. Since, as we noted
previously, $x_2(t)$ is determined entirely by $x_2(0)$, there is no way
in which $x_1(0)$ can be deduced from $y(t)$. Hence the system is not
c.o.

Theorem 2.4   The system $\dot{x}(t) = Ax(t) + Bu(t)$, $y(t) = Hx(t)$ is c.o./
c.rec., or the system $x(k + 1) = Ax(k + 1) + Bu(k)$, $y(k) = Hx(k)$ is
c.o., if and only if the rn $\times$ n *observability matrix* $0(A, H)$ having
block rows

$$H, HA, HA^2, \ldots, HA^{n-1} \tag{2.44}$$

has rank n.

Proof.   We shall only give the argument for the discrete-time
system

$$x(k + 1) = Ax(k), \quad y(k) = Hx(k) \tag{2.45}$$

since that for the continuous-time case can be developed along
analogous lines. It is easy to see that

$$y(i) = HA^i x(0), \quad i = 0, 1, 2, \ldots \tag{2.46}$$

We can write (2.46) in the combined form

$$0(A, H)x(0) = [y(0), y(1), \ldots, y(n - 1)]^T \tag{2.47}$$

If $0(A, H)$ has rank n, then equation (2.47) can always be solved to
give unique values for the elements of $x(0)$, so the system is c.o.

Conversely, if the system is c.o., we shall show that the
assumption rank $0 < n$ produces a contradiction. For this condition
would imply that there exists a nonzero column n-vector p, say,
such that $0p = 0$, i.e.,

$$Hp = 0, \quad HAp = 0, \quad \ldots, \quad HA^{n-1}p = 0 \tag{2.48}$$

Substituting (2.48) into (2.46) shows that when the initial state is x(0) = p, the corresponding output is identically zero, i.e., y(k) = 0, k ⩾ 0. This is the *same* as the output produced by x(0) = 0, and so contradicts the fact that the system is c.o. ∎

Example 4 (continued)   We have

$$A = \begin{bmatrix} 2 & 3 \\ 0 & -2 \end{bmatrix}, \quad H = [0, 1], \quad \mathit{O}(A, H) = \begin{bmatrix} 0 & 1 \\ 0 & -2 \end{bmatrix}$$

so O has rank one, confirming that the system is not c.o.

Remarks on Theorem 2.4   (1) For the discrete case, (2.47) gives a direct way of computing x(0) from the given output data; compare this with Remark 1 on Theorem 2.3.

(2) Notice also for this discrete case that when A is singular, the rank condition is not a necessary one for recon- structibility. This is easily seen from a specific example. In (2.45) let

$$A = \begin{bmatrix} 1 & 2 \\ 2 & 4 \end{bmatrix}, \quad H = [1, 2], \quad \mathit{O}(A, H) = \begin{bmatrix} 1 & 2 \\ 5 & 10 \end{bmatrix}$$

so the system is not c.o., since O(A, H) is singular. However, the system equations give

$$x_1(k) = x_1(k - 1) + 2x_2(k - 1) = y(k - 1)$$
$$x_2(k) = 2x_1(k - 1) + 4x_2(k - 1) = 2y(k - 1)$$

so we can determine the state at any time k from a knowledge of y(k - 1). The system is thus c.rec.

In fact, it can be shown that a necessary and sufficient condition for (2.45) to be c.rec. is that the *nullspace* or (*kernel*) of O(A, H) [i.e., the set of all vectors v such that O(A, H)v = 0] is contained within the kernel of A^n; or, equivalently, that rank $[\lambda_i I - A^T, H^T]$ = n for each nonzero eigenvalue λ_i of A.

(3) A comparison of (2.23) and (2.44) reveals that

$$C(A, B) = [\mathit{O}(A^T, B^T)]^T, \quad \mathit{O}(A, H) = [C(A^T, H^T)]^T \qquad (2.49)$$

Thus in view of Theorem 2.1, the conditions for controllability/
reachability and observability/reconstructibility are directly
related. In fact, we see from (2.49) that any result involving
one pair of properties can be immediately converted into a
corresponding result concerning the other pair. This important
property is known as the *principle of duality*, and can be stated
formally as:

<u>Theorem 2.5</u>   The system

$$\dot{x} = Ax + Bu, \quad y = Hx \tag{2.50}$$

is c.c. (c.o.) if and only if the *dual system*

$$\dot{x} = -A^T x + H^T u, \quad y = B^T x \tag{2.51}$$

is c.o. (c.c.).

Proof.   The observability matrix for (2.51) satisfies, after
transposition,

$$[\mathcal{O}(-A^T, B^T)]^T = [B, -AB, A^2 B, \ldots, (-1)^{n-1} A^{n-1} B]$$

The negative terms do not affect the rank, so rank $\mathcal{O}(-A^T, B^T) =$
rank $\mathcal{C}(A, B)$. Hence, by Theorem 2.1, (2.51) is c.o. if and only if
(2.50) is c.c. Similarly, the controllability matrix for (2.51)
satisfies rank $\mathcal{C}(-A^T, H^T) = $ rank $\mathcal{O}(A, H)$. ∎

<u>Remarks on Theorem 2.5</u>   (1) Consider the duality principle applied
to Theorem 2.2. We obtain the result that $\mathcal{O}(A, H)$ has rank n if
and only if rank $[\lambda_i I - A^T, H^T] = n$, for each eigenvalue λ_i of A.
Similarly, the result in Remark 2 on Theorem 2.4 is the dual of
that in Remark 2 on Theorem 2.3.

(2) The reader may be puzzled as to why the dual system (2.51)
contains $-A^T$, instead of A^T, which would clearly not affect the
proof. In fact, the reasons for choosing (2.51) lie in the theory
of optimal control and filtering (see also Problem 2.8).

Problems

2.9 Return to the predator-prey model in Problem 2.1, and check
     that the system is c.c. under the chosen method of population
     control. Suppose that it is possible to count only the *total*
     number of animals. Is the system c.o. — i.e., is it possible
     to deduce the individual numbers of predators and prey from
     this total?

2.10 Show that the oven model in Problem 2.2 is c.c. — i.e., it is
     possible to control both the oven interior and jacket tempera-
     ture by means of the heating coil in the jacket. Can the oven
     interior temperature be deduced by measuring only the jacket
     (excess) temperature?

2.11 Consider the cattle ranching model in Problem 2.3. If it is
     practicable to count only the total number of cattle, is it
     possible to deduce the numbers of cattle in each of the three
     age groups?

2.12 Verify that the system

$$\dot{x} = \begin{bmatrix} 0 & 3 & 0 \\ 2 & 0 & 0 \\ -1 & 1 & 2 \end{bmatrix} x + \begin{bmatrix} 0 & 0 \\ -2 & 0 \\ 1 & 1 \end{bmatrix} u$$

$$y = \begin{bmatrix} 1 & 2 & 0 \\ 0 & 1 & 1 \end{bmatrix} x$$

     is c.c. and c.o. Determine a basis for the subspace of final
     states which are attainable using u_1 only.

2.13 Describe for what range of values of the parameter θ the
     system

$$\dot{x} = \begin{bmatrix} 0 & 0 & -6 \\ 1 & 0 & 5 \\ 0 & 1 & 2 \end{bmatrix} x + \begin{bmatrix} \theta - 1 \\ \theta \\ 1 \end{bmatrix} u$$

     is c.c.

2.14 Two rods of unit lengths and negligible masses are suspended
by their ends and connected by a spring at a distance ℓ from
the points of suspension. Equal and opposite control forces
u(t) are applied to two particles, each of mass m, which are
fixed to the lower ends of the rods. It can be shown that the
equations of motion for *small* oscillations, in which θ_1^2, θ_2^2
and higher powers can be neglected, are

$$m\ddot{\theta}_1 = -u - mg\theta_1 - k\ell^2(\theta_1 - \theta_2)$$

$$m\ddot{\theta}_2 = u - mg\theta_2 + k\ell^2(\theta_1 - \theta_2)$$

where k is the spring constant. Write the equations in the
form (2.2) by taking as state variables $x_1 = \theta_1 + \theta_2$,
$x_2 = \theta_1 - \theta_2$, $x_3 = \dot{x}_1$, $x_4 = \dot{x}_2$. Hence show that the system is
not c.c. Would this conclusion be altered if the two control
forces were independent?

2.15 The linearized equations describing the equatorial motion of a
satellite in a circular orbit around the earth take the form
(2.2) with

$$A = \begin{bmatrix} 0 & 1 & 0 & 0 \\ 3\omega^2 & 0 & 0 & 2\omega \\ 0 & 0 & 0 & 1 \\ 0 & -2\omega & 0 & 0 \end{bmatrix}, \quad B = \begin{bmatrix} 0 & 0 \\ 1 & 0 \\ 0 & 0 \\ 0 & 1 \end{bmatrix}$$

The state variables represent the position and velocity of the
satellite in two dimensions; $u_1(t)$ is the radial thrust and
$u_2(t)$ is the tangential thrust. Show that the system is
completely controllable. Does it remain so if either the
radial thruster or tangential thruster fails? If not,
determine a basis for the subspace of initial and final states
for which transferability is possible.

2.16 Show that the condition of Theorem 2.2 is equivalent to the
following: $C(A, B)$ has rank n if and only if there is no left
eigenvector η of A such that $\eta B = 0$. Hence deduce that if
(A, B) is c.c., then so is (A + αI, B) for any scalar α.

2.17 If $B = B_1 B_1^T$, where B_1 ($\neq 0$) is an n × k matrix (n $\geqslant$ k $\geqslant$ 1),
such that rank B_1 = k (= rank B), show that

$$C(A, B) = [C(A, B_1)][I_n \; \theta \; B_1^T]$$

where θ denotes Kronecker product, and hence deduce that
rank $C(A, B)$ = rank $C(A, B_1)$. [Use the result that if X is
m × p, Y is p × q, then rank X + rank Y - p $\leqslant$ rank XY $\leqslant$
min(rank X, rank Y).]

2.18 An equivalent form of Theorem 2.2 is that the pair (A, B) is
c.c. if and only if for a row n-vector p,

$$p[\lambda I - A, B] = 0 \quad \text{implies that} \quad p = 0$$

for all complex numbers λ. Use this result to prove that if
the pair (A, B) is c.c. then so are the pairs:
(i)   (A + BK, B) for any m × n matrix K
(ii) (A - PC, B + PCP), where B and C are n × n and positive
      semidefinite, and P is symmetric.

2.19 Show that the transformation $\hat{x}$ = Sx, where S is an arbitrary
nonsingular n × n matrix, applied to (2.2) and (2.3) produces
the system

$$\frac{d\hat{x}}{dt} = A_1\hat{x} + B_1 u, \quad y = H_1\hat{x} \tag{2.52}$$

where

$$A_1 = SAS^{-1}, \quad B_1 = SB, \quad H_1 = HS^{-1} \tag{2.53}$$

The systems are said to be *algebraically equivalent*.
Show also that

$$C(A_1, B_1) = SC(A, B), \quad O(A_1, H_1) = O(A, H)S^{-1} \tag{2.54}$$

and hence deduce that (2.52) is c.c. and c.o. if and only if
the system described by (2.2) and (2.3) is c.c. and c.o.

2.20 Consider the system (2.2) in which u is a scalar, i.e., B is a
column n-vector. Use the result of the preceding problem to
show that if A is similar to $\mathrm{diag}[\lambda_1, \lambda_2, \ldots, \lambda_n]$, in which
some of the eigenvalues are repeated, then the system is not
c.c.

2.21 Suppose that A has distinct eigenvalues, and that in (2.53) S
is chosen so that $A_1 = \mathrm{diag}[\lambda_1, \lambda_2, \ldots, \lambda_n]$. Denote the rows
of the corresponding matrix B_1 by $\hat{b}_1, \ldots, \hat{b}_n$. Show that

$$C(A, B) = S^{-1} \begin{bmatrix} \hat{b}_1 & 0 & \cdots & 0 \\ 0 & \hat{b}_2 & \cdots & 0 \\ \multicolumn{4}{c}{\cdots\cdots\cdots} \\ 0 & 0 & \cdots & \hat{b}_n \end{bmatrix} (V_n^T \otimes I_m)$$

where V_n is the Vandermonde matrix in (1.23). Hence deduce
that the pair (A, B) is c.c. if and only if each row $\hat{b}_i$
contains at least one nonzero element.

2.22 Suppose that for the system (2.2) we have $t_0 = 0$ and
$x(t_0) = 0$. Use (2.11) to show that if u(t) satisfies the
condition $C(A, B)u(t) = 0$, $t \geqslant 0$, then $x(t) \equiv 0$, $t \geqslant 0$.
Similarly, if $u(t) \equiv 0$, $t \geqslant 0$, and x(0) instead satisfies the
condition $0(A, H)x(0) = 0$, show that $y(t) \equiv 0$, $t \geqslant 0$.

2.23 Let

$$C_k = [B, AB, A^2B, \ldots, A^kB], \quad k = 0, 1, 2, \ldots$$

and suppose that when $k = \ell$, rank $C_{\ell+1}$ = rank C_ℓ. Show that
this implies that

rank $C_{\ell+i}$ = rank C_ℓ,   for all $i \geqslant 2$

Hence deduce that if rank B = p, the controllability matrix
(2.23) can be replaced in Theorem 2.1 by C_{n-p}.

2.3 RELATIONSHIPS WITH POLYNOMIALS

We now investigate the form of the controllability/observability
matrices when the A matrix is in companion form (1.26) or (1.30).
We begin with the latter, and recall for convenience that

$$
C^T = \begin{bmatrix} 0 & \vdots & -a_n \\ - - - & \vdots & - - - \\ & \vdots & -a_{n-1} \\ I_{n-1} & \vdots & \vdots \\ & \vdots & -a_1 \end{bmatrix} \tag{2.55}
$$

where

$$
\det(\lambda I - C^T) = \lambda^n + a_1\lambda^{n-1} + \cdots + a_{n-1}\lambda + a_n \triangleq a(\lambda) \tag{2.56}
$$

We saw in Section 1.2 that

$$
b(C^T) \triangleq b_0(C^T)^m + b_1(C^T)^{m-1} + \cdots + b_{m-1}C^T + b_m I \tag{2.57}
$$

with $m < n$, takes the form in (1.60), which we write here as

$$
b(C^T) = [\underset{\sim}{b}, \; C^T\underset{\sim}{b}, \; (C^T)^2\underset{\sim}{b}, \; \ldots, \; (C^T)^{n-1}\underset{\sim}{b}] \tag{2.58}
$$

where

$$
\underset{\sim}{b} = [b_m, b_{m-1}, \ldots, b_1, b_0, 0, \ldots, 0]^T \tag{2.59}
$$

We now recognize the matrix on the right in (2.58) as having
precisely the same form as the controllability matrix defined in
(2.23), i.e.,

$$
C(C^T, \underset{\sim}{b}) = b(C^T) \tag{2.60}
$$

We established in Theorem 1.3 that $b(C^T)$ is nonsingular if and only
if $a(\lambda)$ in (2.56) and

$$
b(\lambda) = b_0\lambda^m + b_1\lambda^{m-1} + \cdots + b_m \tag{2.61}
$$

are relatively prime. Equation (2.60) can therefore be interpreted
in the light of Theorem 2.1 in the following way:

Theorem 2.6   The polynomials $a(\lambda)$ in (2.56) and $b(\lambda)$ in (2.61) are
relatively prime if and only if the pair $(C^T, \underset{\sim}{b})$ is c.c., [or,
equivalently, the pair $(C, \underset{\sim}{b}^T)$ is c.o.].

Proof.   The second part of the theorem follows immediately from
(2.49).

Remarks on Theorem 2.6   (1) We thus have an interesting, and
unexpected, connection between controllability/observability and
relative primeness of polynomials.   The relationship (2.60) can be
used to express results in Chapter 1 involving $b(C^T)$ in
controllability/observability matrix form (see Problem 2.25).
This idea will be developed in some detail for the bezoutian
matrix, in Section 2.4.

   (2) The result can be interpreted in a different way by
returning to the transfer function defined in Section 2.1.   The
continuous-time system equations relative to the controllability
part of Theorem 2.6 are

$$\dot{x}(t) = C^T x(t) + \underset{\sim}{b}u(t) \tag{2.62}$$

where $u(t)$ is a *scalar* control variable.   Taking the Laplace
transform of both sides of (2.62) and assuming that $x(0) = 0$ gives,
as in (2.14),

$$\bar{x}(s) = (sI - C^T)^{-1}\underset{\sim}{b}\bar{u}(s) \tag{2.63}$$

Suppose that the output is here equal to the last component of the
state vector, i.e., $y(t) = x_n(t) = d^T x(t)$, where

$$d^T = [0, 0, \ldots, 0, 1] \tag{2.64}$$

Recalling that the transfer function is equal to $\bar{y}(s)/\bar{u}(s)$, we
obtain from (2.63)

$$g(s) = d^T(sI - C^T)^{-1}\underset{\sim}{b} \tag{2.65}$$

$$= [\text{last row of } (sI - C^T)^{-1}]\underset{\sim}{b} \tag{2.66}$$

It is easy to show that the last row of $(sI - C^T)^{-1}$ is

$$[1, s, s^2, \ldots, s^{n-1}]/\det(sI - C^T)$$

$$= [1, s, \ldots, s^{n-1}]/a(s) \tag{2.67}$$

where $a(s)$ is defined in (2.56). Substituting (2.67) into (2.66) gives

$$g(s) = \frac{b(s)}{a(s)} \tag{2.68}$$

Equation (2.68) states that the single-input, single-output system described by (2.62), with $y = d^T x$, has transfer function equal to the ratio of $b(s)$ in (2.61) to $a(s)$ in (2.56). Since $g(s)$ is a scalar, we can transpose (2.65) to give

$$g(s) = \underset{\sim}{b}^T(sI - C)^{-1}d \tag{2.69}$$

and the right-hand side of (2.69) shows that $g(s)$ is also the transfer function for the system

$$\dot{x} = Cx + du, \quad y = \underset{\sim}{b}^T x \tag{2.70}$$

We can therefore interpret Theorem 2.6 in terms of $g(s)$ as follows:

Theorem 2.7   The following three statements are equivalent.
(i)   The system $\dot{x} = C^T x + \underset{\sim}{b}u$, $y = d^T x$, is c.c.
(ii)   The system described by (2.70) is c.o.
(iii) The numerator and denominator of the transfer function (2.68) are relatively prime.

Remarks on Theorem 2.7   (1) In the engineering literature the roots of $b(s)$ and $a(s)$ are called the *zeros* and *poles*, respectively, of $g(s)$, and a linear common divisor produces a "pole-zero cancellation." If a unit impulse is applied to the system (2.62), so that $\bar{u}(s) = 1$, then the Laplace transform of the output is $\bar{y}(s) \equiv g(s)$. A cancellation means that the "mode" in the solution $y(t)$, corresponding to that particular root of $a(s)$, is no longer present.

(2) By Theorem 1.4, the degree of the g.c.d. of b(s) and a(s) in (2.68) is equal to the rank defect of $C(C^T, \underline{b})$ [or $0(C, \underline{b}^T)$]; and using Theorem 1.5, the g.c.d. can itself be obtained from this matrix.

(3) As was seen in Section 2.1, corresponding results for discrete-time systems can be obtained merely by using the z, instead of the Laplace, transform. Thus, for example, part (i) of Theorem 2.7 is replaced by:

The system described by

$$x(k + 1) = C^T x(k) + \underline{b}u(k), \quad y(k) = d^T x(k)$$

is c.r. if and only if the numerator and denominator of the transfer function g(z) = b(z)/a(z) are relatively prime.

Note, however, that this holds only for strictly discrete systems; for equations which arise by sampling continuous systems, the sampling time must also be taken into account.

We can similarly reinterpret Theorem 1.6 as follows:

Theorem 2.8 The following three statements are equivalent.

(i)   The h + 1 polynomials consisting of $a(\lambda)$ in (2.56) and

$$b_i(\lambda) = b_{i1}\lambda^{n-1} + \cdots + b_{in}, \quad i = 1, 2, \ldots, h$$

are relatively prime.

(ii)   The pair (C^T, B) is c.c., where C^T is the companion matrix (2.55), and B is the n × h matrix having rows
$[b_{1j}, b_{2j}, \ldots, b_{hj}]$, j = n, n - 1, ..., 1.

(iii) The transfer function matrix

$$G(s) = d^T(sI - C^T)^{-1}B$$

$$= [b_1(s), b_2(s), \ldots, b_h(s)]/a(s) \qquad (2.71)$$

for the system $\dot{x} = C^T x + Bu$, $y = d^T x$, has no common factors between numerators and denominator.

Example 5 Let the system in part (iii) of Theorem 2.8 be

$$\dot{x} = \begin{bmatrix} 0 & 0 & -3 \\ 1 & 0 & -2 \\ 0 & 1 & 4 \end{bmatrix} x + \begin{bmatrix} 7 & 0 \\ -3 & -11 \\ 1 & 2 \end{bmatrix} u$$

$$y = [0, 0, 1]x$$

It is easily checked that $C(C^T, B)$ has rank 3, and this is
equivalent to the polynomials

$$a(\lambda) = \lambda^3 - 4\lambda^2 + 2\lambda + 3, \quad b_1(\lambda) = \lambda^2 - 3\lambda + 7,$$

$$b_2(\lambda) = 2\lambda^2 - 11\lambda$$

being relatively prime.

Our results in this section have so far only applied to linear
control systems whose state matrix is in companion form. We now
give a result for the single-input system

$$\dot{x} = Ax + bu \tag{2.72}$$

where A is an *arbitrary* n × n matrix, and b an *arbitrary* column
n-vector. As in (2.63) the vector transfer function for (2.72) is

$$e(s) = \frac{\bar{x}(s)}{\bar{u}(s)} = (sI - A)^{-1}b \tag{2.73}$$

$$\triangleq \frac{1}{k(s)} \begin{bmatrix} \beta_1(s) \\ \vdots \\ \beta_n(s) \end{bmatrix} \tag{2.74}$$

where $k(s)$ is the characteristic polynomial in (2.20) of A.

Theorem 2.9 The polynomials $k(s)$, $\beta_1(s)$, ..., $\beta_n(s)$ in (2.74) are
relatively prime if and only if the pair (A, b) is c.c.

Proof. From the expression for $(sI - A)^{-1}$ in (2.18), we have

$$(sI - A)^{-1}b = \frac{1}{k(s)} (s^{n-1}I + s^{n-2}B_1 + s^{n-3}B_2 + \cdots + B_{n-1})b \tag{2.75}$$

where from Problem 2.5

$$B_i = A^i + k_1 A^{i-1} + \cdots + k_i I, \quad i = 1, 2, \ldots, n - 1 \qquad (2.76)$$

Substituting (2.76) into (2.75), and collecting terms, produces

$$(sI - A)^{-1} b = \frac{1}{k(s)} (\theta_0(s)b + \theta_1(s)Ab + \theta_2(s)A^2 b + \cdots$$

$$+ \theta_{n-1}(s)A^{n-1}b) \qquad (2.77)$$

where the $\theta_i(s)$ are themselves polynomials, with $\theta_{n-1} = 1$. The numerator on the right-hand side of (2.77) is the vector $[\beta_1(s), \ldots, \beta_n(s)]^T$ in (2.74).

If the polynomials $k(s)$, $\beta_1(s)$, $\ldots$, $\beta_n(s)$ have a common factor, this means that there is some value of s, say s_0, which is a root of both numerator and denominator in (2.77). That is, $k(s_0) = 0$, and

$$\theta_0(s_0)b + \theta_1(s_0)Ab + \theta_2(s_0)A^2 b + \cdots + \theta_{n-1}(s_0)A^{n-1}b = 0$$
$$(2.78)$$

and since $\theta_{n-1} \neq 0$, (2.78) implies that $b, Ab, A^2 b, \ldots, A^{n-1}b$ are linearly dependent, i.e., rank $C(A, b) < n$. Hence, if the pair (A, b) is c.c., the polynomials must be relatively prime.

Conversely, we have to show that if (A, b) is not c.c., then the polynomials $k(s)$, $\beta_1(s)$, $\ldots$, $\beta_n(s)$ are not relatively prime. On this hypothesis, since rank $C(A, b) < n$, there exist scalars γ_i, not all zero, such that

$$\gamma_0 b + \gamma_1 Ab + \gamma_2 A^2 b + \cdots + \gamma_{n-1}A^{n-1}b = 0 \qquad (2.79)$$

From (2.73) we have

$$b = (sI - A)e \qquad (2.80)$$

and substituting for b from (2.80) into (2.79) gives

$$(sI - A)(\gamma_0 e + \gamma_1 Ae + \gamma_2 A^2 e + \cdots + \gamma_{n-1}A^{n-1}e) = 0 \qquad (2.81)$$

using the fact that $sI - A$ commutes with all powers of A. By setting s equal to any value for which $sI - A$ is nonsingular,

(2.81) therefore implies that

$$\sum_{i=0}^{n-1} \gamma_i A^i e = 0 \tag{2.82}$$

From (2.80) we also have

$$se = Ae + b, \quad s^2 e = s(Ae + b) = A^2 e + Ab + sb$$

$$s^i e = A^i e + A^{i-1} b + s A^{i-2} b + s^2 A^{i-3} b + \cdots + s^{i-1} b, \quad i \geqslant 3 \tag{2.83}$$

Using the expressions in (2.83) we can therefore write

$$(\gamma_0 + \gamma_1 s + \cdots + \gamma_{n-1} s^{n-1}) e = (\gamma_0 + \gamma_1 A + \gamma_2 A^2 + \cdots$$

$$+ \gamma_{n-1} A^{n-1}) e + \gamma_1 b + \gamma_2 (Ab + sb)$$

$$+ \cdots + \gamma_{n-1} (A^{n-2} b + s A^{n-3} b +$$

$$\cdots + s^{n-2} b) \tag{2.84}$$

and by (2.82) the first term on the right-hand side of (2.84) is zero. Hence (2.84) reduces to

$$e = v(s)/(\gamma_0 + \gamma_1 s + \cdots + \gamma_{n-1} s^{n-1}) \tag{2.85}$$

where $v(s)$ is a column n-vector having degree at most $n - 2$ in s. Clearly, the denominator in (2.85) has degree at most $n - 1$. Comparison of (2.85) with (2.74), in which the denominator $k(s)$ has degree n, therefore reveals that the polynomials in (2.74) must have a common factor. ∎

Example 6   Let the matrices in (2.72) be

$$A = \begin{bmatrix} -1 & 2 \\ 4 & -3 \end{bmatrix}, \quad b = \begin{bmatrix} b_1 \\ b_2 \end{bmatrix}$$

It is trivial to obtain det $C(A, b) = 2(2b_1 + b_2)(b_1 - b_2)$, and

$$(sI - A)^{-1} b = \frac{1}{(s + 5)(s - 1)} \begin{bmatrix} (s + 3)b_1 + 2b_2 \\ 4b_1 + (s + 1)b_2 \end{bmatrix} \tag{2.86}$$

If $b_2 = -2b_1$ then (A, b) is not c.c., and in (2.86) the vector becomes

$$\begin{bmatrix} (s - 1)b_1 \\ -2(s - 1)b_1 \end{bmatrix}$$

showing that there is a common factor $(s - 1)$ on the right-hand side. Similarly, if $b_2 = b_1$, there is a common factor $(s + 5)$.

Remark on Theorem 2.9 Be clear about the difference from Theorem 2.8. In (2.71) the denominator is the characteristic polynomial of C, and the coefficients of the polynomials $b_i(s)$ are obtained directly from the elements of B; in (2.74), k(s) is the characteristic polynomial of A, but the $\beta_i(s)$ are only obtained indirectly from b, a knowledge of the elements of $(sI - A)^{-1}$ also being required.

Problems

2.24 Interpret Problem 2.13 in terms of polynomials, using Theorem 2.6.

2.25 Use Theorem 2.8 to write down the pair (C^T, B) corresponding to the three polynomials in Example 11, Chapter 1.

2.26 For each of the following systems determine for what values of the parameter θ the system is not c.c. Confirm Theorem 2.9 in each case by finding the polynomials in (2.74).

(a)
$$\dot{x} = \begin{bmatrix} -1 & 1 & -1 \\ 0 & -1 & \theta \\ 0 & 1 & 3 \end{bmatrix} x + \begin{bmatrix} 0 \\ 2 \\ 1 \end{bmatrix} u$$

$$\dot{x} = \begin{bmatrix} -1 & 0 & 2 \\ 0 & 1 & 2 \\ 2 & 2 & 0 \end{bmatrix} x + \begin{bmatrix} 1 \\ \theta \\ 0 \end{bmatrix} u$$

2.27 Rewrite equation (1.86) using (2.60), so as to relate the Sylvester matrix to the controllability matrix $C(C_{III}, J_n \underset{\sim}{b})$, where C_{III} is the companion form defined in (1.33).

2.4 CANONICAL FORMS: SINGLE INPUT AND OUTPUT

In so-called classical control theory, a continuous-time linear
system with a *single* input is often described by the single
equation

$$z^{(n)} + a_1 z^{(n-1)} + a_2 z^{(n-2)} + \cdots + a_n z = u(t) \tag{2.87}$$

where $z(t)$, $u(t)$ are scalars, $z^{(j)} \triangleq d^j z/dt^j$, and the a_i are
constants. It is trivial to convert (2.87) into state space form;
simply set

$$w_1 = z, \ w_2 = z^{(1)}, \ \ldots, \ w_n = z^{(n-1)} \tag{2.88}$$

so that

$$\dot{w}_i = w_{i+1}, \quad i = 1, 2, \ldots, n - 1 \tag{2.89}$$

and from (2.88) and (2.87)

$$\dot{w}_n = z^{(n)} = -a_n w_1 - a_{n-1} w_2 \cdots - a_1 w_n + u \tag{2.90}$$

Equations (2.89) and (2.90) are equivalent to the single-input
system

$$\dot{w} = Cw + du \tag{2.91}$$

where $w = [w_1, w_2, \ldots, w_n]^T$, d was defined in (2.64) as

$$d = [0, 0, \ldots, 0, 1]^T \tag{2.92}$$

and C is the companion form in (1.26), having last row
$[-a_n, \ldots, -a_2, -a_1]$. Thus (2.91) is the state space form of
(2.87). The controllability matrix for this system is

$$C(C, d) = \begin{bmatrix} 0 & 0 & 0 & . & 0 & 1 \\ 0 & 0 & 0 & . & 1 & \theta_1 \\ \vdots & \vdots & \vdots & \vdots & \vdots & \vdots \\ 0 & 0 & 1 & . & \theta_{n-4} & \theta_{n-3} \\ 0 & 1 & \theta_1 & . & \theta_{n-3} & \theta_{n-2} \\ 1 & \theta_1 & \theta_2 & . & \theta_{n-2} & \theta_{n-1} \end{bmatrix} \tag{2.93}$$

where the θ_i are easily determined in terms of the a_i (see Problem 2.30). We see from its triangular form that the matrix (2.93) is nonsingular, showing that (2.91) is c.c. That is, if a system is modeled by the classical equation (2.87), then this is automatically c.c.; in other words, if only scalar differential equations like (2.87) are considered, then the concept of systems which are not c.c. will never be suspected. This is one of the chief reasons why the notion of controllability has been appreciated only relatively recently. It also provides an important justification for the use of state space methods. These do *not* consist of mere reformulation in matrix-vector terms, as they in fact reveal new concepts and properties.

The following result shows that an arbitrary single-input system can be transformed into the classical form (2.87), or its state space equivalent (2.91), if and only if it is c.c.

Theorem 2.10   Consider the system

$$\dot{x} = Ax + bu \tag{2.94}$$

where A is an arbitrary n × n matrix, and b an arbitrary nonzero column n-vector. Then there exists a nonsingular transformation

$$w = Px \tag{2.95}$$

which transforms (2.94) into (2.91) if and only if rank $C(A, b) = n$. In this case $P = C(C, d)[C(A, b)]^{-1}$,

Proof.   Substitution of (2.95) into (2.94) produces

$$\dot{w} = PAP^{-1}w + Pbu \tag{2.96}$$

If a suitable P exists, then comparison of (2.91) and (2.96) reveals that $C = PAP^{-1}$, $d = Pb$. It easily follows that $C^i = PA^iP^{-1}$, $i = 2, 3, \ldots$, so that we can write

$$C(C, d) = [d, Cd, C^2d, \ldots, C^{n-1}d]$$

$$= [Pb, (PAP^{-1})Pb, \ldots, (PA^{n-1}P^{-1})Pb]$$

$$= P[b, Ab, \ldots, A^{n-1}b]$$

$$= PC(A, b) \hspace{4cm} (2.97)$$

Since we noted earlier that $C(C, d)$ has rank n, and since by
assumption P is nonsingular, (2.97) shows that rank $C(A, b) = n$.
The stated expression for P then follows from (2.97). Conversely,
suppose that (A, b) is c.c. Define a unique row n-vector t by
$tC(A, b) = d^T$, which on expansion produces

$$tb = 0, \quad tAb = 0, \quad \ldots, \quad tA^{n-2}b = 0, \quad tA^{n-1}b = 1 \hspace{1cm} (2.98)$$

We establish this part of the theorem by showing that P in (2.95)
is given by

$$P = O(A, t) \hspace{4cm} (2.99)$$

First, we show that P in (2.99) is nonsingular by considering a
linear combination of its rows. If

$$\varepsilon_1 t + \varepsilon_2 tA + \varepsilon_3 tA^2 + \cdots + \varepsilon_n tA^{n-1} = 0 \hspace{2cm} (2.100)$$

for some scalars ε_i, postmultiply (2.100) by b and use (2.98) to
obtain $\varepsilon_n = 0$. Similarly, postmultiply (2.100) by Ab, A^2b, $\ldots$,
$A^{n-1}b$, in turn, to show that $\varepsilon_{n-1} = 0$, $\varepsilon_{n-2} = 0$, $\ldots$, $\varepsilon_1 = 0$.
Thus the rows of P are linearly independent.

Next, it follows immediately from (2.98) and (2.99) that
$Pb = d$, so it only remains to confirm that $PAP^{-1} = C$. Denote the
columns of P^{-1} by $s_1, \ldots, s_n$. Then from the identity $PP^{-1} = I_n$,
it follows that

$$\left. \begin{array}{l} tA^i s_j = 1, \quad j = i + 1 \\[2mm] \hspace{1.3cm} = 0, \quad \text{otherwise} \end{array} \right\} \quad i = 0, 1, \ldots, n - 1 \hspace{1cm} (2.101)$$

However,

$$PAP^{-1} = \begin{bmatrix} tA \\ tA^2 \\ \vdots \\ tA^n \end{bmatrix} [s_1, \ldots, s_n] \tag{2.102}$$

and the reader can check that substitution of (2.101) into (2.102) shows that the latter does indeed have the required companion form, with $a_i = -tA^n s_{n-i+1}$. ∎

Example 7 Consider the system

$$\dot{x} = \begin{bmatrix} 2 & 7 \\ -1 & 3 \end{bmatrix} x + \begin{bmatrix} 2 \\ -1 \end{bmatrix} u \tag{2.103}$$

whose controllability matrix is

$$C(A, b) = \begin{bmatrix} 2 & -3 \\ -1 & -5 \end{bmatrix}$$

which is nonsingular, so the system is c.c. From (2.98), with $t = [t_1, t_2]$, we obtain the equations

$$2t_1 - t_2 = 0, \quad -3t_1 - 5t_2 = 1$$

which have solution $t_1 = 1/13$, $t_2 = 2/13$. Hence from (2.99)

$$P = \begin{bmatrix} t \\ tA \end{bmatrix} = \frac{1}{13} \begin{bmatrix} -1 & -2 \\ 0 & -13 \end{bmatrix}$$

It is easy to compute

$$P^{-1} = \begin{bmatrix} -13 & 2 \\ 0 & -1 \end{bmatrix}, \quad PAP^{-1} = \begin{bmatrix} 0 & 1 \\ -13 & 5 \end{bmatrix} = C \tag{2.104}$$

and the characteristic polynomial of A is the same as that of C, i.e., $\lambda^2 - 5 + 13$.

Remarks on Theorem 2.10   (1) Since any c.c. single-input system can
be put into the form (2.91), the latter is termed the *controllable
canonical form* for (2.94).   Suppose that associated with (2.91) is
an arbitrary scalar output y = hw.   Then by duality, the system

$$\dot{v} = C^T v + h^T u, \qquad y = d^T v \tag{2.105}$$

is the *observable* canonical form, to which any single-input,
single-output system can be transformed if and only if it is c.o.
(see Problem 2.33).

 (2) To investigate further the form of the transformation
matrix P, invert both sides of (2.97) to give

$$P^{-1} = C(A, b)[C(C, d)]^{-1} \tag{2.106}$$

It is straightforward to use the expression for the elements of
$C(C, d)$ in (2.93) as given in Problem 2.30 to obtain

$$[C(C, d)]^{-1} = \begin{bmatrix} a_{n-1} & a_{n-2} & \cdots & a_1 & 1 \\ a_{n-2} & a_{n-3} & \cdots & 1 & 0 \\ \cdot & \cdot & & \cdot & \\ \cdot & \cdot & & & \\ a_1 & 1 & \cdot & & 0 \\ 1 & 0 & & & \end{bmatrix} \tag{2.107}$$

and we recognize the matrix on the right in (2.107) as none other
than T in (1.35), a matrix which transforms C into C^T.   Thus,
knowing A, b and the characteristic polynomial of A, (2.106) and
(2.107) provide an alternative way of determining P^{-1}; its disadvantage is the need to predetermine the characteristic polynomial.
Notice, incidentally, that (2.106) shows that P is unique.

Example 7 (continued)   We found that a_1 = -5, and A, b are given in
(2.103), so (2.106) and (2.107) imply that

$$P^{-1} = \begin{bmatrix} 2 & -3 \\ -1 & -5 \end{bmatrix} \begin{bmatrix} -5 & 1 \\ 1 & 0 \end{bmatrix}$$

$$= \begin{bmatrix} -13 & 2 \\ 0 & -1 \end{bmatrix}, \quad \text{as in (2.104)}$$

Consider now the controllable and observable canonical forms in (2.91) and (2.105), rewritten here for convenience:

$$\dot{w} = Cw + du, \quad y = b^T w \quad (I) \tag{2.108}$$

$$\dot{v} = C^T v + bu, \quad y = d^T v \quad (II) \tag{2.109}$$

where b is defined in (2.59), i.e., $b = [b_m, b_{m-1}, \ldots]^T$. Suppose that (2.108) is *also* c.o., and that (2.109) is *also* c.c., and denote by C_I, 0_I, C_{II}, 0_{II} their respective controllability and observability matrices. It follows immediately by duality [see (2.49)] that

$$C_I = 0_{II}^T, \quad 0_I = C_{II}^T \tag{2.110}$$

Recall that in equation (1.120) we showed that the bezoutian matrix Z associated with $a(\lambda)$ (the characteristic polynomial of C) and $b(\lambda)$ in (2.61) satisfies

$$Z = Tb(C) \tag{2.111}$$

Furthermore, we saw in (2.60) that $b(C^T) = C_{II}$, i.e.,

$$b(C) = C_{II}^T \ (= 0_I) \tag{2.112}$$

and since by (2.107), $T = C_I^{-1}$, the expressions (2.111) and (2.112) produce a controllability/observability matrix expression for Z:

$$Z = C_I^{-1} 0_I \tag{2.113}$$

However, there are three other such expressions. First, using the fact that Z is symmetric, we have

$$Z = Z^T = O_I^T (C_I^T)^{-1} \tag{2.114}$$

$$= C_{II} O_{II}^{-1}, \quad \text{using (2.110)} \tag{2.115}$$

Next, C_I^{-1} in (2.107) is symmetric, so C_I is also symmetric. Thus, transposing the first part of (2.110) shows that $C_I = O_{II}$. Substituting this into (2.113) gives

$$Z = O_{II}^{-1} O_I \tag{2.116}$$

Finally, transposing (2.116) and again applying (2.110) similarly shows that

$$Z = C_{II} C_I^{-1} \tag{2.117}$$

We can combine these facts, together with a further interesting property of Z.

Theorem 2.11   The bezoutian matrix Z associated with the poly-
nomials $a(\lambda)$ in (2.56) and $b(\lambda)$ in (2.61) can be expressed in terms of controllability and observability matrices according to (2.113) and (2.115) to (2.117). In addition, when $a(\lambda)$ and $b(\lambda)$ are relatively prime, Z transforms the controllable canonical form (2.108) into the observable form (2.109).

Proof.   Only the last part remains to be established. The trans-
formation matrix P in Theorem 2.10 sends the arbitrary system (2.94) into (2.108). Setting A in (2.97) equal to C^T thus gives the transformation from the observable form (2.109) to the controllable form (2.108). We seek the *inverse* of this transformation, which from (2.106) with $A \equiv C^T$ is

$$P^{-1} = C(C^T, b) [C(C, d)]^{-1}$$

$$= C_{II} C_I^{-1} = Z, \quad \text{by (2.117)} \blacksquare$$

Example 8   Consider the polynomials

$$a(\lambda) = \lambda^3 - 3\lambda^2 + \lambda + 5, \quad b(\lambda) = 2\lambda^2 - 3\lambda + 1$$

Their bezoutian matrix was given in (1.126), and it is easy to determine Z^{-1}, this being simplified by the fact (Theorem 1.14) that it is orthosymmetric, so

$$Z = \begin{bmatrix} 16 & -13 & 1 \\ -13 & 8 & -3 \\ 1 & -3 & 2 \end{bmatrix}, \quad Z^{-1} = \frac{-1}{156} \begin{bmatrix} 7 & 23 & 31 \\ 23 & 31 & 35 \\ 31 & 35 & -41 \end{bmatrix} \quad (2.118)$$

The system (2.108) is

$$\dot{w} = \begin{bmatrix} 0 & 1 & 0 \\ 0 & 0 & 1 \\ -5 & -1 & 3 \end{bmatrix} w + \begin{bmatrix} 0 \\ 0 \\ 1 \end{bmatrix} u, \quad y = [1, -3, 2]w$$

and the reader can readily check that $v = Zw$, with Z given by (2.118), does indeed transform this into (2.109).

Remarks on Theorem 2.11   (1) It is intriguing that the bezoutian matrix, which originally arose in the theory of resultants, should play such a role in linear control system representations. Indeed, we shall encounter further evidence of the usefulness of the bezoutian in Chapter 3.

(2) Recall from Theorem 1.12 (Remark 1) that relative primeness of $a(\lambda)$ and $b(\lambda)$ is equivalent to nonsingularity of Z. This is equivalent to each of (2.108) and (2.109) being both c.c. and c.o.

Problems

2.28 Determine the tranformation (2.95) and the controllable canonical form for the system

$$\dot{x} = \begin{bmatrix} 1 & 1 & -2 \\ -1 & 2 & 1 \\ 0 & 1 & -1 \end{bmatrix} x + \begin{bmatrix} 1 \\ 0 \\ 2 \end{bmatrix} u$$

2.29 Consider the controllable canonical form obtained in the
     preceding problem, and let the system output be
     $y = -w_1 + 2w_2 + w_3$. Verify that the system is also c.o. By
     determining the appropriate bezoutian matrix using a suitable
     method from Section 1.5, carry out the transformation of
     Theorem 2.11 into the observable canonical form.

2.30 Show that the elements of $C(C, d)$ in (2.93) are given by

$$\theta_i = - \sum_{j=0}^{i-1} a_{j+1}\theta_{i-j-1}, \quad i = 1, 2, \ldots, n - 1; \quad \theta_0 = 1 \quad (2.119)$$

Hence verify the expression for $[C(C, d)]^{-1}$ in (2.107). Show
also that

$$\frac{1}{\det(sI - A)} = \frac{1}{s^n} + \frac{\theta_1}{s^{n+1}} + \frac{\theta_2}{s^{n+2}} + \cdots + \frac{\theta_{n-1}}{s^{2n-1}} + \cdots$$

2.31 Deduce from Theorem 2.10 that a *necessary* condition for (2.94)
     to be c.c. is that A is nonderogatory (defined in Section
     1.2). An alternative characterization of a nonderogatory
     matrix is that in its Jordan form there is only one Jordan
     block associated with each distinct eigenvalue (see the proof
     of Theorem 1.4). Thus the result of this problem is an
     extension of that in Problem 2.20.

2.32 Show that the matrix P^{-1} defined by (2.106) and (2.107) can be
     written as

$$[B_{n-1}b, B_{n-2}b, \ldots, B_1b, b]$$

where the matrices B_i are defined in (2.76) with $k_i \equiv a_i$, for
all i [since A is similar to C, $\det(sI - A) \equiv \det(sI - C)$].

2.33 Show that the system $\dot{x} = Ax$, $y = b^Tx$, if c.o., is tranformed
     into $\dot{v} = C^Tv$, $y = d^Tv$, by means of $x = Qv$, where Q is given by
     $\mathcal{O}(A, b^T)Q = \mathcal{O}(C^T, d^T)$.

2.5 REALIZATION OF TRANSFER FUNCTIONS

We now consider the problem of determining a state variable
description corresponding to a known transfer function g(s), which
in practice can be identified from experimentally determined data.
The formal definition is: Given *any* scalar rational transfer
function g(s) (assumed *proper*, i.e., the degree of its numerator is
less than that of its denominator) then any *triple* R = {F, h, j}
consisting of an N × N matrix F, a column n-vector h and a row
N-vector j such that

$$g(s) = j(sI - F)^{-1}h \qquad (2.120)$$

is called a state space *realization* of g(s) of *order* N, since the
system

$$\dot{x} = Fx + hu, \quad y = jx \qquad (2.121)$$

has transfer function g(s). There is no unique triple satisfying
(2.120), but among all realizations there will be some in which N
takes its smallest value. Such realizations are termed *minimal*,
and are important since they correspond to a set of state equations
(2.121) in which the number of state variables is as small as
possible. In other words, in a nonminimal realization of a
transfer function, some of the state variables are redundant. In
particular, (2.108) and (2.109) are, respectively, a c.c. and a
c.o. realization of g(s) in (2.68), but need not necessarily be
minimal, as the following illustrates.

Example 9 Consider a transfer function

$$g(s) = \frac{s^2 + 7s + 12}{s^3 + 7s^2 + 14s + 8} \qquad (2.122)$$

The c.c. realization in (2.108) of g(s) is

$$C = \begin{bmatrix} 0 & 1 & 0 \\ 0 & 0 & 1 \\ -8 & -14 & -7 \end{bmatrix}, \quad d = \begin{bmatrix} 0 \\ 0 \\ 1 \end{bmatrix}, \quad b^T = [12, 7, 1] \qquad (2.123)$$

However, there is a common factor $(s + 4)$ between numerator and denominator of (2.122), and after removing this the reader can readily confirm that two alternative realizations of $g(s)$ are

$$R_1 = \left\{ \begin{bmatrix} 0 & 1 \\ -2 & -3 \end{bmatrix}, \begin{bmatrix} 0 \\ 1 \end{bmatrix}, [3, 1] \right\} \tag{2.124}$$

$$R_2 = \left\{ \begin{bmatrix} -2 & 0 \\ 0 & -1 \end{bmatrix}, \begin{bmatrix} -1 \\ 2 \end{bmatrix}, [1, 1] \right\} \tag{2.125}$$

so (2.123) is certainly not minimal, since the order of both R_1 and R_2 is two.

Example 9 suggests that minimality of a realization is related to cancellation of factors between numerator and denominator of $g(s)$. We now establish this in a precise form:

Theorem 2.12 Let $g(s) = p(s)/q(s)$ be an arbitrary proper transfer function, with $\delta q = N$, and let F be any $N \times N$ matrix such that $\det(sI - F) = q(s)$. Then a realization $R = \{F, h, j\}$ of $g(s)$ is minimal if and only if $p(s)$ and $q(s)$ are relatively prime.

Proof. Suppose first that R is minimal. If $p(s)$ and $q(s)$ are not relatively prime, then we can remove their g.c.d. to give $p/q \equiv \hat{p}/\hat{q}$, where $\delta\hat{q} < N$. We can then write down a realization, using (2.108) or (2.109), of $\hat{p}/\hat{q}$ in which C is the companion matrix of $\hat{q}$, and so has dimension less than N. This contradicts the assumption that R is minimal, so p and q must be relatively prime.

Conversely, given that $p(s)$ and $q(s)$ are relatively prime, we must show that R is minimal. Suppose that it is not, i.e., there exists a realization $R' = \{F', h', j'\}$ of $g(s)$ in which F' has dimension less than N. Now

$$\frac{p(s)}{q(s)} = \frac{j'\,\text{adj}(sI - F')h'}{\det(sI - F')} \tag{2.126}$$

and since the denominator on the right-hand side of (2.126) has degree less than N, this implies that p and q have a nontrivial g.c.d. — a contradiction, so R is minimal. ∎

It is interesting to interpret Theorem 2.12 in terms of the canonical form realizations (2.108) and (2.109). In view of Theorem 2.7 we can state:

Theorem 2.13 The realizations $\{C, d, b^T\}$ in (2.108) and $\{C^T, b, d^T\}$ in (2.109) of $b(s)/a(s)$ in (2.68) are minimal if and only if they are both c.c. and c.o.

We can go further, however, and extend this characterization of minimality to *any* realization:

Theorem 2.14 A realization of a given transfer function is minimal if and only if it is c.c. and c.o.

Proof. This will be given in Chapter 4 for the general case when $G(s)$ is the transfer function matrix in (2.16) (Theorem 4.13).

Example 10 It is easily confirmed that

$$F = \begin{bmatrix} 1 - a & 1 \\ a - 3 & -3 \end{bmatrix}, \quad h = \begin{bmatrix} 0 \\ 1 \end{bmatrix}, \quad j = [0, 1]$$

is a realization for

$$g(s) = \frac{s + a - 1}{s^2 + (a + 2)s + 2a} = \frac{s + a - 1}{(s + 2)(s + a)}$$

By Theorem 2.12, the realization is minimal if and only if numerator and denominator are relatively prime, i.e., $a \neq 3$. Furthermore, it is easy to obtain det $C(F, h) = -1$, det $0(F, j) = 3 - a$, so the realization is always c.c., but is c.o. only provided that $a \neq 3$, thus agreeing with Theorem 2.14.

Example 11 Return to the transfer function (2.122). The reader can check that the observability matrix for the c.c. realization (2.123) is singular, so this realization is not minimal. It is, however, easily verified that (2.124) and (2.125) are minimal realizations of (2.122), since each is both c.c. and c.o.

In general, a simple procedure for obtaining a minimal realization of b(s)/a(s), where a(s), b(s) are defined in (2.56) and (2.61), respectively, is as follows:

(a) Let C be the companion matrix having last row $[-a_n, \ldots, -a_1]$, b = $[b_m, b_{m-1}, \ldots]^T$, $d^T = [0, 0, \ldots, 0, 1]$. If $C(C^T, b)$ has full rank, then from (2.109) $\{C^T, b, d^T\}$ is a minimal realization.

(b) If rank $C(C^T, b) < n$, determine the g.c.d. e(s) of a(s) and b(s) using Theorem 1.5 [remembering that $b(C^T) = C(C^T, b)$; see (2.60)]. After removing the g.c.d. from numerator and denominator, a minimal realization can be written down as in (a). In particular, the order of minimal realizations is $n - \delta e$, which by Theorem 1.4 is equal to rank $b(C^T)$ = rank $C(C^T, b)$.

Example 12   We use the polynomials in Example 10c, Chapter 1, so

$$g(s) = \frac{s^4 - 3s^3 + s^2 + 3s - 2}{s^5 + s^4 - 9s^3 - 5s^2 + 16s + 12}$$

The matrix $C(C^T, b)$ was given in (1.76), and was found to have rank equal to 3, so this is the order of minimal realizations. It was also found from (1.76) that e(s) = $s^2 - s - 2$, and on removing this g.c.d. from g(s), the reduced transfer function is $(s^2 - 2s + 1)/(s^3 + 2s^2 - 5s - 6)$. Applying (2.109) to this gives as a minimal realization of g(s):

$$\left\{ \begin{bmatrix} 0 & 0 & 6 \\ 1 & 0 & 5 \\ 0 & 1 & -2 \end{bmatrix}, \begin{bmatrix} 1 \\ -2 \\ 1 \end{bmatrix}, [0, 0, 1] \right\}$$

It follows from Theorems 2.11 and 2.13 that when each of the canonical forms (2.108) and (2.109) is both c.c. and c.o., then the bezoutian matrix transforms one into the other. However, alternative minimal realizations can be obtained (see Problem 2.34), and it is interesting that any two minimal realizations are simply related:

Theorem 2.15 If $R_1 = \{A, \beta, \gamma\}$ is a minimal realization of $g(s)$, then $R_2 = \{\hat{A}, \hat{\beta}, \hat{\gamma}\}$ is also minimal if and only if the systems are *algebraically equivalent*, as defined in Problem 2.19.

Proof. From (2.53), algebraic equivalence requires that there exists a nonsingular matrix S such that

$$\hat{A} = SAS^{-1}, \quad \hat{\beta} = S\beta, \quad \hat{\gamma}S = \gamma \qquad (2.127)$$

If R_1 is a minimal realization, and (2.127) holds, then

$$\hat{\gamma}(sI - \hat{A})^{-1}\hat{\beta} = \gamma S^{-1}(sI - SAS^{-1})^{-1}S\beta$$

$$= \gamma(sI - A)^{-1}\beta$$

so R_2 is indeed a realization of $g(s)$. Since $\hat{A}$ obviously has the same dimensions as A, R_2 has the same order as R_1, and so is also minimal.

Conversely, if R_1 and R_2 are minimal, we shall show that (2.127) holds by demonstrating that a suitable S is given explicitly by

$$S = \hat{0}^{-1}0 \qquad (2.128)$$

where $\hat{0} \triangleq 0(\hat{A}, \hat{\gamma})$, $0 \triangleq 0(A, \gamma)$. First, since R_1 and R_2 are minimal, then by Theorem 2.14 both 0 and $\hat{0}$ are nonsingular, and hence so is S. Equation (2.128) is equivalent to $\hat{0}S = 0$, or

$$\begin{bmatrix} \hat{\gamma} \\ \hat{\gamma}\hat{A} \\ \vdots \end{bmatrix} S = \begin{bmatrix} \gamma \\ \gamma A \\ \vdots \end{bmatrix}$$

whence $\hat{\gamma}S = \gamma$, showing that the third relationship in (2.127) is satisfied. Next, since R_1 and R_2 are realizations of the same transfer function, we have

$$\gamma(sI - A)^{-1}\beta = \hat{\gamma}(sI - \hat{A})^{-1}\hat{\beta} \qquad (2.129)$$

Taking the inverse Laplace transform of both sides of (2.129) shows (see Table 2.1) that

$$\gamma e^{At} \beta = \hat{\gamma} e^{\hat{A}t} \hat{\beta}$$

which on using the series expansion for e^{At} in (2.9) gives

$$\gamma(I + tA + \frac{t^2}{2!} A^2 + \cdots)\beta = \hat{\gamma}(I + t\hat{A} + \frac{t^2}{2!} \hat{A}^2 + \cdots)\hat{\beta} \qquad (2.130)$$

Equating coefficients of powers of t in (2.130) produces

$$\gamma A^i \beta = \hat{\gamma}\hat{A}^i\hat{\beta}, \quad i = 0, 1, 2, \ldots \qquad (2.131)$$

which implies that

$$0\beta = \hat{0}\hat{\beta} \qquad (2.132)$$

Substituting from (2.128) reduces (2.132) to $S\beta = \hat{\beta}$, showing that the second relationship in (2.127) is satisfied. Finally, it is left for the reader to verify that $SA = \hat{A}S$ (Problem 2.37). ∎

Problems

2.34 Show that if the roots $\lambda_1, \lambda_2, \ldots, \lambda_n$ of a(s) are distinct, and

$$g(s) = \frac{b(s)}{a(s)} = \sum_{i=1}^{n} \frac{g_i}{s - \lambda_i} \quad (\text{all } g_i \neq 0)$$

then a minimal realization of g(s) is

$$\{\text{diag}[\lambda_1, \lambda_2, \ldots, \lambda_n], [1, 1, \ldots, 1, 1]^T, [g_1, g_2, \ldots, g_n]\}$$

2.35 Determine a minimal realization for each of the three transfer functions whose denominator and numerator are given by the polynomials in Problem 1.21.

2.36 Use (2.128) to obtain a transformation between the minimal realizations (2.124) and (2.125) of g(s) in (2.122). Verify that your expression is correct.

2.37 Use (2.131) to prove that

$$O(A, \gamma)A^k C(A, \beta) = O(\hat{A}, \hat{\gamma})\hat{A}^k C(\hat{A}, \hat{\beta})$$

for k = 0, 1. Hence show that S in (2.128) is equal to $C(\hat{A}, \hat{\beta})[C(A, \beta)]^{-1}$, and therefore establish that $SA = \hat{A}S$.

2.38 If $\{A_1, b_1, h_1\}$, $\{A_2, b_2, h_2\}$ are realizations of $g_1(s)$ and $g_2(s)$, respectively, show that

$$\left\{ \begin{bmatrix} A_1 & b_1 h_2 \\ 0 & A_2 \end{bmatrix}, \begin{bmatrix} 0 \\ b_2 \end{bmatrix}, [h_1, 0] \right\}$$

is a realization of $g_1(s)g_2(s)$.

2.39 Use the result of Problem 2.20 to show that if $\{A, b, h\}$ is a minimal realization and $\det(sI - A)$ has a repeated root, then A cannot be diagonalized by a similarity transformation.

2.40 If $\{A, b_1, h\}$, $\{A, b_2, h\}$ are two realizations of the same transfer function, show that $hA^i b_1 = hA^i b_2$, i = 0, 1, 2, ... Hence deduce that if each realization is c.o., then $b_1 \equiv b_2$.

2.6 SCALAR LINEAR STATE FEEDBACK

We return to the concept of feedback, introduced in Section 2.1. We again restrict ourselves to single-input systems, described by either

$$\dot{x}(t) = Ax(t) + bu(t) \tag{2.133}$$

or

$$x(k + 1) = Ax(k) + bu(k) \tag{2.134}$$

and consider the situation where *linear state feedback* is applied, i.e., the control function is taken to be a linear combination of the state variables

$$u = fx \tag{2.135}$$

where $f = [f_1, f_2, \ldots, f_n]$ is the *feedback vector*. Substitution of (2.135) into (2.133) or (2.134) produces the respective *closed-loop systems*

$$\dot{x}(t) = (A + bf)x(t), \quad x(k + 1) = (A + bf)x(k) \qquad (2.136)$$

so in both cases the *closed-loop matrix* is

$$A_c = A + bf \qquad (2.137)$$

Our aim is to choose f so as to specify the behavior of the solutions of the closed-loop systems (2.137), as functions of time. Note that by (2.9) and (2.10), these solutions can be written, respectively, as

$$x(t) = \exp(A_c t)x(0), \quad x(k) = (A_c)^k x(0)$$

and the transition matrices $\exp(A_c t)$, $(A_c)^k$ are determined by the eigenvalues and eigenvectors of A_c. In fact, we have control only over the former, as the following key result makes clear.

Theorem 2.16   Let the characteristic polynomial of $A_c = A + bf$ be

$$\phi(\lambda) = \det(\lambda I_n - A_c)$$

$$= \lambda^n + \phi_1 \lambda^{n-1} + \cdots + \phi_n \qquad (2.138)$$

Then provided that the pair (A, b) is c.c., there exists a unique vector f such that the coefficients ϕ_i in (2.138) are equal to any set of n numbers.

Proof. Since the pair (A, b) is c.c., by Theorem 2.10 there exists a nonsingular transformation $w = Px$ in (2.95) which transforms (2.133) into the controllable canonical form (2.91), namely

$$\dot{w} = Cw + du$$

Applying linear feedback with respect to w, i.e.,

$$u = \hat{f}w \quad (= \hat{f}Px = fx) \qquad (2.139)$$

produces

$$\dot{w} = (C + d\hat{f})w$$

Since $d = [0, 0, \ldots, 0, 1]^T$, it follows that the matrix $C + d\hat{f}$ has the same companion form as C, but with last row

$$[-a_n + \hat{f}_n, \ -a_{n-1} + \hat{f}_{n-1}, \ \ldots, \ -a_1 + \hat{f}_1]$$

where

$$\hat{f} = [\hat{f}_n, \ \hat{f}_{n-1}, \ \ldots, \ \hat{f}_1] \tag{2.140}$$

This implies that the characteristic polynomial of $C + d\hat{f}$ is

$$\lambda^n + (a_1 - \hat{f}_1)\lambda^{n-1} + \cdots + (a_{n-1} - \hat{f}_{n-1})\lambda + (a_n - \hat{f}_n) \tag{2.141}$$

In order to make (2.141) identical with the desired polynomial (2.138), we therefore select

$$\hat{f}_i = a_i - \phi_i, \quad i = 1, 2, \ldots, n \tag{2.142}$$

Moreover, from Theorem 2.10 we have

$$A = P^{-1}CP, \quad b = P^{-1}d \tag{2.143}$$

so

$$A_c = A + bf$$

$$= P^{-1}CP + P^{-1}d\hat{f}P, \quad \text{by (2.142)}$$

$$= P^{-1}(C + d\hat{f})P \tag{2.144}$$

showing that A_c is similar to $C + d\hat{f}$, and thus has the same characteristic polynomial (2.138). The uniqueness of $f = \hat{f}P$ follows from that of $\hat{f}$, displayed in (2.142), and P (Remark 2 on Theorem 2.10). ∎

Example 13   Return to the system (2.103), in Example 7. We found that the coefficients in the characteristic polynomial of A were $a_1 = -5$, $a_2 = 13$. Suppose that the closed-loop characteristic polynomial is required to be

126   Basic Properties of Control Systems

$$\phi(\lambda) = \lambda^2 + 11\lambda + 1$$

From (2.142) $\hat{f}_1 = -5 - 11$, $\hat{f}_2 = 13 - 1$, so $\hat{f} = [12, -16]$. Hence by (2.139) the desired feedback vector is

$$f = \hat{f}P = \frac{1}{13}[-12, 184]$$

using the matrix P found in Example 7. As a check, compute

$$A_c = A + bf = \frac{1}{13}\begin{bmatrix} 2 & 459 \\ -1 & -145 \end{bmatrix}$$

which does indeed have the stated characteristic polynomial.

<u>Remarks on Theorem 2.16</u>   (1) The result is often called the "pole-assignment" theorem, since we are preselecting the poles of the closed-loop transfer function, corresponding to some output $y = hx$. This transfer function is

$$g_c(s) = h(sI - A_c)^{-1}b$$
$$= h \text{ adj}(sI - A_c)b/\phi(s) \tag{2.145}$$

for (2.133), with s replaced by z for (2.134). Notice, however, that if A, b, and the ϕ_i are all real (as is usually the case in practice), then so is f, and hence also A_c. Thus, in particular, any complex eigenvalues of A_c will occur in conjugate pairs. It is preferable, therefore, to describe the result as one on assignment of the characteristic polynomial.

(2) If w is an eigenvector of $C + d\hat{f}$, then from the similarity relationship (2.144) it follows that $P^{-1}w$ is an eigenvector of A_c. However, the eigenvectors of the companion form matrix $C + d\hat{f}$ are entirely determined by its eigenvalues which are those of A_c (see Problem 1.11). In other words, once $\phi(\lambda)$ has been chosen, this automatically fixes the eigenvectors of A_c.

(3) We can also deduce from (2.144) that since A_c is similar to the companion form matrix $C + d\hat{f}$, it is nonderogatory. This implies that the Jordan form of A cannot be made arbitrary (see the

comment after Problem 2.31), and hence that the solutions of
(2.136) as functions of time cannot be completely prespecified.

(4) A converse of Theorem 2.16 holds, namely that the closed-
loop characteristic polynomial can be assigned *only* if the system
is c.c., but we leave the proof of this result until the general
multi-input case in Chapter 4 (Theorem 4.1).

(5) Using the expression for P in Theorem 2.10,
$u = \hat{f}C(C, d)[C(A, b)]^{-1}x$. However, an alternative formula, which
does not require a knowledge of the characteristic polynomial of A,
is as follows:

Theorem 2.17   The feedback vector f in Theorem 2.16 can be
expressed as

$$f = -d^T[C(A, b)]^{-1}\phi(A) \tag{2.146}$$

where $\phi(\lambda)$ is the desired closed-loop polynomial (2.138), and
$d^T = [0, 0, \ldots, 0, 1]$.

Proof.   From (2.142) $\phi_i = a_i - \hat{f}_i$, so by (2.138)

$$\phi(A) = A^n + (a_1 - \hat{f}_1)A^{n-1} + \cdots + (a_n - \hat{f}_n)I$$

$$= (A^n + a_1 A^{n-1} + \cdots + a_n I) - (\hat{f}_1 A^{n-1} + \cdots + \hat{f}_n I) \tag{2.147}$$

$$= -(\hat{f}_1 A^{n-1} + \hat{f}_2 A^{n-2} + \cdots + \hat{f}_n I) \tag{2.148}$$

where the expression within the first set of parentheses in (2.147)
vanishes by virtue of the Cayley-Hamilton theorem. Postmultiplying
both sides of (2.148) by b gives

$$\phi(A)b = -(\hat{f}_1 A^{n-1}b + \hat{f}_2 A^{n-2}b + \cdots + \hat{f}_n b)$$

$$= -C(A, b)\hat{f}^T \tag{2.149}$$

where $\hat{f}$ is defined in (2.140). Since (A, b) is c.c. we can obtain
from (2.149)

$$\hat{f}^T = - [C(A, b)]^{-1}\phi(A)b \qquad (2.150)$$

Now $\hat{f}d = d^T\hat{f}^T$, since both are scalars, so (2.150) implies that

$$-d^T[C(A, b)]^{-1}\phi(A)b = \hat{f}d$$

$$= fP^{-1}d, \quad \text{by (2.139)}$$

$$= fb, \quad \text{by (2.143)} \qquad (2.151)$$

Since (2.151) must hold for all possible vectors b for which (A, b) is c.c., it follows that f must equal the stated expression in (2.146). ■

Example 14 We repeat Example 13, using Theorem 2.17. The matrix A and vector b are given in (2.103). Equation (2.146) gives

$$f = -[0, 1][C(A, b)]^{-1}(A^2 + 11A + I)$$

$$= -[0, 1]\begin{bmatrix} 2 & -3 \\ -1 & -5 \end{bmatrix}^{-1}\begin{bmatrix} 20 & 112 \\ -16 & 36 \end{bmatrix}$$

$$= \frac{1}{13}[-12, 184]$$

agreeing with our previous calculation.

Problems

2.41 Determine linear state feedback for the system in Problem 2.28 such that the resulting closed-loop system has eigenvalues -2, -3, -4.
  (a) Using the canonical form found previously.
  (b) Using Theorem 2.17.
  Check your result by calculating A_c and determining its characteristic polynomial.

2.42 A simple model for the national income of a state is based on the following assumptions:
  (i)   National income I(k) in year k
        = (consumer expenditure) + (private investment)
        + [government expenditure, g(k)]

(ii) Consumer expenditure is proportional to national income in the previous year.

(iii) Private investment is proportional to the change in consumer spending over the previous year.

This leads to the equation

$$I(k + 2) = \alpha(1 + \beta) I(k + 1) - \alpha\beta I(k) + g(k + 2)$$

where $0 < \alpha < 1$, $0 < \beta \leqslant 1$. Choose as state variables $x_1(k) = I(k)$, $x_2(k) = I(k + 1)$, and determine linear feedback in the form

$$g(k + 2) = \theta_1 x_1(k) + \theta_2 x_2(k)$$

such that the resulting closed-loop system has characteristic polynomial $\lambda^2 - \alpha\beta\lambda + 1$.

2.43 Suppose that b is a right eigenvector of A corresponding to a nonzero eigenvalue λ. Show that $\lambda + fb$ and b are, respectively, a corresponding eigenvalue and eigenvector of $A_c = A + bf$. If instead f is a left eigenvector of A corresponding to λ, and e is a right eigenvector of A corresponding to an eigenvalue μ different from λ, show that μ and e are also, respectively, an eigenvalue and eigenvector of A_c.

2.44 Let q be a left eigenvector of A corresponding to an eigenvalue λ. Show that if b is orthogonal to q, then $A_c = A + bf$ also has an eigenvalue λ. Hence deduce from Theorem 2.16 that the pair (A, b) is not c.c.

2.45 Let ω be an arbitrary column n-vector, and ν an arbitrary row n-vector, and define the $(n + 1) \times (n + 1)$ matrix

$$X = \begin{bmatrix} I_n & \omega \\ \nu & 1 \end{bmatrix}$$

By considering the product

$$\begin{bmatrix} I_n & 0 \\ -\nu & 1 \end{bmatrix} X \qquad\qquad (2.152)$$

show that $\det X = 1 - \nu\omega$. By reversing the order of the factors in (2.152), show also that $\det(I_n - \omega\nu) = 1 - \nu\omega$.

2.46 With ω, ν as in the preceding problem, and W any $n \times n$ matrix such that $1 + \nu W^{-1}\omega \neq 0$, prove that

$$(W + \omega\nu)^{-1} = W^{-1} - \frac{W^{-1}\omega\nu W^{-1}}{1 + \nu W^{-1}\omega}$$

Use this result to show that if $\{A, b, h\}$ is a realization of a transfer function $g(s) - 1$, then $\{A - bh, b, h\}$ is a realization of $1 - 1/g(s)$.

2.47 Use the result in Problem 2.45 to prove that

$$\frac{\det(sI - A_c)}{\det(sI - A)} = 1 - f(sI - A)^{-1}b$$

where $A_c = A + bf$.

2.48 Let $g_0(s) = h(sI - A)^{-1}b$ be the open-loop transfer function corresponding to an arbitrary output $y = hx$ for (2.133). Use the result of the preceding problem to show that the closed-loop transfer function in (2.145) is given by

$$g_c(s) = \frac{g_0(s) \det(sI - A)}{\det(sI - A_c)}$$

2.49 By using Theorem 2.16 and its converse (Remark 4), prove that the pair (A, b) is c.c. if and only if the pair $(A + br, b)$ is c.c., for an arbitrary row n-vector r.

BIBLIOGRAPHICAL NOTES

Section 2.0. Revision material on the Laplace transform can be conveniently found in Spiegel (1965), and a standard text on the z-transform is that by Jury (1964). An interesting general treatment of differential equations which emphasized their

applications is by Braun (1978); and for state space modeling, see
McClamroch (1980).

Section 2.1. From the large number of recent books on state space
control theory, two particularly readable introductions are by
Elgerd (1967) and Auslander et al. (1974). On linear systems the
choice is even wider, but noteworthy treatments are by Chen (1970),
Fortmann and Hitz (1977), Fossard and Guéguen (1977), Rugh (1975),
and Kailath (1980), the latter being very comprehensive and nearest
in spirit to the present book. Jacquot (1981), Kuo (1977), and
Strejc (1981) give good accounts of discrete-time systems. The
introduction to mathematical control theory by Barnett (1975a)
lists further textbooks. For numerical evaluation of the
exponential of a matrix, see Moler and Van Loan (1978); and of
transfer function matrices, see Varga and Sima (1981).

Section 2.2. The concepts of controllability and observability
were first formally introduced by Kalman (1962, 1963), and a full
treatment will be found in the books listed for Section 2.1; those
by Fortmann and Hitz, Kailath, and Rugh also cover reachability and
reconstructibility. A good survey was given by Willems and Mitter
(1971), and the book by Kwakernaak and Sivan (1972) is also worth
consulting for applications in the field of optimal control. The
proof of sufficiency in Theorem 2.1 follows Hartwig (1982), and
Theorem 2.2 is due to Hautus (1969). Some definitions of other
types of controllability can be found in Rosenbrock (1970),
including a proof of the result in Remark 5 on Theorems 2.1 and
2.2. A review of conditions for controllability and reconstructibility of discrete-time systems was given by Grasselli (1980)
[see also Inouye (1982)]. A study of numerical aspects of testing
controllability and observability has been made by Paige (1981)
(see also Remark 5 on Theorem 4.1, Chapter 4). Controllability has
been linked with the solution of linear algebraic equations by
de Souza and Bhattacharyya (1981). Problem 2.16 is taken from
Hautus (1969), Problem 2.21 is based on a result of Power (1979),

and Problem 2.23 on work by Chen et al. (1966).

Section 2.3. Theorem 2.6 first appeared in Kalman (1963), and the generalization in Theorem 2.8 is due to Barnett (1971b). The result in Theorem 2.9 was obtained by Butman and Sivan (1964).

Section 2.4. A good discussion of the canonical forms given here, together with others, is in Kailath (1980). The results of Theorem 2.11 were first formally presented by Sidhu (1977). For a general survey of single-input, single-output canonical forms, see Maroulas and Barnett (1978a), where further references can be found. The results in Problem 2.30, together with a number of related extensions, were given by Bass and Gura (1966).

Section 2.5. Again Kailath's book is a good source, although the basic results, Theorems 2.14 and 2.15, will be found in all the standard texts.

Section 2.6. Theorem 2.16 was developed by Rissanen (1960), and formula (2.146) by Ackermann (1972). An interesting interpretation of this characteristic polynomial assignment theorem in terms of an optimal control problem has been devised by Heymann (1979). An efficient algorithm for applying linear state feedback has been proposed by Miminis and Paige (1982).

3

Root Location and Stability

3.0 INTRODUCTION

The basic motivation for this chapter arises in the study of
systems of linear differential or difference equations in state
space form. Stability of the solution means, in intuitive terms,
that initial perturbations from an equilibrium state either lead
to subsequent motions which are not in some sense "too large," or
(preferably) result in the system eventually returning to its
equilibrium condition. Some precise definitions of stability are
given in Section 3.1, together with statements of the two
fundamental theorems for the continuous- and discrete-time cases
in terms of the eigenvalues of the system state matrices. In this
book, however, we focus our attention on investigating the
stability nature of a linear system in terms of the coefficients
of its characteristic polynomial. For the continuous-time case
this requires determination of the numbers of zeros of the
polynomial in the halves of the complex plane to the left and
right of the imaginary axis; the corresponding regions for
discrete-time systems are the interior and exterior of the disk
centered at the origin, with unit radius. There are a very large
number of results available in this area, and in Section 3.2 we
have attempted to present a fairly comprehensive survey. At this

stage only statements of the theorems are given, first for real
polynomials — the continuous-time case in Section 3.2.1, the
discrete-time case in Section 3.2.2 — and then extended to complex
polynomials in Section 3.2.3. It would take a great deal of space
to give separate proofs of all the individual results, and indeed
this would not be the most instructive way of proceeding. Instead,
following the theme of this book, we concentrate on a method
involving appropriate matrix equations. There are several
advantages of following this approach: (i) it provides purely
algebraic proofs; (ii) it provides a unifying tool for applying to
both the continuous- and discrete-time theorems, and exposes many
interrelationships; (iii) it is a generalization of the linear
case of Liapunov's method of stability analysis, itself an
important idea used in control theory; and (iv) it makes use of
the companion matrix representation of polynomials. A very recent
simple proof of the basic complex discrete-time theorem is given in
Section 3.3.3. This is also purely algebraic, but uses a
completely different method. The classical procedure using the
Cauchy index is briefly described in Section 3.4, since this is
particularly important in dealing with critical cases when some
roots lie actually on the imaginary axis, or unit circle. The
bilinear mapping, and root location relative to nonstandard
regions, are briefly considered in Section 3.5. Throughout the
chapter we attempt, wherever possible, to highlight connections
between the various results and methods, and to point out
connections with the controllability/observability expressions of
Chapter 2.

3.1 STABILITY OF LINEAR SYSTEMS

We consider systems described by sets of linear differential
equations, or linear difference equations, of the form

$$\dot{x}(t) = Ax(t), \quad t \geqslant 0 \tag{3.1}$$

and

$$x(k + 1) = A_1 x(k), \quad k = 0, 1, 2, \ldots \tag{3.2}$$

respectively, where as in Chapter 2, $x(t)$ or $x(k)$ denote the n-dimensional state vector. The representations (3.1) and (3.2) can be regarded as arising from the linear control systems in (2.2) and (2.4) where linear feedback in the form

$$u = Ky \tag{3.3}$$

has been applied, where y denotes the output vector. The continuous-time equations (2.2) and (2.3) are

$$\dot{x} = Ax + Bu, \quad y = Hx \tag{3.4}$$

so substitution of (3.3) into (3.4) produces the closed-loop system

$$\dot{x}(t) = (A + BKH)x(t) \tag{3.5}$$

Thus for a linear control system the "A" matrix in (3.1) is actually equal to $A + BKH$; it is easy to check using (2.6) that the same expression also arises for the discrete-time case. However, we shall leave the implications of linear feedback until Chapter 4, and concentrate in this chapter on the closed-loop forms (3.1) and (3.2), it being understood that "A" or "A_1" may well refer to the system closed-loop matrix after feedback has been applied.

The concept of stability is a fundamental one for dynamic systems, and there are a variety of types of stability which have been defined. Clearly, in (3.1) or (3.2), if $x(0) = 0$, then $x(\tau) = 0$ for all subsequent time $\tau > 0$. In other words, if the system is initially at the origin, it will remain there. Our interest thus lies in what happens if an initial perturbation, represented by the initial condition $x(0) = x_0 \neq 0$, is applied. The definitions we shall need are as follows:

The *equilibrium* (or *"critical"*) state $x = 0$ is *stable* (*in the sense of Lipaunov*) if for any positive scalar ε, there exists a positive scalar δ such that $|x_0| < \delta$ implies that the solution x of (3.1) [or (3.2)] satisfies $|x(t)| < \varepsilon$ (or $|x(k)| < \varepsilon$), where

$$|x| \triangleq \left[x_1^2 + x_2^2 + \cdots + x_n^2 \right]^{\frac{1}{2}} \tag{3.6}$$

If in addition

$$\lim_{t \to \infty} x(t) = 0 \quad \text{or} \quad \lim_{k \to \infty} x(k) = 0$$

then the origin is called *asymptotically stable*.

The definition of a stable system means that we can specify the maximum deviation ε of the norm of $x(t)$ or $x(k)$, as measured by (3.6), and there will always be a corresponding maximum allowable perturbation δ of x from the equilibrium position $x = 0$. Asymptotic stability has the additional property, desirable in real-life applications, that eventually the system returns to its equilibrium state.

The origin is *unstable* if it is not stable; however large ε, and however small the initial perturbation δ, there exists some $|x_0| < \delta$ and time τ_1 (> 0) such that $|x(\tau_1)| \geqslant \varepsilon$. If this holds for every initial state x_0, the equilibrium is *completely unstable*.

Example 1   Discussion of the elementary case $n = 2$ will help to elucidate the concepts.   Suppose first that

$$A = \begin{bmatrix} a & 0 \\ 0 & b \end{bmatrix} \tag{3.7}$$

so that the solution of (3.1) is simply

$$x_1(t) = x_1(0) e^{at}, \quad x_2(t) = x_2(0) e^{bt}$$

Clearly,

$$x_1(t) \to 0, \quad x_2(t) \to 0 \quad \text{as } t \to \infty$$

if and only if $Re(a) < 0$, $Re(b) < 0$, and the origin is then an asymptotically stable equilibrium point.   If, however, a (say) has a positive real part, then $x_1(t) \to \infty$ as $t \to \infty$, so the origin is unstable, [although if $x_1(0) = 0$, $x_2(0) \neq 0$, a divergent motion will not in fact be activated]. If both a and b have positive real parts, then the solution diverges whatever the (nonzero) values of $x_1(0)$ and $x_2(0)$, so we have complete instability. Finally, if (say) $a = i\alpha$, then

$$x_1(t) = x_1(0) (\cos \alpha t + i\sin \alpha t)$$

which represents oscillatory motion. Thus if $Re(b) \leqslant 0$, the origin is stable, but not asymptotically stable.

This example has been made more straightforward by the assumption, implicit in (3.7), that A has a diagonal Jordan form. Thus if we consider instead

$$A = \begin{bmatrix} a & 1 \\ 0 & b \end{bmatrix} \tag{3.8}$$

which has the same eigenvalues as (3.7), then when $a = b$ the corresponding solution of (3.1) takes the form

$$x_1(t) = \left(x_1(0) + tx_2(0)\right)e^{at}$$

$$x_2(t) = x_2(0)e^{at}$$

Clearly, if $Re(a) = 0$, then $x_1(t) \to \infty$ as $t \to \infty$ whenever $x_2(0) \neq 0$, so the origin is unstable, in contrast with the situation discussed above with respect to A in (3.7).

This example illustrates the fact that the stability nature of the equilibrium point depends upon the signs of the real parts of the eigenvalues of A. It also shows that when some of these eigenvalues are purely imaginary, knowledge of the structure of the Jordan form of A is required.

For completeness we state in full the appropriate stability theorems for (3.1) and (3.2). Proofs are not given, however, since these are very standard, depending upon consideration of $\exp(Jt)$, where J is the Jordan form of A (see Problem 3.2).

Theorem 3.1 The origin of (3.1) is:

(i)   Asymptotically stable if all the eigenvalues λ_1, ..., λ_n of A
       have negative real parts

(ii)   Stable, but not asymptotically stable, if $Re(\lambda_k) \leqslant 0$, k = 1,
       2, ..., n, and for each λ_k such that $Re(\lambda_k) = 0$ (of which
       there must be at least one) the corresponding Jordan block in
       the Jordan form of A has order one

(iii) Unstable if at least one of the conditions (i), (ii) is
       violated

(iv)   Completely unstable if $Re(\lambda_k) > 0$, k = 1, 2, ..., n

Remarks on Theorem 3.1   (1) Our elementary discussion in Example 1
is readily confirmed. In particular, when there are two equal
imaginary eigenvalues, then (3.8) is a 2 × 2 Jordan block for this
eigenvalue, whereas (3.7) contains two 1 × 1 blocks.

     (2) An equivalent form of the stability criterion (ii) is that
any λ_i with $Re(\lambda_i) = 0$ must be a simple zero of the minimum
polynomial of A.

     (3) The asymptotic stability criterion (i) is conveniently
described in geometric terms as requiring that all the eigenvalues
λ_i lie in the open left half of the complex λ-plane, viz.,
$Re(\lambda) < 0$. A matrix having this property is often called a
stability matrix.

     (4) If the asymptotic stability condition (i) is satisfied,
then for *every* x_0 we have $x(t) \to 0$ as $t \to \infty$, so the origin is in
fact *globally asymptotically stable (or asymptotically stable in
the large)*. Notice also that since the solution of (3.1) can be
written as $x(t) = \exp(At)x_0$, this implies that $\exp(At) \to 0$ as
$t \to \infty$.

     As we noted in Section 2.1, exponential terms $\exp(\lambda_i t)$ in the
solution of the continuous-time equation (3.1) are replaced by
terms $(\mu_i)^k$ for the discrete-time case (3.2), where μ_1, ..., μ_n
are the eigenvalues of A_1. This leads to:

Theorem 3.2   The origin of (3.2) is:

(i)   Asymptotically stable if $|\mu_k| < 1$, $k = 1, 2, \ldots, n$

(ii)   Stable, but not asymptotically stable, if $|\mu_k| \leqslant 1$, $k = 1, 2,$
       $\ldots, n$, and for each μ_k such that $|\mu_k| = 1$ (of which there
       must be at least one) the corresponding Jordan block in the
       Jordan form of A_1 has order one

(iii) Unstable if at least one of the conditions (i), (ii) is
       violated

(iv)   Completely unstable if $|\mu_k| > 1$, $k = 1, 2, \ldots, n$

Remarks on Theorem 3.2   (1) The correspondence between Theorems 3.1
and 3.2 is obtained by substituting

$$\text{Re}(\lambda_i) < 0 \Leftrightarrow |\mu_i| < 1 \tag{3.9}$$

and this achieved by the standard *bilinear mapping* (or
transformation)

$$\lambda = \frac{\mu + 1}{\mu - 1}, \quad \mu = \frac{\lambda + 1}{\lambda - 1} \tag{3.10}$$

It is easy to confirm that (3.10) is equivalent to taking

$$A = (A_1 + I)(A_1 - I)^{-1}, \quad A_1 = (A - I)^{-1}(A + I) \tag{3.11}$$

The relationships (3.11) can thus be used to convert the stability
problem for discrete time into that for continuous time, and vice
versa. The transformations (3.10) will be studied further in
Section 3.5.1.

       (2) The asymptotic stability criterion is now that all the
eigenvalues of A_1 must lie inside the unit disk $|\mu| < 1$, in the
complex μ-plane. If A_1 has this property it is termed a *convergent*
matrix, because $A_1^k \to 0$ as $k \to \infty$.

    A direct method of testing a given system for stability is
obviously to apply one of the powerful standard computer algorithms
for determining eigenvalues. Discussion of such a procedure lies
within the realm of numerical analysis, and is outside the scope of

this book. We therefore now proceed instead to consideration of
the stability problems in terms of the characteristic polynomial of
A or A_1. Indeed, we can solve the more general situation of
finding the numbers of roots inside and outside the appropriate
stability regions in the complex λ- and μ-planes.

Problems

3.1   Prove that when n = 2, the necessary and sufficient conditions
      for:
      (a) A to be a stability matrix are det A > 0, tr(A) < 0
      (b) A_1 to be a convergent matrix are $|\det A_1| < 1$,
          $|tr(A_1)| < 1 + \det A_1$.

3.2   Let

$$J_i = \begin{bmatrix} \lambda_i & 1 & 0 & & & \\ & \lambda_i & 1 & & 0 & \\ & & \ddots & \ddots & & \\ & & & \lambda_i & 1 \\ 0 & & & & \lambda_i \end{bmatrix}$$

      be an r × r Jordan block corresponding to an eigenvalue λ_i of
      A.   Show that

$$\exp(J_i t) = \begin{bmatrix} 1 & t & t^2/2! & \cdots & t^{r-1}/(r-1)! \\ 0 & 1 & t & \cdots & t^{r-2}/(r-2)! \\ \cdot & \cdot & \cdot & \cdots & \cdot \\ \cdot & \cdot & \cdot & \cdots & 1 \end{bmatrix} e^{\lambda_i t}$$

      Hence deduce the behavior of $\exp(\lambda_i t)$ as $t \to \infty$.

3.3   If A is a stability matrix, show that the solution x(t) of the
      system

$$\dot{x}(t) = Ax(t) - b$$

      where b is a constant vector, satisfies the condition

$$\lim_{t \to \infty} x(t) = A^{-1}b$$

3.4 Use Gershgorin's theorem, quoted in Problem 1.15, to show that if A is *diagonal dominant*, i.e., $|a_{ii}| > \rho_i$, for all i, and in addition all the diagonal elements of A are real and negative, then A is a stability matrix.

3.5 Let A and B be real n × n matrices, and denote by λ_k, k = 1, ..., n, the eigenvalues of the complex matrix $A_c = A + iB$. By applying suitable block transformations to

$$\begin{bmatrix} A_c & 0 \\ 0 & \bar{A}_c \end{bmatrix}$$

transform it into

$$A_r = \begin{bmatrix} A & B \\ -B & A \end{bmatrix} \tag{3.12}$$

and hence show that the eigenvalues of A_r are λ_k together with their conjugates $\bar{\lambda}_k$. Notice that this result shows that the location of the eigenvalues of A_c, relative to the imaginary axis or the unit circle, can be determined by solving the same problem for the 2n × 2n real matrix A_r.

3.2 ROOT LOCATION AND STABILITY CRITERIA

In this section we give a number of theorems on determining the location of the roots of polynomials relative to the imaginary axis or the unit circle in the complex plane. In view of the extremely large body of literature in this area it is impossible to be exhaustive, so the emphasis is on the more important and useful results. It is helpful to identify each result with the name of its originator(s), where possible. We use (SC) and (RL) to denote the *stability criterion* and *root location* parts of the theorems.

3.2.1 Real Continuous-Time Case
Consider the polynomial

$$a(\lambda) = a_0 \lambda^n + a_1 \lambda^{n-1} + \cdots + a_{n-1}\lambda + a_n, \qquad a_0 > 0 \tag{3.13}$$

where the a_i are all real. By a mild abuse of definitions we shall call $a(\lambda)$ asymptotically stable if all its roots have negative real parts; the term *Hurwitz* polynomial is also sometimes used. We begin with what is perhaps the best known, and certainly one of the most useful tests, that involving the Routh array described in Section 1.6.1 for determining greatest common divisor.

<u>Theorem 3.3 (Routh)</u> Construct the Routh array $\{r_{ij}\}$ having initial two rows

$$\{r_{01}, r_{02}, r_{03}, \ldots\} = \{a_0, a_2, a_4, \ldots\}$$

$$\{r_{11}, r_{12}, r_{13}, \ldots\} = \{a_1, a_3, a_5, \ldots\} \tag{3.14}$$

and subsequent rows defined by

$$r_{ij} = - \begin{vmatrix} r_{i-2,1} & r_{i-2,j+1} \\ r_{i-1,1} & r_{i-1,j+1} \end{vmatrix} \Big/ r_{i-1,1}, \quad i = 2, 3, \ldots, n \tag{3.15}$$

(SC) The polynomial $a(\lambda)$ in (3.13) is asymptotically stable if and only if all the first column elements r_{i1}, $i = 0, 1, 2, \ldots,$ n, are positive.

(RL) Provided that the array is regular, i.e., $r_{i1} \neq 0$, for all i, then $a(\lambda)$ has no purely imaginary roots, and there are k and n - k roots with positive and negative real parts, respectively, where

$$k = V(r_{01}, r_{11}, r_{21}, \ldots) \tag{3.16}$$

$\overset{\Delta}{=}$ number of variations in sign in the sequence of first column elements

<u>Example 2</u> Let

$$a(\lambda) = \lambda^8 + 2\lambda^7 - 3\lambda^6 + 5\lambda^5 + 5\lambda^4 + \lambda^3 + 7\lambda^2 + 3\lambda + 2 \tag{3.17}$$

so that from (3.14) and (3.15) the Routh array is

r_{0j}	1	-3	5	7	2
r_{1j}	2	5	1	3	
r_{2j}	$-\dfrac{11}{2}$	$\dfrac{9}{2}$	$\dfrac{11}{2}$	2	
r_{3j}	$\dfrac{73}{11}$	3	$\dfrac{41}{11}$		
r_{4j}	$\dfrac{510}{73}$	$\dfrac{627}{73}$	2		
r_{5j}	$-\dfrac{2631}{510}$	$\dfrac{932}{510}$			
r_{6j}	$\dfrac{29,109}{2631}$	2			
r_{7j}	$\dfrac{80,341}{29,109}$				
r_{8j}	2				

There are four variations in sign in the first column (from r_{11} to r_{21}, from r_{21} to r_{31}, etc.), so $a(\lambda)$ has four roots with positive real parts and $8 - 4 = 4$ roots with negative real parts.

The array above should be compared with that in Example 18, Chapter 1. The polynomial (3.17) was deliberately chosen so that the initial two rows are the same as in the earlier example.

<u>Remarks on Theorem 3.3</u> (1) Notice that at any stage a row can be multiplied by any convenient positive scaling factor without affecting the signs of the first column elements. Alternatively, the optimal fraction-free array of Section 1.6.3 could be used. The reader is urged to construct this array $\{t_{ij}\}$ defined in (1.148) for the polynomial (3.17). It will then be seen the number of sign changes in the sequence of first column elements t_{i1} is *not* the same as for the elements r_{i1} in Example 2. In fact, it can be shown that

$$k = V\left(t_{01}, t_{11}, \frac{t_{21}}{t_{11}}, \frac{t_{31}}{t_{21}}, \ldots, \frac{t_{n1}}{t_{n-1,1}}\right) \tag{3.18}$$

and we shall see the reason for this in Remark 3 on our next result, Theorem 3.4.

(2) Since $a(\lambda)$ is real, any complex roots will occur in
conjugate pairs $\alpha \pm i\beta$, corresponding to a factor

$$(\lambda - \alpha - i\beta)(\lambda - \alpha + i\beta) = \lambda^2 - 2\alpha\lambda + \alpha^2 + \beta^2$$

Hence if $a(\lambda)$ is asymptotically stable, it will be expressible as a
product of factors

$$\Pi(\lambda + \gamma)\Pi(\lambda^2 - 2\alpha\lambda + \alpha^2 + \beta^2)$$

with $\alpha < 0$, $\gamma > 0$. Since all the coefficients in this product are
positive, it follows that a simple *necessary* condition for $a(\lambda)$ to
be asymptotically stable is $a_i > 0$, for all i. In other words, if
any coefficient of $a(\lambda)$ is negative or zero, then $a(\lambda)$ cannot be
asymptotically stable (as is illustrated by Example 2). Moreover,
it can be shown that this necessity of having positive coefficients
extends to *all* the elements of the Routh array $\{r_{ij}\}$, so if it is
required only to test for stability, the procedure can be stopped
on encountering a nonpositive element. This fact is seldom pointed
out in textbooks, although it is contained in Routh's original
paper.

(3) When a zero first column element is encountered the
procedure breaks down, since (3.15) cannot be continued. The
theorem shows that in this case $a(\lambda)$ is not asymptotically stable,
but if it is desired to determine the location of the roots the
Routh algorithm must be modified. There are two distinct types of
such *singular cases* to consider. Suppose that a singularity occurs
in the $(p + 1)$th row, i.e., $r_{p1} = 0$, but the complete row is not
zero. Then (exactly as for the g.c.d. array) the elements are
shifted bodily to the left until a nonzero element is obtained in
the first position. Construction of the array is then continued,
with further shiftings if necessary.

Singular case: Type (i). If no complete row of zeros is obtained
throughout the procedure described above, this implies that $a(\lambda)$
has no purely imaginary roots. A modified Routh array is then
constructed in which r_{p1} is replaced by a parameter ε (assumed

small), and the array is continued in the usual way. If a further
zero first-column element is encountered later, replace this by a
second parameter η, and so on. The value of k is given by (3.16)
applied to this modified array.

Example 3   Let

$$a(\lambda) = \lambda^4 + \lambda^3 + 4\lambda^2 + 4\lambda + 3 \tag{3.19}$$

so that the array is

    1   4   3
    1   4
    0   3

Replacing r_{21} by ε gives

    1   4   3
    1   4
    ε   3
    $4 - \dfrac{3}{\varepsilon}$
    3

so that from (3.16) we have

$$k = V(1, 1, \varepsilon, 4 - \frac{3}{\varepsilon}, 3)$$

$$= 2$$

showing that (3.18) has two roots with positive real parts, and
4 - 2 = 2 with negative real parts. Notice that the sign of ε is
irrelevant, provided that it is assumed fixed: for when ε > 0, then
4 - (3/ε) < 0 since ε can be as small as we wish; and when ε < 0,
then 4 - (3/ε) > 0. Thus in either case there are two variations
in sign in the sequence.

    The preceding idea is based on the fact that introducing the
parameter(s) is equivalent to making perturbations in some or all
of the coefficients of a(λ). It is known that the roots of a
polynomial are continuous functions of its coefficients.

Therefore, the modified polynomial will have the same number of roots in each half-plane as the original one, provided that the perturbations are sufficiently small, and provided also that there are no purely imaginary roots. In the latter case these would move off the imaginary axis under the perturbations, thus altering the counts. We therefore need to use a different argument in this case, as is now outlined.

Singular Case: Type (ii). We now encounter at some stage in the array (with shiftings as described above, if necessary) a row consisting entirely of zeros. As we saw in Section 1.6.1, this means that the polynomials which we can associate with the initial two rows of the array have a nontrivial g.c.d. $\delta(\lambda)$, say. It can be shown that any purely imaginary root of $a(\lambda)$ is also a root of $\delta(\lambda)$, and we can write

$$a(\lambda) = \delta(\lambda)a_r(\lambda) \tag{3.20}$$

where $a_r(\lambda)$ has no imaginary roots. Thus the number of right-half-plane roots k_r of $a_r(\lambda)$ can be determined using Routh's algorithm, modified if necessary for type (i) singularities. The required number k of right-half-plane roots of $a(\lambda)$ is then given by

$$k = k_r + k_\delta \tag{3.21}$$

where k_δ is the number of such roots of $\delta(\lambda)$. It can also be shown that $\delta(\lambda)$ is a polynomial in λ^2, and so has the same numbers of roots to the right and left of the imaginary axis. Thus if $\delta(\lambda)$ has degree q, and s purely imaginary roots, it follows that $q = 2k_\delta + s$, i.e.,

$$k_\delta = \frac{1}{2}(q - s) \tag{3.22}$$

In order to determine s, we note that it is equal to the number of *real* roots of $\delta(i\omega)$, which is a real polynomial in ω. This number s can be calculated using Sturm's theorem, which is given later in Section 3.4.

Example 4   Consider the polynomial

$$a(\lambda) = \lambda^6 + \lambda^5 + 3\lambda^4 + 3\lambda^3 + 3\lambda^2 + 2\lambda + 1 \qquad (3.23)$$

The Routh array of Theorem 3.3 is

r_{0j} 1   3   3   1

r_{1j} 1   3   2

r_{2j} 0   1   1

and because $r_{21} = 0$ we shift the third row to the left, to produce a modified array

r_{1j} 1   3   2

r_{2j} 1   1

r_{3j} 2   2

r_{4j} 0   0

Since a complete row of zeros has arisen, we have a singularity of type (ii). The g.c.d. $\delta(\lambda)$ is provided by the last nonvanishing row of the modified array (as we saw in Section 1.6.1). We have $\delta(\lambda) = 2\lambda^2 + 2$, so that $\delta(\lambda)$ has roots $\pm i$, i.e., $s = 2$, and from (3.22) $k_\delta = \frac{1}{2}(2 - 2) = 0$. Division of $a(\lambda)$ in (3.23) by $\delta(\lambda)$ produces

$$a_r(\lambda) = \lambda^4 + \lambda^3 + 2\lambda^2 + 2\lambda + 1$$

and to determine k_r we construct the array for $a_r(\lambda)$. This has a singularity of type (i), so we obtain the following:

1   2   1

1   2

ε 1

$2 - \dfrac{1}{\varepsilon}$

1

Thus from (3.16)

$$k_r = V(1, 1, \varepsilon, 2 - \frac{1}{\varepsilon}, 1) = 2$$

so finally from (3.21) the number of right-half-plane roots of $a(\lambda)$
is $k = 2$. Since $a(\lambda)$ is of sixth degree, and has the same
imaginary roots $\pm i$ as $\delta(\lambda)$, it follows that $a(\lambda)$ has two roots with
negative real parts.

A simpler way of dealing with singularities of type (ii), in
the case when there are *no* singularities of type (i), is as
follows. Let the row immediately preceding the complete row of
zeros be $\{r_{mj}\}$, and associate with it a polynomial

$$f_m(\omega) = r_{m1}\omega^{n-m} + r_{m2}\omega^{n-m-2} + r_{m3}\omega^{n-m-4} + \cdots \qquad (3.24)$$

Replace the zero row by a new row consisting of the coefficients of
the derivative of $f_m(\omega)$ with respect to ω. Then continue with the
construction and interpretation of the array in the usual way.
This procedure can be applied each time a zero row is encountered.

Example 5   The array for

$$a(\lambda) = \lambda^4 + 4\lambda^3 + 7\lambda^2 + 16\lambda + 12$$

is

$$
\begin{array}{llll}
r_{0j} & 1 & 7 & 12 \\
r_{1j} & 4 & 16 \\
r_{2j} & 3 & 12 \\
r_{3j} & 0
\end{array}
$$

Here there is a singularity of type (ii) with $m = 2$, so from (3.24)

$$f_2(\omega) = 3\omega^2 + 12, \quad df_2/d\omega = 6\omega$$

The revised array is thus

$$
\begin{array}{ccc}
1 & 7 & 12 \\
4 & 16 & \\
3 & 12 & \\
6 & & \\
12 & &
\end{array}
$$

which has no sign variations in the first column, so by (3.16) the number of right-half-plane roots of $a(\lambda)$ is $k = 0$. We see from $\{r_{2j}\}$ in the original array that $\delta(\lambda) = 3(\lambda^2 + 4)$, showing that $\delta(\lambda)$, and therefore also $a(\lambda)$, have imaginary roots $\pm 2i$.

(4) There are a number of other ways of dealing with a singularity of type (i). One method is to replace $a(\lambda)$ by $(\lambda + c)a(\lambda)$, where c is a positive constant which can always be chosen so that $r_{p1} \neq 0$. Another approach is to apply the algorithm to the *reverse* polynomial $\hat{a}(\lambda) = \lambda^n a(1/\lambda)$, whose roots are the reciprocals of those of $a(\lambda)$. It is left as an easy exercise for the reader to show that the root distribution of $\hat{a}(\lambda)$ with respect to the imaginary axis is the same as that of $a(\lambda)$.

Example 6   Return to the polynomial in (3.19). Multiplying by $(\lambda + 2)$ produces

$$
\lambda^5 + 3\lambda^4 + 6\lambda^3 + 12\lambda^2 + 11\lambda + 6
$$

whose Routh array is now regular:

$$
\begin{array}{ccc}
1 & 6 & 11 \\
3 & 12 & 6 \\
2 & 9 & \\
-\dfrac{3}{2} & 6 & \\
17 & &
\end{array}
$$

The new polynomial has the same number of roots with positive real parts, namely

$$k = V(1, 3, 2, -\frac{3}{2}, 17) = 2$$

agreeing with what was found before.

Alternatively, $\hat{a}(\lambda)$ is formed simply by reversing the order of the coefficients in (3.19):

$$\hat{a}(\lambda) = 3\lambda^4 + 4\lambda^3 + 4\lambda^3 + \lambda + 1$$

The corresponding Routh array is regular, and again produces k = 2, as the reader should confirm.

(5) In applications it is often required that the system (3.1) should be asymptotically stable, and that in addition the solution x(t) should decay at least as fast as exp(-σt), where σ is a given positive constant. This is equivalent to requiring that all the roots of a(λ) should lie to the left of a line λ = -σ in the complex λ-plane. To test this condition, simply set λ = λ' - σ, so that Re(λ') < 0 if and only if Re(λ) < -σ, and apply the Routh method to a(λ').

Theorem 3.4 (Hurwitz) Construct the n × n *Hurwitz* matrix

$$H = \begin{bmatrix} a_1 & a_3 & a_5 & \cdots & a_{2n-1} \\ a_0 & a_2 & a_4 & \cdots & a_{2n-2} \\ 0 & a_1 & a_3 & \cdots & a_{2n-3} \\ 0 & a_0 & a_2 & \cdots & a_{2n-4} \\ 0 & 0 & a_1 & \cdots & \cdot \\ \cdot & \cdot & \cdot & \cdots & \cdot \\ 0 & 0 & 0 & \cdots & a_n \end{bmatrix} \qquad (3.25)$$

where $a_r = 0$, r > n.

(SC) The polynomial a(λ) in (3.13) is asymptotically stable if and only if all the leading principal minors H_1, H_2, ..., H_n of H are positive.

(RL) Provided that $H_i \neq 0$, for all i, then $a(\lambda)$ has k and n - k
roots with positive and negative real parts, respectively,
where

$$k = V\left(a_0, H_1, \frac{H_2}{H_1}, \frac{H_3}{H_2}, \ldots, \frac{H_n}{H_{n-1}}\right)$$

$$= V(a_0, H_1, H_3, \ldots) + V(1, H_2, H_4, \ldots) \tag{3.26}$$

Notice that the first two rows of H are the same as the second
and first rows, respectively, in the Routh array in (3.14);
subsequent rows of H are then formed by repeatedly shifting this
pair one position to the right.

Example 7   The Hurwitz matrix for

$$a(\lambda) = \lambda^4 + \lambda^3 - 12\lambda^2 - 28\lambda - 16 \tag{3.27}$$

is

$$H = \begin{bmatrix} 1 & -28 & & 0 \\ 1 & -12 & -16 & 0 \\ 0 & 1 & -28 & 0 \\ 0 & 1 & -12 & -16 \end{bmatrix} \tag{3.28}$$

and the leading principal minors (shown within dashed lines) are

$$H_1 = 1, \quad H_2 = \begin{vmatrix} 1 & -28 \\ 1 & -12 \end{vmatrix} = 16 \tag{3.29}$$

$$H_3 = \begin{vmatrix} 1 & -28 & 0 \\ 1 & -12 & -16 \\ 0 & 1 & -28 \end{vmatrix} = -432, \quad H_4 = \det H = 6912$$

Thus from (3.26)

$$k = V(1, 1, 16, -27, -16) = 1$$

showing that $a(\lambda)$ has, respectively, one and three roots with
positive and negative real parts.

<u>Remarks on Theorem 3.4</u> (1) Although Hurwitz's work was published independently of Routh's, and 18 years later, in 1895, the two results are often linked together as the "Routh-Hurwitz conditions". This description is in fact reasonably appropriate, since the two methods are indeed equivalent, as the following development illustrates for the case n = 3: It is easy to verify that reduction of H to upper triangular form by elementary rows operations is expressible as

$$
\begin{bmatrix} 1 & 0 & 0 \\ -\dfrac{a_0}{a_1} & 1 & 0 \\ \dfrac{-a_1 a_0}{a_3 a_0 - a_1 a_2} & \dfrac{a_1^2}{a_3 a_0 - a_1 a_2} & 1 \end{bmatrix}
\begin{bmatrix} a_1 & a_3 & 0 \\ a_0 & a_2 & 0 \\ 0 & a_1 & a_3 \end{bmatrix}
=
\begin{bmatrix} r_{11} & r_{12} & r_{13} \\ 0 & r_{21} & r_{22} \\ 0 & 0 & r_{31} \end{bmatrix}
$$

$$(3.30)$$

where the r_{ij} in (3.30) are precisely the elements in the Routh stability array for $a(\lambda)$. It follows immediately from (3.30) that

$$ H_1 = r_{11}, \qquad H_2 = r_{11} r_{21}, \qquad H_3 = r_{11} r_{21} r_{31} $$

The argument is easily extended to the general case, showing that provided $H_i \neq 0$, for all i, the elements on the principal diagonal of the matrix obtained by reducing H to upper triangular form are just the elements in the first column of the Routh array. Since the elementary operations performed on H do not alter the values of its principal minors, we have

$$ H_i = r_{11} r_{21} r_{31} \cdots r_{i1}, \qquad i = 1, 2, \ldots, n \tag{3.31} $$

from which

$$ r_{11} = H_1, \qquad r_{i1} = \frac{H_i}{H_{i-1}}, \qquad i = 2, \ldots, n \tag{3.32} $$

Example 7 (continued)   The Routh array for $a(\lambda)$ in (3.27) is

r_{0j} 1   -12   -16

r_{1j} 1   -28

r_{2j} 16   -16

r_{3j} -27

r_{4j} -16

and (3.31) gives

$$H_1 = 1, \quad H_2 = 1.16, \quad H_3 = 1.16(-27), \quad H_4 = 1.16(-27)(-16)$$

agreeing with the previous calculations.

(2) In the last column of H in (3.25), only the (n, n) entry is nonzero, so expanding det H by this column shows that

$$H_n = a_n H_{n-1} \qquad\qquad (3.33)$$

It is also worth noting that *Orlando's formula* states that

$$H_{n-1} = \varepsilon(n) a_0^{n-1} \prod_{1 \leqslant i < j \leqslant n} (\lambda_i + \lambda_j) \qquad\qquad (3.34)$$

where the λ_i are the roots of $a(\lambda)$, and $\varepsilon(n)$ is defined in (1.22). In particular, (3.34) shows that $H_{n-1} = 0$ if and only if $a(\lambda)$ has at least one pair of roots λ_i and $-\lambda_i$.

(3) As stated in Remark 1, it is easy to show that (3.30) can be generalized to give

$$TH = \begin{bmatrix} r_{11} & r_{12} & r_{13} & \cdots & r_{1n} \\ 0 & r_{21} & r_{22} & \cdots & \cdot \\ 0 & 0 & r_{31} & \cdots & \cdot \\ \cdot & \cdot & \cdot & \cdots & \cdot \\ 0 & 0 & 0 & \cdots & r_{n1} \end{bmatrix} \qquad\qquad (3.35)$$

where T is lower triangular with all elements on the principal
diagonal equal to unity. By considering the minors of the first i
rows on both sides of (3.35) it follows that

$$H_{ij} = r_{11}r_{21} \cdots r_{i-1,1}r_{i,j-i+1}$$

$$= H_{i-1}r_{i,j-i+1}, \quad j = i, \ i+1, \ \ldots, \ n \tag{3.36}$$

where we have introduced the notation H_{ij} to stand for the minor
formed from rows 1, 2, ..., i - 1, i and columns 1, 2, ..., i - 1,
j of H (we shall subsequently apply this notation to *any* matrix;
notice that in particular $H_{ii} \overset{\Delta}{=} H_i$). It is interesting to relate
(3.36) to the optimal fraction-free form $\{t_{ij}\}$ of the Routh array,
introduced in (1.148). It was noted in Problem 1.38 that the two
arrays are related by

$$r_{ij} = \frac{t_{ij}}{t_{i-1,1}}, \quad i \geqslant 2$$

When this is substituted into (3.36) we obtain

$$H_{ij} = t_{11} \frac{t_{21}}{t_{11}} \frac{t_{31}}{t_{21}} \cdots \frac{t_{i-1,1}}{t_{i-2,1}} \frac{t_{i,j-i+1}}{t_{i-1,1}}$$

$$= t_{i,j-i+1} \tag{3.37}$$

showing that the elements of the optimal fraction-free array are
precisely equal to the minors of the Hurwitz matrix. In
particular, setting j = i in (3.37) shows that the principal minors
H_i are equal to the first column elements t_{i1}, and hence (3.18) is
equivalent to (3.26).

Example 7 (continued) For H in (3.28) we have, for example,

$$H_{23} = \begin{vmatrix} 1 & 0 \\ 1 & -16 \end{vmatrix}$$

$$= -16 = r_{11}r_{22}$$

which verifies (3.36) in this case.

The fraction-free array (1.148) for $a(\lambda)$ in (3.27) is

t_{0j} 1   -12   -16

t_{1j} 1   -28

t_{2j} 16   -16

t_{3j} -432

t_{4j} 6912

The first column elements are equal to the H_i found earlier, and $t_{23} = -16 = H_{23}$.

(4) The theory for singular cases when some of the H_i are zero can also be developed. However, this will not be given since Routh's algorithm is to be preferred as a computational procedure, as only 2 × 2 determinants have to be calculated.

(5) It was noted in Remark 2 on Theorem 3.3 that a necessary condition for $a(\lambda)$ to be asymptotically stable is $a_i > 0$, for all i. However, this condition is of course not sufficient, as is illustrated by (3.19) in Example 3. Nevertheless, if a certain positivity condition is satisfied, then only about half the Hurwitz determinants need to be calculated, as the following reduced form of the stability criterion of Theorem 3.4 shows:

Theorem 3.5 (Liénard-Chipart)
(SC) The polynomial $a(\lambda)$ in (3.13) is asymptotically stable if and only if one of the sequences Q_1, Q_2 and one of Q_3, Q_4, has all its members positive, where

$$Q_1 = \{a_n, a_{n-2}, a_{n-4}, \cdots\} \tag{3.38}$$

$$Q_2 = \{a_n, a_{n-1}, a_{n-3}, \cdots\} \tag{3.39}$$

$$Q_3 = \{H_1, H_3, H_5, \cdots\} \tag{3.40}$$

$$Q_4 = \{H_2, H_4, H_6, \cdots\} \tag{3.41}$$

Example 8 Consider

$$a(\lambda) = \lambda^4 + 2\lambda^3 + 9\lambda^2 + \lambda + 4 \tag{3.42}$$

whose coefficients are all positive. According to (3.40) we need
only to calculate H_1 and H_3 to test for stability, and these are

$$H_1 = 2, \quad H_3 = \begin{vmatrix} 2 & 1 & 0 \\ 1 & 9 & 4 \\ 0 & 2 & 1 \end{vmatrix} = 1$$

showing that $a(\lambda)$ is asymptotically stable.

It is now appropriate to point out that if the rows of the
Hurwitz matrix in (3.25) are rearranged by taking rows 1, 3, 5, ...,
followed by rows 2, 4, 6, ..., then the following matrix is
obtained:

$$\begin{bmatrix} a_1 & a_3 & a_5 & a_7 & \cdots \\ 0 & a_1 & a_3 & a_5 & \cdots \\ 0 & 0 & a_1 & a_3 & \cdots \\ & \cdots & & \cdots \\ & \cdots & & \cdots \\ a_0 & a_2 & a_4 & a_6 & \cdots \\ 0 & a_0 & a_2 & a_4 & \cdots \\ 0 & 0 & a_0 & a_2 & \cdots \end{bmatrix} \tag{3.43}$$

The matrix (3.43) has the form of a Sylvester resultant matrix,
defined in (1.84) (the numbers of rows of each type in (3.43)
depend upon whether n is even or odd). We saw in Chapter 1
(Theorem 1.7) that a Sylvester matrix can be related to a
polynomial in an appropriate companion matrix. It is therefore to
be expected that the Hurwitz matrix H, and hence in particular
minors of H, can be expressed in a similar fashion. In fact, since
the polynomials associated with the rows in (3.43) have degrees
only about $\frac{1}{2}n$, it turns out that the determinants involved will

have orders only about half those in the Hurwitz and Liénard-
Chipart theorems.

We now set out the details of this reduction. Assume that
$a(\lambda)$ in (3.13) is monic ($a_0 = 1$), and define polynomials associated
with the rows in (3.43):

$$f(\lambda) = a_n + a_{n-2}\lambda + a_{n-4}\lambda^2 + \cdots$$

$$g(\lambda) = a_{n-1} + a_{n-3}\lambda + a_{n-5}\lambda^2 + \cdots \qquad (3.44)$$

When n is even, let F denote the companion matrix in the form
(1.26) of $f(\lambda)$, which is then monic and has degree $\frac{1}{2}n$. When n is
odd, let G denote the companion matrix in the form (1.26) of $g(\lambda)$,
which is then monic and has degree $\frac{1}{2}(n - 1)$. Furthermore, define

$$L^{(1)} = g(F)J_{n/2}, \qquad L^{(2)} = FL^{(1)} \qquad (3.45)$$

$$M_1^{(1)} = f(G)J_{(n-1)/2} \qquad (3.46)$$

$$M^{(2)} = \begin{bmatrix} a_1 & a_3 & \cdots & a_n \\ & & & 0 \\ & M^{(1)} & & \vdots \\ & & & 0 \end{bmatrix} \qquad (3.47)$$

and let $L_i^{(1)}$ denote the $i \times i$ leading principal minor of $L^{(1)}$, and
so on.

Theorem 3.6 (Barnett)

(SC/RL) The Hurwitz minors in Theorems 3.4 and 3.5 are given by:

n even: $H_{2i-1} = L_i^{(1)}$, $i = 1, 2, \ldots, \frac{1}{2}n$

$H_{2i} = (-1)^i L_i^{(2)}$, $i = 1, 2, \ldots, \frac{1}{2}n - 1$ $\qquad (3.48)$

n odd: $H_{2i-1} = M_i^{(2)}$

$H_{2i} = (-1)^i M_i^{(1)}$, $i = 1, 2, \ldots, \frac{1}{2}(n - 1)$ $\qquad (3.49)$

Example 9 Return to $a(\lambda)$ in (3.27), Example 7. Here $n = 4$, and from (3.44)

$$f(\lambda) = -16 - 12\lambda + \lambda^2, \quad g(\lambda) = -28 + \lambda$$

and

$$F = \begin{bmatrix} 0 & 1 \\ 16 & 12 \end{bmatrix}$$

Thus from (3.45)

$$L^{(1)} = (F - 28I)J_2 \quad L^{(2)} = FL^{(1)}$$

$$= \begin{bmatrix} 1 & -28 \\ -16 & 16 \end{bmatrix} \quad = \begin{bmatrix} -16 & 16 \\ -176 & -256 \end{bmatrix}$$

and (3.48) gives

$$H_1 = L_1^{(1)} = 1, \quad H_3 = L_2^{(1)} = -432$$

$$H_2 = (-1)L_1^{(2)} = 16, \quad H_4 = (-1)^2 L_2^{(2)} = 6912$$

agreeing with the values found directly in (3.29).

Remarks on Theorem 3.6 (1) As Example 9 illustrates, the largest minor to be evaluated in the Hurwitz or Liénard-Chipart theorems using (3.48) or (3.49) has order only $\frac{1}{2}n$ or $\frac{1}{2}(n - 1)$. This reduction is to some extent offset by the extra effort involved in determining the matrices L and M. When required, H_n is best obtained from (3.33).

(2) Construction of $g(F)$ and $f(G)$ in formulas (3.45) and (3.46) can be achieved as explained in Section 1.2. Notice that because F is in companion form, the first $\frac{1}{2}n - 1$ rows of $L^{(2)}$ are rows 2, 3, ..., $\frac{1}{2}n$ of $L^{(1)}$. Note also that in view of (2.60), $g(F)$ and $f(G)$ can be expressed as controllability/observability-type matrices.

(3) The results (3.48) and (3.49) can be extended to cover all the minors of H, so that in view of (3.36), all the elements of the

Routh array [or the fraction-free form, via (3.37)] can be expressed in terms of minors of $L^{(1)}$, $L^{(2)}$, $M^{(1)}$, $M^{(2)}$.

Another approach which is also equivalent to computing the Hurwitz minors, and which also involves matrices of approximately half the order, is now outlined. We begin by writing

$$a(\lambda) = a_{ev}(z) + \lambda a_{od}(z), \quad z = \lambda^2 \tag{3.50}$$

so that a_{ev} and a_{od} contain, respectively, the even and odd powers in $a(\lambda)$. It is assumed that $a_{ev}(z)$ and $a_{od}(z)$ are relatively prime, which implies that $a(\lambda)$ has no purely imaginary (or zero) root, as was seen in the discussion of the second type of Routh singularity. The formula (3.34) thus implies that $H_{n-1} \neq 0$, and since $a_n \neq 0$, equation (3.33) shows that $H_n \neq 0$. The following expansion is then obtained:

$$\frac{a_{od}(z)}{a_{ev}(z)} = s_{-1} + \frac{s_0}{z} - \frac{s_1}{z^2} + \frac{s_2}{z^3} - \frac{s_3}{z^4} + \cdots \tag{3.51}$$

where the numbers s_0, s_1, s_2, $\ldots$, s_{n-1} (n even); s_{-1}, s_0, s_1, $\ldots$, s_{n-2} (n odd) are the *Markov parameters* of $a(\lambda)$.

Theorem 3.7 (Chebyshev-Markov-Gantmacher)

(SC) The polynomial $a(\lambda)$ in (3.13) is asymptotically stable if and only if the two real symmetric matrices

$$
S^{(1)} = \begin{bmatrix}
s_0 & s_1 & \cdots & s_{m-1} \\
s_1 & s_2 & \cdots & s_m \\
\cdot & \cdot & \cdots & \cdot \\
s_{m-1} & s_m & \cdots & s_{2m-2}
\end{bmatrix}
$$

$$
S^{(2)} = \begin{bmatrix}
s_1 & s_2 & \cdots & s_m \\
s_2 & s_3 & \cdots & s_{m+1} \\
\cdot & \cdot & \cdots & \cdot \\
s_m & s_{m+1} & \cdots & s_{2m-1}
\end{bmatrix}
\tag{3.52}
$$

are positive definite, where $m = \frac{1}{2}n$ or $\frac{1}{2}(n - 1)$ according as n is even or odd; in the latter case there is an additional condition, $s_{-1} > 0$.

Example 10 For the polynomial in (3.42) we have from (3.50)

$$a_{ev}(\lambda^2) = \lambda^4 + 9\lambda^2 + 4, \qquad a_{od}(\lambda^2) = 2\lambda^2 + 1$$

so

$$\frac{a_{od}(z)}{a_{ev}(z)} = \frac{2z + 1}{z^2 + 9z + 4} \qquad\qquad (3.53)$$

By setting the right side of (3.53) equal to the expansion in (3.51) and comparing coefficients, we obtain

$$s_{-1} = 0, \quad s_0 = 2, \quad s_1 = 17, \quad s_2 = 145, \quad s_3 = 1237$$

Since $m = 2$, (3.52) gives

$$S^{(1)} = \begin{bmatrix} 2 & 17 \\ 17 & 145 \end{bmatrix}, \quad S^{(2)} = \begin{bmatrix} 17 & 145 \\ 145 & 1237 \end{bmatrix} \qquad (3.54)$$

which are easily seen to be positive definite, showing that $a(\lambda)$ is asymptotically stable, as found earlier in Example 8.

Remarks on Theorem 3.7 (1) Notice that the matrices in (3.52) have orthosymmetric (or Hankel) form (see Problem 1.16). As for Theorem 3.6, the largest minor to be evaluated is only about half that in the Hurwitz criterion. This time there is the added advantage that $S^{(1)}$ and $S^{(2)}$ are symmetric, which is to be balanced against the effort required to determine the s_i.

   (2) The test for positive definiteness requires that all the leading principal minors of the matrices in (3.52) be positive. This can be conveniently carried out by triangularizing the matrices using elementary operations, after which all the diagonal elements (i.e., the pivots) should be positive.

   (3) If the condition on positive definiteness of $S_1^{(1)}$ and $S_2^{(2)}$ is expressed in the form

$$S_1^{(1)} > 0, \quad S_1^{(2)} > 0, \quad S_2^{(1)} > 0, \quad S_2^{(2)} > 0, \quad \ldots, \quad S_m^{(2)} > 0$$

$$(3.55)$$

where $S_j^{(i)}$ denotes the $j \times j$ leading principal minor of $S^{(i)}$, then it can be shown that the sequence (3.55) is just

$$H_1, \ H_2, \ \ldots, \ H_n \quad (n \text{ even})$$

$$\frac{H_2}{a_1^2}, \ \frac{H_3}{a_1^3}, \ \ldots, \ \frac{H_n}{a_1^n} \quad (n \text{ odd})$$

$$(3.56)$$

Example 10 (continued)   The Hurwitz minors for $a(\lambda)$ in (3.42) are

$$H_1 = 2, \quad H_2 = 17, \quad H_3 = 1, \quad H_4 = 4$$

and for $S^{(1)}$ and $S^{(2)}$ in (3.52) the sequence (3.55) is

$$S_1^{(1)} = 2, \quad S_1^{(2)} = 17, \quad S_2^{(1)} = \det S^{(1)} = 1,$$

$$S_2^{(2)} = \det S^{(2)} = 4$$

which confirms (3.56) for this example.

A complete solution to the root location problem was published by Hermite in 1856, twenty-one years before Routh's paper.

Theorem 3.8 (Hermite)

(SC) The polynomial $a(\lambda)$ in (3.13) is asymptotically stable if and only if the symmetric $n \times n$ *Hermite matrix* $H^e = [h_{ij}]$ defined by

$$\left. \begin{array}{l} h_{ij} = \displaystyle\sum_{k=0}^{i-1} (-1)^{k+i-1} a_k a_{i+j-k-1}, \quad i \leqslant j \\[4mm] \quad\ = 0, \quad i + j \text{ odd}; \quad i, \ j = 1, \ 2, \ \ldots, \ n \end{array} \right\}$$

$$(3.57)$$

is positive definite.

(RL) Provided that the leading principal minors satisfy $H_i^e \neq 0$, for all i, then $a(\lambda)$ has k and n - k roots with positive and negative real parts, where

$$k = V(1, H_1^e, H_2^e, \ldots, H_n^e) \qquad (3.58)$$

Example 11 Using (3.57) and the fact that $h_{ji} = h_{ij}$, the Hermite matrix for n = 5 is easily established to be

$$H^e = \begin{bmatrix} a_0a_1 & 0 & a_0a_3 & 0 & a_0a_5 \\ 0 & a_1a_2 - a_0a_3 & 0 & a_1a_4 - a_0a_5 & 0 \\ a_0a_3 & 0 & \begin{matrix}a_0a_5 - a_1a_4 \\ + a_2a_3\end{matrix} & 0 & a_2a_5 \\ 0 & a_1a_4 - a_0a_5 & 0 & a_3a_4 - a_2a_5 & 0 \\ a_0a_5 & 0 & a_2a_5 & 0 & a_4a_5 \end{bmatrix}$$

$$(3.59)$$

Example 12 Return to $a(\lambda)$ in (3.27). Since this has degree 4, we can obtain H^e by deleting the last row and column in (3.59), and setting $a_5 = 0$. This gives

$$H^e = \begin{bmatrix} 1 & 0 & -28 & 0 \\ 0 & 16 & 0 & -16 \\ -28 & 0 & 352 & 0 \\ 0 & -16 & 0 & 448 \end{bmatrix}$$

and

$$H_1^e = 1, \quad H_2^e = 16, \quad H_3^e = -6912, \quad H_4^e = 2,985,984 \qquad (3.60)$$

so from (3.58) k = 1, agreeing with what was found previously in Example 7.

Remarks on Theorem 3.8   (1) If the order of the rows and columns in H^e is reversed to give a matrix $H^e_{(1)}$, i.e.,

$$H^e_{(1)} = JH^e J \qquad (3.61)$$

then the elements $h^{(1)}_{ij}$ of $H^e_{(1)}$ are given by a bezoutian-type expression [see (1.110)] of the form

$$\frac{a(\lambda)a(-\mu) - a(\mu)a(-\lambda)}{\lambda - \mu} \equiv 2 \sum_{i,j=1}^{n} (-1)^{j-1} h^{(1)}_{ij} \lambda^i \mu^j \qquad (3.62)$$

In particular, it therefore follows at once from Theorem 1.12 that H^e is nonsingular if and only if $a(\lambda)$ has no pair of roots $+\lambda_i$, $-\lambda_i$ (thus there must be no zero or purely imaginary root). It also follows from Theorem 1.12 that $H^e_{(1)}$ could be expressed in terms of $a(-C)$, where C is the companion matrix of $a(\lambda)$, a slight modification being necessary because of the factor $(-1)^{j-1}$ in (3.62). Furthermore, (1.117) and (1.118) can be easily modified to give an easy way of constructing $H^e_{(1)}$.

(2) Because of the symmetrical pattern of zero elements in H^e, a simple permutation of rows and columns of H^e can be used to express it as the direct sum of two symmetric matrices. For example, with H^e as in (3.59), we have

$$PH^e P^T = \begin{bmatrix} H^{e1} & 0 \\ 0 & H^{e2} \end{bmatrix} \qquad (3.63)$$

where P is a permutation matrix,

$$H^{e1} = \begin{bmatrix} a_0 a_1 & a_0 a_3 & a_0 a_5 \\ a_0 a_3 & \begin{matrix} a_0 a_5 - a_1 a_4 \\ + a_2 a_3 \end{matrix} & a_2 a_5 \\ a_0 a_3 & a_2 a_5 & a_4 a_5 \end{bmatrix} \qquad (3.64)$$

$$H^{e2} = \begin{bmatrix} a_1 a_2 - a_0 a_3 & a_1 a_4 - a_0 a_5 \\ a_1 a_4 - a_0 a_5 & a_3 a_4 - a_2 a_5 \end{bmatrix} \qquad (3.65)$$

In general, H^{e1} has order $\frac{1}{2}n$ or $\frac{1}{2}(n+1)$ according to whether n is even or odd, and consists of the elements h_{ij} for which i and j are both odd; H^{e2} has order $\frac{1}{2}n$ or $\frac{1}{2}(n-1)$ and comprises the h_{ij} for which both i and j are even; and the permutation matrix in (3.63) is

$$P = \begin{bmatrix} P_1 \\ P_2 \end{bmatrix}, \quad P_1 = \begin{bmatrix} e_1 \\ e_3 \\ e_5 \\ \vdots \end{bmatrix}, \quad P_2 = \begin{bmatrix} e_2 \\ e_4 \\ \vdots \end{bmatrix}$$

where e_i denotes the ith row of I_n.

(3) The leading principal minors of H^e are directly related to the Hurwitz minors via the relationships

$$H_1^e = a_0 H_1, \quad H_2^e = a_0 H_1 H_2, \quad H_3^e = a_0 H_2 H_3, \quad \ldots, \quad H_n^e = a_0 H_{n-1} H_n$$

$$(3.66)$$

This establishes the connection between the Hurwitz and Hermite theorems. For example, the Hermite minors in (3.60) and the Hurwitz minors in (3.29), for the polynomial in (3.27), can readily be seen to satisfy (3.66).

(4) It can be shown that the matrices $S^{(1)}$, $S^{(2)}$ in (3.52) are related to the so-called reduced-order Hermite matrices H^{e1}, H^{e2} in (3.63) by

$$S^{(i)} = WH^{ei}W^T, \quad i = 1, 2$$

$$(3.67)$$

where W is a triangular matrix with unit diagonal elements. This establishes the equivalence of the Hermite and Chebyshev-Markov-Gantmacher stability criteria, since clearly H^e is positive definite if and only if H^{e1}, H^{e2} are positive definite.

(5) The preceding remark can be used to obtain the following reduced form of the Hermite stability criterion. This is a symmetric version of the result obtained when the reduced-order expressions in Theorem 3.6 (using appropriate companion matrices) are applied to the Liénard-Chipart criterion.

Theorem 3.9 (Anderson)

(SC) The polynomial $a(\lambda)$ in (3.13) is asymptotically stable if and only if *either* H^{e1} *or* H^{e2} is positive definite, and *one* of the sequences Q_1, Q_2 defined in (3.38) and (3.39) has all its members positive.

Example 13   Return again to $a(\lambda)$ in (3.42), which has all its coefficients positive. We need only determine H^{e2}, which is obtained from (3.65) (setting $a_5 = 0$) to be

$$H^{e2} = \begin{bmatrix} 17 & 4 \\ 4 & 4 \end{bmatrix}$$

Clearly this is positive definite, confirming that $a(\lambda)$ is asymptotically stable. This approach can be contrasted with that in Example 8, using the Liénard-Chipart theorem.

3.2.2 Real Discrete-Time Case

We now refer to the real polynomial

$$a_1(\mu) = \alpha_0 \mu^n + \alpha_1 \mu^{n-1} + \cdots + \alpha_{n-1}\mu + \alpha_n, \qquad \alpha_0 > 0 \qquad (3.68)$$

and present some analogues of the results of the preceding section, beginning with a tabular array corresponding to Routh's. We shall call $a_1(\mu)$ *convergent* if all its roots have modulus less than one.

Theorem 3.10 (Jury-Marden)   Construct an array having initial rows

$$\{c_{11}, c_{12}, \ldots, c_{1,n+1}\} = \{\alpha_0, \alpha_1, \ldots, \alpha_n\}$$
$$\{d_{11}, d_{12}, \ldots, d_{1,n+1}\} = \{\alpha_n, \alpha_{n-1}, \ldots, \alpha_0\}$$
$$(3.69)$$

and subsequent rows defined by

$$c_{ij} = \begin{vmatrix} c_{i-1,1} & c_{i-1,j+1} \\ d_{i-1,1} & d_{i-1,j+1} \end{vmatrix}, \qquad i = 2, 3, \ldots, n + 1 \qquad (3.70)$$

$$d_{ij} = c_{i,n-j-i+3} \qquad (3.71)$$

(SC) The polynomial $a_1(\mu)$ in (3.68) is convergent if and only if
the first column elements satisfy $d_{21} > 0$, $d_{i1} < 0$, $i = 3$, 4,
..., $n + 1$.

(RL) Provided that the array is *regular*, i.e., $d_{i1} \neq 0$, for all i,
then $a_1(\mu)$ has no roots with unit modulus, and there are k and
n - k roots inside and outside the unit disk $|\mu| < 1$, where k
is the number of negative products in the sequence

$$P_k = (-1)^k d_{21} d_{31} \cdots d_{k+1,1}, \quad k = 1, 2, \ldots, n \qquad (3.72)$$

Example 14   Let

$$a_1(\mu) = 2\mu^3 + 4\mu^2 - 5\mu + 3 \qquad (3.73)$$

so that from (3.69), (3.70), and (3.71) the array is

$$
\begin{array}{lrrrr}
c_{1j} & 2 & 4 & -5 & 3 \\
d_{1j} & 3 & -5 & 4 & 2 \\
c_{2j} & -22 & 23 & -5 & \\
d_{2j} & -5 & 23 & -22 & \\
c_{3j} & -391 & 459 & & \\
d_{3j} & 459 & -391 & & \\
c_{41} & -57,800 & & & \\
= d_{41} & & & &
\end{array}
\qquad (3.74)
$$

Thus in (3.72), $P_1 > 0$, $P_2 < 0$, $P_3 < 0$, so $a_1(\mu)$ has two roots
inside the unit disk and one outside.

Remarks on Theorem 3.10   (1) As Example 14 illustrates, (3.71)
simply states that for each i, the row $\{d_{ij}\}$ consists of the row
$\{c_{ij}\}$ in reverse order; this point has been emphasized by linking
the rows in pairs in (3.74). The array as defined is a little
different from that found in many standard texts. However, it has
been deliberately formulated so as to be as similar as possible to

the Routh scheme: the 2×2 determinants (3.15) and (3.70) all
contain the first-column elements and the other columns are
selected from left to right.   Notice also that (3.70) is identical
to the first part of (1.144), showing that the Jury-Marden array is
a special case of the second tabular g.c.d. algorithm in Section
1.6.2 applied to $a_1(\mu)$ and $\mu^n a_1(1/\mu)$, just as the Routh array is a
special case of the first tabular scheme (Section 1.6.1) applied to
the even and odd parts of $a(\lambda)$.

(2) In line with the comments in Remark 1 on Theorem 3.3
(Routh), the array (3.70) can be modified if so desired to include
division by first-column elements; and any linked pair of rows can
be scaled by a constant positive factor without affecting the signs
of the first-column elements d_{i1}.   It should be appreciated that
there is no real reason why the Routh array is usually defined to
include divisions by first-column elements, whereas the Jury-Marden
array is not.   This has simply become the accepted convention, and
either array can be expressed with or without divisions.

(3) When an element d_{i1} equal to zero is encountered, then the
theorem shows that $a_1(\mu)$ is not convergent.   As for the Routh
algorithm, there are two types of singular cases to be considered
if it is required to determine the location of the roots.

Singular Case: Type (i).   Here no complete row of zeros is
obtained, implying that $a_1(\mu)$ has no roots on the unit circle.   As
for the Routh array, we introduce a small perturbation.   This is
done by replacing μ in $a_1(\mu)$ by $(1 + \varepsilon)\mu$, and using the fact that

$$[(1 + \varepsilon)\mu]^k \triangleq \mu^k (1 + k\varepsilon) \tag{3.75}$$

By applying the theorem to the perturbed polynomial, it is thus
possible to determine the number of roots inside the disk of radius
$1 + \varepsilon$, and hence to deduce the number inside the unit disk itself
(note that we are again appealing to the continuity of the roots
with respect to the polynomial coefficients).

Example 15   Let

$$a_1(\mu) = \mu^3 + 3\mu^2 + 2\mu + 1$$

so the array of Theorem 3.10 is

$$
\begin{array}{llll}
c_{1j} & 1 & 3 & 2 & 1 \\
d_{1j} & 1 & 2 & 3 & 1 \\
c_{2j} & -1 & 1 & 0 \\
d_{2j} & 0 & -1 & 1
\end{array}
$$

Replace $a_1(\mu)$ by $a_1[(1 + \varepsilon)\mu]$, which by (3.75) leads to

$$(1 + 3\varepsilon)\mu^3 + 3(1 + 2\varepsilon)\mu^2 + 2(1 + \varepsilon)\mu + 1$$

The array for this polynomial is constructed using (3.69) to (3.71), to give

$$
\begin{array}{lllll}
c_{1j} & 1 + 3\varepsilon & 3(1 + 2\varepsilon) & 2(1 + \varepsilon) & 1 \\
d_{1j} & 1 & 2(1 + \varepsilon) & 3(1 + 2\varepsilon) & 1 + 3\varepsilon \\
c_{2j} & 2\varepsilon - 1 & 13\varepsilon + 1 & 6\varepsilon \\
d_{2j} & 6\varepsilon & 13\varepsilon + 1 & 2\varepsilon - 1 \\
c_{3j} & -17\varepsilon - 1 & -4\varepsilon + 1 \\
d_{3j} & -4\varepsilon + 1 & -17\varepsilon - 1 \\
c_{41} & 26\varepsilon \\
= d_{41}
\end{array}
$$

Notice that in the construction of the array, terms in ε^2 have been ignored at each step. For this modified array, the sequence (3.72) is

$$P_1 = -6\varepsilon, \quad P_2 = 6\varepsilon(1 - 4\varepsilon), \quad P_3 = -6\varepsilon(1 - 4\varepsilon)26\varepsilon \qquad (3.76)$$

Provided that ε is sufficiently small, then *irrespective of its sign*, the sequence (3.76) contains two negative terms. We can therefore conclude from Theorem 3.10 that $k = 2$, i.e., that $a_1(\mu)$ has two roots inside and one outside the unit disk.

Singular Case: Type (ii). Here at some stage in the array a
complete row of zeros is obtained. This implies that there are
some roots on the unit circle, and/or roots such that $\mu_i \bar{\mu}_j = 1$.
The perturbation method just described can still be used. Suppose
that with ε assumed positive, we use the array to find that there
are p_1 and $n - p_1$ roots inside and outside the disk of radius
$1 + \varepsilon$. By changing the sign of ε, we can similarly find that there
are p_2 and $n - p_2$ roots inside and outside the disk of radius
$1 - \varepsilon$. It therefore follows that there are $p_1 - p_2$ roots actually
on the unit circle, with p_2 inside and $n - p_1$ outside.

Example 16   It is easy to check that for

$$a_1(\mu) = \mu^3 + (2 - \sqrt{2})\mu^2 + (1 - 2\sqrt{2})\mu + 2 \tag{3.77}$$

we obtain $c_{31} = c_{32} = 0$. We therefore construct $a_1[(1 + \varepsilon)\mu]$ using
(3.75) to obtain the following modified array:

c_{1j}	$1 + 3\varepsilon$	$(2 - \sqrt{2})(1 + 2\varepsilon)$	$(1 - 2\sqrt{2})(1 + \varepsilon)$	2
d_{1j}	2	$(1 - 2\sqrt{2})(1 + \varepsilon)$	$(2 - \sqrt{2})(1 + 2\varepsilon)$	$1 + 3\varepsilon$
c_{2j}	$-3 - 4(1 + \sqrt{2})\varepsilon$	$3\sqrt{2} + (8 - \sqrt{2})\varepsilon$	$-3 + 6\varepsilon$	
d_{2j}	$-3 + 6\varepsilon$	$3\sqrt{2} + (8 - \sqrt{2})\varepsilon$	$-3 - 4(1 + \sqrt{2})\varepsilon$	
c_{3j}	$-6\varepsilon(4 + \sqrt{2})$	$12\varepsilon(5 + 2\sqrt{2})$		
d_{3j}	$12\varepsilon(5 + 2\sqrt{2})$	$-6\varepsilon(4 + \sqrt{2})$		
c_{41}	$-72\varepsilon^2(57 + 36\sqrt{2})$			

$= d_{41}$

Notice that because the singularity occurs in the third pair of
rows of the original array, we must retain terms in ε^2 in subsequent rows of the modified array. We compute the signs of P_k in
(3.72) as follows:

$$\begin{array}{lcc} & \varepsilon > 0 & \varepsilon < 0 \\ P_1 = 3 - 6\varepsilon & + & + \\ P_2 = 12(5 + 2\sqrt{2})\varepsilon(-3 + 6\varepsilon) & - & + \\ P_3 = 864(57 + 36\sqrt{2})\varepsilon^3(-3 + 6\varepsilon) & - & + \end{array}$$

Thus by Theorem 3.10 we have $p_1 = 2$, $p_2 = 0$, so $a_1(\mu)$ in (3.77) has two roots on the unit circle and one outside. In fact, the roots of $a_1(\mu)$ are -2, $(1 \pm i)/\sqrt{2}$.

(4) If it is required that the solution $x(k)$ of (3.2) should converge to zero at least as fast as $(\tau)^k$, for some given $\tau < 1$, then this is equivalent to requiring that all the roots of $a_1(\mu)$ should lie inside the disk $|\mu| < \tau$. To test for this, set $\mu = \mu'\tau$, so $|\mu'| < 1$ if and only if $|\mu| < \tau$, and apply the Jury-Marden criterion to $a_1(\mu')$.

<u>Theorem 3.11 (Schur-Cohn)</u> Construct the $2n \times 2n$ matrix

$$\Delta = \begin{bmatrix} \Delta_1 & \Delta_2 \\ \Delta_2 & \Delta_1 \end{bmatrix} \tag{3.78}$$

$$\Delta_1 = \begin{bmatrix} \alpha_0 & \alpha_1 & \alpha_2 & \cdots & \alpha_{n-1} \\ 0 & \alpha_0 & \alpha_1 & \cdots & \alpha_{n-2} \\ 0 & 0 & \alpha_0 & \cdots & \alpha_{n-3} \\ & \mathbf{0} & & \ddots & \alpha_0 \end{bmatrix} \tag{3.79}$$

$$\Delta_2 = \begin{bmatrix} & & 0 & 0 & \alpha_n \\ & \mathbf{0} & 0 & \alpha_n & \alpha_{n-1} \\ \vdots & & \alpha_n & \alpha_{n-1} & \alpha_{n-2} \\ \alpha_n & \cdots & \alpha_2 & \alpha_1 \end{bmatrix} \tag{3.80}$$

and denote by $\Delta^{(2n-2k)}$ the $2(n-k) \times 2(n-k)$ centrally situated submatrix of Δ obtained by deleting rows and columns numbers 1, 2, ..., k; 2n, 2n - 1, ..., 2n - k + 1 (with $\Delta^{(2n)} \triangleq \Delta$).

(SC) The polynomial $a_1(\mu)$ in (3.68) is convergent if and only if
$|\Delta^{(2i)}| > 0$, $i = 1, 2, \ldots, n$ (Δ is then said to be *positive innerwise*)

(RL) If $|\Delta^{(2i)}| \neq 0$, for all i, then $a_1(\mu)$ has k and n - k roots
inside and outside the unit disk, where

$$k = n - V(1, |\Delta^{(2)}|, |\Delta^{(4)}|, \ldots, |\Delta^{(2n)}|) \tag{3.81}$$

Example 17   When n = 4, the matrix defined by (3.78) to (3.80) is

$$\Delta = \Delta^{(8)} = \begin{bmatrix} \alpha_0 & \alpha_1 & \alpha_2 & \alpha_3 & 0 & 0 & 0 & \alpha_4 \\ 0 & \alpha_0 & \alpha_1 & \alpha_2 & 0 & 0 & \alpha_4 & \alpha_3 \\ 0 & 0 & \alpha_0 & \alpha_1 & 0 & \alpha_4 & \alpha_3 & \alpha_2 \\ 0 & 0 & 0 & \alpha_0 & \alpha_4 & \alpha_3 & \alpha_2 & \alpha_1 \\ 0 & 0 & 0 & \alpha_4 & \alpha_0 & \alpha_1 & \alpha_2 & \alpha_3 \\ 0 & 0 & \alpha_4 & \alpha_3 & 0 & \alpha_0 & \alpha_1 & \alpha_2 \\ 0 & \alpha_4 & \alpha_3 & \alpha_2 & 0 & 0 & \alpha_0 & \alpha_1 \\ \alpha_4 & \alpha_3 & \alpha_2 & \alpha_1 & 0 & 0 & 0 & \alpha_0 \end{bmatrix} \tag{3.82}$$

The matrices within the dashed lines, in sequence of reducing
dimensions, are $\Delta^{(6)}$, $\Delta^{(4)}$ and $\Delta^{(2)}$, and are the *inners* of Δ,
defined in Remark 3 on Theorem 1.7.

Remarks on Theorem 3.11   (1) Notice that in the equation (3.82), if
the last four columns are reversed in order, and the last four are
then also reversed in order, a matrix of Sylvester resultant type
(1.84) is obtained, associated with $a_1(\mu)$ and the reverse polynomial
$\hat{a}_1(\mu) = \mu^4 a_1(1/\mu)$. This holds for all n, and suggests that the
relationship between Sylvester matrices and companion matrices
(Theorem 1.7) can be used to derive a companion matrix version of
Theorem 3.11. This will indeed turn out to be the case, but we

shall delay this development for application to a different
Schur-Cohn result (see Theorem 3.13), which is based on the
bezoutian form of the resultant.

(2) In the continuous-time case it was seen (Remark 1 on
Theorem 3.4) that the Hurwitz determinants can be expressed in
terms of the first column elements of the Routh array according to
(3.31). A precisely analogous result is that the Schur-Cohn
determinants are related to the first-column elements in the
Jury-Marden array (3.70) according to

$$|\Delta^{(2)}| = d_{21}, \qquad |\Delta^{(4)}| = -d_{31},$$

$$|\Delta^{(2i)}| = \frac{\varepsilon^{(i)} d_{i+1,1}}{(d_{21})^{i-2}(d_{31})^{i-3} \cdots d_{i-1,1}}, \qquad i = 3, \ldots, n \qquad (3.83)$$

where $\varepsilon(i)$ is defined in (1.22).

Example 18   Consider $a_1(\mu)$ in (3.73), where $n = 3$, so (3.78) to
(3.80) give

$$\Delta = \begin{bmatrix} 2 & 4 & -5 & 0 & 0 & 3 \\ 0 & 2 & 4 & 0 & 3 & -5 \\ 0 & 0 & 2 & 3 & -5 & 4 \\ 0 & 0 & 3 & 2 & 4 & -5 \\ 0 & 3 & -5 & 0 & 2 & 4 \\ 3 & -5 & 4 & 0 & 0 & 2 \end{bmatrix} \qquad (3.84)$$

Evaluating the determinants of the inners, indicated by the dashed
lines, and comparing with the array computed in (3.74), we have

$$|\Delta^{(2)}| = \begin{bmatrix} 2 & 3 \\ 3 & 2 \end{bmatrix} = -5 = d_{21}$$

$$|\Delta^{(4)}| = -459 = -d_{31}$$

$$|\Delta^{(6)}| = |\Delta| = -11,560$$

$$= \frac{\varepsilon(3)d_{41}}{d_{21}}$$

thus agreeing with (3.83).

The correspondence with the Routh-Hurwitz result (3.31) can be made even closer by defining a modified Jury-Marden array which, like the standard Routh array, involves division by first column elements. We replace (3.70) by

$$c'_{ij} = \begin{vmatrix} c'_{i-1,1} & c'_{i-1,j+1} \\ d'_{i-1,1} & d'_{i-1,j+1} \end{vmatrix} \bigg/ d'_{i-2,1}, \quad i = 4, 5, \ldots \qquad (3.85)$$

the first three rows being unaltered, and again each row of d's consists of the c's in reverse order. It then turns out that

$$\Delta^{(2i)} = \varepsilon(i)d'_{i+1,1} \qquad (3.86)$$

and it is interesting to compare (3.86) with the formula $H_i = t_{i1}$, obtained from (3.37), expressing the Hurwitz determinants in terms of the optimal fraction-free Routh array.

(3) Since the α_i in (3.68) are all real, then as for the continuous-time case (Remark 2 on Theorem 3.3), we can write $a_1(\mu)$ as a product of factors

$$a_1(\mu) = \Pi(\mu + \gamma)\Pi(\mu^2 - 2\alpha\mu + \alpha^2 + \beta^2)$$

$$= \Pi(\mu + \gamma)\Pi[(\mu - \alpha)^2 + \beta^2] \qquad (3.87)$$

where α, β, γ are real. Clearly, the quadratic terms in (3.87) are always positive, and if $a_1(\mu)$ is convergent, then we have $|\gamma| < 1$, which implies that $1 + \gamma > 0$. It therefore follows from (3.87) that

$$a_1(1) > 0 \qquad (3.88)$$

Similar reasoning shows that

$$(-1)^n a_1(-1) > 0 \qquad\qquad\qquad (3.89)$$

and (3.88) and (3.89) are simple *necessary* conditions for $a_1(\mu)$ to
be convergent, partially analogous to the condition that all the
coefficients be positive in the continuous-time case. In the
latter case this enabled the Hurwitz Theorem 3.4 to be expressed in
the reduced Liénard-Chipart form (Theorem 3.5). Similar simplifi-
cations of Theorem 3.11 are also possible, of which one is:

<u>Theorem 3.12 (Jury)</u> Let X denote the principal submatrix of Δ_1 in
(3.79) obtained by deleting the last row and last column; and let Y
denote the submatrix of Δ_2 in (3.80) obtained by deleting the first
row and last column.

(SC) The polynomial $a_1(\mu)$ in (3.68) is convergent if and only if
      (3.88) and (3.89) hold, and the two $(n - 1) \times (n - 1)$ matrices
      $Z_1 = X - Y$, $Z_2 = X + Y$ are positive innerwise.

(RL) If the inners satisfy $|z_j^{(i)}| \neq 0$, for all i, j = 1, 2, then
      define

$$\begin{aligned}
V &\equiv V\left(|z_1^{(n-1)}|,\ |z_1^{(n-3)}|,\ \ldots,\ |z_1^{(1)}|,\ |z_2^{(1)}|,\ \ldots,\right.\\
  &\qquad\qquad\qquad \left. |z_2^{(n-1)}|\right),\qquad n\ \text{even}\\
  &\equiv V\left(|z_1^{(n-1)}|,\ |z_1^{(n-3)}|,\ \ldots,\ |z_1^{(2)}|,\ 1,\ |z_2^{(2)}|,\ \ldots,\right.\\
  &\qquad\qquad\qquad \left. |z_2^{(n-1)}|\right),\qquad n\ \text{odd}
\end{aligned} \qquad (3.90)$$

The numbers of roots of $a_1(\mu)$ inside and outside the unit disk are,
respectively, k and n - k, where

$$k = n - V - \tau \qquad\qquad\qquad (3.91)$$

and

$$\tau = \begin{cases} 1, & (-1)^{n+V} a_1(1) a_1(-1) < 0 \\ 0, & (-1)^{n+V} a_1(1) a_1(-1) > 0 \end{cases} \qquad (3.92)$$

Note that the inner submatrix $Z_j^{(n-1-2k)}$ has order $n - 1 - 2k$, and is obtained from Z_j by deleting rows and columns numbers $1, 2, \ldots, k; n - 1, \ldots, n - k$ (with $Z_j^{(n-1)} \triangleq Z_j$).

Example 19 Consider again $a_1(\mu)$ in (3.73), where from (3.78) and (3.84) we can write down

$$\Delta_1 = \begin{bmatrix} 2 & 4 & -5 \\ 0 & 2 & 4 \\ 0 & 0 & 2 \end{bmatrix}, \qquad X = \begin{bmatrix} 2 & 4 \\ 0 & 2 \end{bmatrix}$$

$$\Delta_2 = \begin{bmatrix} 0 & 0 & 3 \\ 0 & 3 & -5 \\ 3 & -5 & 4 \end{bmatrix}, \qquad Y = \begin{bmatrix} 0 & 3 \\ 3 & -5 \end{bmatrix}$$

Thus

$$Z_1 = X - Y = \begin{bmatrix} 2 & 1 \\ -3 & 7 \end{bmatrix}, \qquad Z_2 = X + Y = \begin{bmatrix} 2 & 7 \\ 3 & -3 \end{bmatrix}$$

and since $n = 3$, we obtain from (3.90)

$$V = V\left(|Z_1^{(2)}|, \ 1, \ |Z_2^{(2)}| \right)$$

$$= V\left(|Z_1|, \ 1, \ |Z_2| \right)$$

$$= V(17, \ 1, \ -27) = 1$$

Since $(-1)^4 a_1(1) a_1(-1) = 4.10 > 0$, (3.92) gives $\tau = 0$. Hence from (3.91) $a_1(\mu)$ has two roots inside (and one outside) the unit disk, as before.

We now give a result involving a symmetric matrix, corresponding to the Hermite theorem for the continuous-time case.

Theorem 3.13 (Schur-Cohn-Fujiwara)

(SC) The polynomial $a_1(\mu)$ in (3.68) is convergent if and only if the symmetric *Schur-Cohn matrix* $K = [k_{ij}]$, i, j = 1, ..., n, defined by

$$k_{ij} = \sum_{r=0}^{i-1} \left(\alpha_{i-1-r} \alpha_{j-1-r} - \alpha_{n+r-i+1} \alpha_{n+r-j+1} \right), \quad i \leq j \qquad (3.93)$$

is positive definite.

(RL) Provided that the leading principal minors satisfy $K_i \neq 0$, for all i, then $a_1(\mu)$ has k and n - k roots, respectively, inside and outside the unit disk $|\mu| < 1$, where

$$k = n - V(1, K_1, K_2, \ldots, K_n) \qquad (3.94)$$

Example 20   Using (3.93) and $k_{ij} = k_{ji}$, the Schur-Cohn matrix for n = 3 is easily found to be

$$K = \begin{bmatrix} \alpha_0^2 - \alpha_3^2 & \alpha_0\alpha_1 - \alpha_2\alpha_3 & \alpha_0\alpha_2 - \alpha_1\alpha_3 \\ \alpha_0\alpha_1 - \alpha_2\alpha_3 & \alpha_0^2 + \alpha_1^2 - \alpha_2^2 - \alpha_3^2 & \alpha_0\alpha_1 - \alpha_2\alpha_3 \\ \alpha_0\alpha_2 - \alpha_1\alpha_3 & \alpha_0\alpha_1 - \alpha_2\alpha_3 & \alpha_0^2 - \alpha_3^2 \end{bmatrix} \qquad (3.95)$$

In particular, when $a_1(\mu)$ is the cubic in (3.73) we obtain

$$K = \begin{bmatrix} -5 & 23 & -22 \\ 23 & -14 & 23 \\ -22 & 23 & -5 \end{bmatrix}$$

and the leading principal minors are

$$K_1 = -5, \qquad K_2 = -459, \qquad K_3 = -11,560$$

The number of roots outside the unit disk is therefore

$$V(1, -5, -459, -11,560) = 1$$

agreeing with what was found in Example 14.

Remarks on Theorem 3.13   (1) As was the symmetric Hermite matrix in
Theorem 3.8, K is directly related to a bezoutian expression.
Specifically, KJ (which is K with its columns reversed in order) is
the matrix defined by (1.111) associated with the polynomials $a_1(\mu)$
and the reverse polynomial $\hat{a}_1(\mu) = \mu^n a_1(1/\mu)$. This is easily
confirmed for the case n = 3 by comparing (3.95) with the matrix
obtained by setting

$$a_i = \alpha_i, \quad b_i = \alpha_{n-i}, \quad i = 0, 1, \ldots, n \qquad\qquad (3.96)$$

in (1.112), with n = 3. It now also follows that a simpler way of
constructing K than using (3.93) is to apply (3.96) to (1.118),
which gives the last row of KJ, and the other rows are then
constructed from the recurrence formula (1.117), in which C is the
companion matrix in the form (1.26) of $a_1(\mu)$ (for convenience we
can assume that $\alpha_0 = 1$).

   (2) In view of the relationship between a bezoutian matrix and
a polynomial in C (stated in Theorem 1.12) we can now give the
promised companion matrix version of the second Schur-Cohn theorem
(again assuming that $\alpha_0 = 1$):

Theorem 3.14 (Barnett)   The leading principal minors K_i in Theorem
3.13 are equal to those of the matrix $\hat{a}_1(C)$, where C is the
companion matrix in the form (1.26) of $a_1(\mu)$ (with $\alpha_0 = 1$).

Example 21   Consider the case n = 3. Since

$$\hat{a}_1(\mu) = \alpha_3\mu^3 + \alpha_2\mu^2 + \alpha_1\mu + 1$$

has the same degree as $a_1(\mu)$, the first row of $\hat{a}_1(C)$ is given by
(1.51) as

$$g_1 = [1 - \alpha_3^2, \ \alpha_1 - \alpha_2\alpha_3, \ \alpha_2 - \alpha_1\alpha_3]$$

From Theorem 1.2, the other two rows are $g_2 = g_1 C$, $g_3 = g_2 C$, where
C has last row $[-\alpha_3, -\alpha_2, -\alpha_1]$. It is left as an easy exercise for

the reader to complete the construction of $\hat{a}_1(C)$, and to verify that the relationship (1.120) is here

$$K = \begin{bmatrix} 1 & 0 & 0 \\ \alpha_1 & 1 & 0 \\ \alpha_2 & \alpha_1 & 1 \end{bmatrix} \hat{a}_1(C) \qquad\qquad (3.97)$$

where K is given by (3.95). The expression (3.97) exposes the identity of the minors of $\hat{a}_1(C)$ and K, and can be extended to the general case (see Problem 3.26).

Remarks on Theorem 3.14   (1) It can again be pointed out that in view of (2.60), the matrix $\hat{a}_1(C)$ can be written in controllability/observability form.

(2) Unlike the Hermite matrix in Theorem 3.8, the matrix K does not contain a pattern of zero elements. However, it can be split up into two matrices, each of order about $\frac{1}{2}n$, to give a result corresponding to Theorem 3.9. Instead of going into full details, we merely illustrate the case n = 3:

$$\begin{bmatrix} 1 & 0 & 1 \\ 0 & 1 & 0 \\ -1 & 0 & 1 \end{bmatrix} K \begin{bmatrix} 1 & 0 & -1 \\ 0 & 1 & 0 \\ 1 & 0 & 1 \end{bmatrix} = 2 \begin{bmatrix} K_1 & 0 \\ 0 & K_2 \end{bmatrix}$$

where

$$K_1 = \begin{bmatrix} \alpha_0^2 - \alpha_3^2 + \alpha_0\alpha_2 - \alpha_1\alpha_3 & \alpha_0\alpha_1 - \alpha_2\alpha_3 \\ \alpha_0\alpha_1 - \alpha_2\alpha_3 & \frac{1}{2}(\alpha_0^2 + \alpha_1^2 - \alpha_2^2 - \alpha_3^2) \end{bmatrix}$$

$$K_2 = [\alpha_0^2 - \alpha_3^2 - \alpha_0\alpha_2 + \alpha_1\alpha_3]$$

and the stability criterion of Theorem 3.13 reduces to requiring K_1 and K_2 to be positive definite. Explicit formulas can be given for the elements of K_1 and K_2 in the general case.

3.2.3 Complex Polynomials

Many of the results of the preceding two sections can be extended
to the case when the polynomials $a(\lambda)$ and $a_1(\mu)$ have complex
coefficients, although some, like the Liénard-Chipart criterion
(Theorem 3.5) are specifically restricted to the real case. We
give here only some of the more useful generalizations.

Assume for this section that the coefficients a_i in (3.13) are
complex, with real and imaginary parts denoted by a_i' and a_i'',
respectively, $i = 1, 2, \ldots, n$; it is convenient to also assume
throughout that $a_0 = 1$. The extension of the Routh array for the
continuous-time case is worthwhile, since it involves only real
arithmetic:

Theorem 3.15 (Complex Routh) Construct a Routh array $\{r_{ij}\}$
according to (3.15) but with initial two rows

$$r_{0j} = \{1, -a_1'', -a_2', a_3'', a_4', -a_5'', -a_6', \ldots\} \qquad (3.98)$$

$$r_{1j} = \{a_1', -a_2'', -a_3', a_4'', a_5', -a_6'', -a_7', \ldots\} \qquad (3.99)$$

(SC) The complex polynomial $a(\lambda)$ is asymptotically stable if and
   only if all the products

$$r_{11}, \quad r_{11}r_{21}r_{31}, \quad r_{11}r_{21}r_{31}r_{41}r_{51}, \quad \cdots, \quad r_{11}r_{21}\cdots r_{2n-1,1}$$
$$(3.100)$$

   are positive.

(RL) Provided that $r_{i1} \neq 0$, for all i, then $a(\lambda)$ has no purely
   imaginary roots, and the number of roots with positive real
   parts is equal to the number of variations in sign in the
   sequence consisting of (3.100), preceded by 1.

Remarks on Theorem 3.15 (1) Construction of a Routh array having
initial rows (3.98) and (3.99) is equivalent to finding leading
principal minors of a Hurwitz-type matrix (see Remark 1 on Theorem
3.4). Theorem 3.15 can thus be shown to be equivalent to forming a
$(2n - 1) \times (2n - 1)$ Hurwitz matrix with initial two rows (3.98) and
(3.99); the sequence (3.100) is replaced by the sequence of

odd-order leading principal minors of this matrix. The stability
criterion thereby obtained is named after *Bilharz*.

(2) Yet another way of expressing the theorem is interesting
because it involves continued fractions. Specifically, if we
express

$$\frac{a_1'\lambda^{n-1} + ia_2''\lambda^{n-2} + a_3'\lambda^{n-3} + ia_4''\lambda^{n-4} + \cdots}{\lambda^n + ia_1''\lambda^{n-1} + a_2'\lambda^{n-2} + ia_3''\lambda^{n-3} + a_4'\lambda^{n-4} + \cdots}$$

as a continued fraction

$$\cfrac{1}{c_1\lambda + k_1 + \cfrac{1}{c_2\lambda + k_2 + \cfrac{\ddots}{ + \cfrac{1}{c_n\lambda + k_n}}}} \tag{3.101}$$

where the c's are real and nonzero, and the k's are purely
imaginary or zero, then the number of roots of $a(\lambda)$ having negative
real parts is equal to the number of positive c's. The algorithm
for obtaining (3.101) is equivalent to constructing the Routh array
in the statement of the theorem, and in particular, reduces to the
standard algorithm of Theorem 3.3 when $a(\lambda)$ is a real polynomial.

(3) Theorem 3.15 has the advantage that only real arithmetic
is involved, but this is tempered by the fact that one is
effectively dealing with a real polynomial of degree 2n. This
increase in degree is avoided when complex arithmetic is allowed,
as the generalization of Hermite's result (Theorem 3.8) shows:

Theorem 3.16 (Complex Hermite) Define a hermitian matrix
$H^e = [h_{ij}]$ as in (3.102).

(SC) The complex polynomial $a(\lambda)$ is asymptotically stable if and
     only if H^e is positive definite.

$$H^e = \begin{bmatrix}
a_1 + \bar{a}_1 & a_2 - \bar{a}_2 & a_3 + \bar{a}_3 & \cdots & \cdots & a_n + (-1)^{n-1}\bar{a}_n \\[4pt]
-a_2 + \bar{a}_2 & a_1\bar{a}_2 + a_2\bar{a}_1 - a_3 - \bar{a}_3 & a_3\bar{a}_1 - a_1\bar{a}_3 & \cdots & \cdots & \cdot \\[4pt]
a_3 + \bar{a}_3 & a_4 - \bar{a}_4 + a_1\bar{a}_3 - a_3\bar{a}_1 & -a_4 + \bar{a}_4 & \cdots & \cdot & \cdot \\[4pt]
 & & a_5 + \bar{a}_5 - a_1 a_4 & & & \cdot \\[4pt]
\cdot & -a_3\bar{a}_1 & -a_4\bar{a}_1 + a_2\bar{a}_3 + a_3\bar{a}_2 & \cdots & \cdot & \\[4pt]
(-1)^{n-2}a_{n-1} + \bar{a}_{n-1} & \cdot & \cdot & & & -a_n\bar{a}_{n-2} + a_{n-2}\bar{a}_n \\[4pt]
(-1)^{n-1}a_n + \bar{a}_n & \cdot & \cdot & & & a_n\bar{a}_{n-1} + a_{n-1}\bar{a}_n
\end{bmatrix}$$

$$(3.102)$$

(RL) Provided that the leading principal minors satisfy $H_i^e \neq 0$, for all i, then $a(\lambda)$ has k and n - k roots with positive and negative real parts, where

$$k = V(1, H_1^e, H_2^e, \ldots, H_n^e) \qquad (3.103)$$

Remarks on Theorem 3.16   (1) When the a_i are all real, H^e in (3.102) reduces to two times the real symmetric matrix defined in (3.57), Theorem 3.8.

(2) The minors H^e can also be evaluated in terms of the first-column elements of a *complex* Routh array having initial two rows

$$\{1, a_1, a_2, \ldots\}, \quad \{1, -\bar{a}_1, \bar{a}_2, -\bar{a}_3, \ldots\}$$

thus providing a direct complex analogue of the Routh-Hurwitz connection (3.31).

(3) It can be shown that

$$\det H^e = (-1)^n \prod_{i,j=1}^{n} (\lambda_1 + \bar{\lambda}_j) \qquad (3.104)$$

which can be regarded as a generalization of Orlando's formula (3.34).

(4) As for the real case, on defining

$$H_{(1)}^e = JH^e J$$

where H^e is the matrix in (3.102), then the elements $h_{ij}^{(1)}$ of $H_{(1)}^e$ are given by a bezoutian expression

$$\frac{\bar{a}(\lambda)a(-\mu) - \bar{a}(\mu)a(-\lambda)}{\lambda - \mu} = \sum_{i,j=1}^{n} (-1)^{j-1} h_{ij}^{(1)} \lambda^i \mu^j$$

which is the complex generalization of (3.62) [note that $\bar{a}(\lambda) = \bar{a}_0 \lambda^n + \bar{a}_1 \lambda^{n-1} + \cdots$]. Again, Theorem 1.12 could be used to express $H_{(1)}^e$ in terms of $a(-C)$, where C is now a complex companion matrix for $\bar{a}(\lambda)$.

(5) Instead of applying this companion matrix approach directly, we shall follow it for a version of Hermite's theorem involving a *real symmetric* matrix. This defined by the similarity transformation

$$H^r = FH^eF*$$ (3.105)

where

$$F = diag[i^{n-1}, i^{n-2}, ..., i^2, i, 1]$$ (3.106)

and it is easy to show that the elements of H^r are given by

$$h_{k\ell}^r = (-1)^k (i)^{k+\ell} h_{k\ell}; \quad k, \ell = 1, 2, ..., n$$ (3.107)

The leading principal minors of H^r can be used instead of those of H^e in the theorem. For either matrix the theory of singular cases can be developed in detail, but this will not be done here.

A simple way of constructing H^r [and hence H^e via (3.105) or (3.107)] is to define

$$\phi_k = Re(i^k a_k), \quad \psi_k = Im(i^k a_k), \quad k = 1, 2, ..., n$$ (3.108)

together with an associated real companion matrix

$$\Phi = \begin{bmatrix} 0 & & \\ \vdots & I_{n-1} & \\ -\phi_n & \cdots & -\phi_1 \end{bmatrix}$$ (3.109)

Then the rows $h_1^r, ..., h_n^r$ of H^r are given by

$$h_1^r = 2[\psi_1, \psi_2, ..., \psi_n]$$ (3.110)

$$h_k^r = \phi_{k-1}h_1 + h_{k-1}J_n\phi J_n, \quad k = 2, ..., n$$ (3.111)

Furthermore, by using once again the bezoutian-companion matrix link (1.120), H^r can be expressed as the product of a triangular matrix and a polynomial in Φ, leading to:

Theorem 3.17 (Barnett)   The leading principal minors in Theorem
3.16 can be replaced by those of the matrix

$$\psi(\Phi)J_n \triangleq (\psi_1\Phi^{n-1} + \psi_2\Phi^{n-2} + \cdots + \psi_n I_n)J_n \qquad (3.112)$$

where Φ is defined in (3.109).

Remarks on Theorem 3.17   (1) The matrix $\psi(\Phi)$ can be constructed
from Theorem 1.2.   It can be expressed in controllability/
observability form using (2.60).

(2) A slightly different version of the theorem is possible
using a complex companion matrix associated with $i^n a(-i\lambda)$.

Example 22   We apply the results given so far in this section to

$$a(\lambda) = \lambda^3 + (2 + 2i)\lambda^2 + 7i\lambda - 5 + 5i \qquad (3.113)$$

$$\quad\quad\quad a_1' \ a_1'' \qquad a_2'' \quad a_3' \ a_3''$$

(a) The Routh array of Theorem 3.15, commencing with (3.98)
and (3.99), is

$$
\begin{array}{llll}
r_{0j} & 1 & -2 & 0 & -5 \\
\end{array}
$$

r_{0j}	1	-2	0	-5
r_{1j}	2	-7	5	
r_{2j}	$\frac{3}{2}$	$-\frac{5}{2}$	-5	
r_{3j}	$-\frac{11}{3}$	$\frac{35}{3}$		
r_{4j}	$\frac{25}{11}$	-5		
r_{51}	$\frac{18}{5}$			

From (3.100) the number of roots with positive real parts is

$$k = V(1, \ 2, \ -11, \ -90) = 1$$

and there are thus $3 - 1 = 2$ roots with negative real parts.

(b) The complex Hermite matrix (3.102) is

$$H^e = \begin{bmatrix} 4 & 14i & -10 \\ -14i & 38 & 40i \\ -10 & -40i & 70 \end{bmatrix} \qquad (3.114)$$

After calculating the leading principal minors of H^e, (3.103) gives

$$k = V(1, 4, -44, -2080) = 1$$

(c) Instead of using (3.107) to construct the real Hermite matrix, we apply (3.110) and (3.111) to illustrate their use. From (3.108)

$$\phi_1 = -2, \quad \phi_2 = 0, \quad \phi_3 = 5$$

$$\psi_1 = 2, \quad \psi_2 = -7, \quad \psi_3 = 5$$

so in (3.110) we have $h_1^r = 2[2, -7, 5]$, and the other rows of H^r are given by

$$h_k^r = \phi_{k-1} h_1^r + h_{k-1}^r J_3 \Phi J_3, \qquad k = 2, 3$$

where

$$\Phi = \begin{bmatrix} 0 & 1 & 0 \\ 0 & 0 & 1 \\ -5 & 0 & 2 \end{bmatrix} \qquad (3.115)$$

This leads to

$$H^r = \begin{bmatrix} 4 & -14 & 10 \\ -14 & 38 & -40 \\ 10 & -40 & 70 \end{bmatrix} \qquad (3.116)$$

and it is easily verified that the leading principal minors of (3.116) are equal to those of H^e in (3.114). The reader can also check that the relationship (3.107) holds between the elements in (3.114) and (3.116).

(d) In Theorem 3.17, the matrix $\psi(\Phi)$ has first row

$$r_1 = [\psi_3,\ \psi_2,\ \psi_1] = [5,\ -7,\ 2]$$

and subsequent rows $r_2 = r_1\Phi$, $r_3 = r_2\Phi$. Using (3.115), this gives

$$\psi(\Phi)J = \begin{bmatrix} 2 & -7 & 5 \\ -3 & 5 & -10 \\ -1 & -10 & 15 \end{bmatrix} \tag{3.117}$$

Using the leading principal minors of (3.117) produces

$$k = V[1,\ 2,\ -11,\ -260] = 1$$

Notice that each pth-order minor is $(\tfrac{1}{2})^p$ times the corresponding minor of H^r.

We now turn to the discrete-time problem, and again assume throughout that the polynomial $a_1(\mu)$ in (3.68) has complex coefficients, but is *monic*.

Theorem 3.18 When $a_1(\mu)$ in (3.68) is complex, the results of Section 3.2.2 still apply, subject to the following modifications:
(i)   Theorem 3.10 (Jury-Marden)
       The second row in (3.69) is replaced by

$$\{d_{11},\ \ldots,\ d_{1,n+1}\} = \{\bar{a}_n,\ \bar{a}_{n-1},\ \ldots,\ \bar{a}_1,\ 1\} \tag{3.118}$$

and (3.71) by

$$d_{ij} = \bar{c}_{i,n-j-i+3} \tag{3.119}$$

(ii)   Theorem 3.11 (Schur-Cohn)
       The matrix in (3.78) is replaced by

$$\Delta = \begin{bmatrix} \Delta_1 & \Delta_2 \\ \bar{\Delta}_2 & \bar{\Delta}_1 \end{bmatrix} \tag{3.120}$$

(iii) Theorem 3.13 (Schur-Cohn-Fujiwara)

The matrix K is now hermitian, and the formula (3.93) is replaced by

$$k_{ij} = \sum_{r=0}^{i-1} \left(\alpha_{i-1-r}\bar{\alpha}_{j-1-r} - \bar{\alpha}_{n+r-i+1}\alpha_{n+r-j+1}\right), \quad i \leq j \quad (3.121)$$

(iv) Theorem 3.14 (Barnett)

The polynomial $\hat{a}_1(\mu)$ is now defined by

$$\hat{a}_1(\mu) = \mu^n \bar{a}_1(1/\mu)$$

$$= \bar{\alpha}_n \mu^n + \bar{\alpha}_{n-1}\mu^{n-1} + \cdots + 1 \quad (3.122)$$

Example 23 We apply the four parts of Theorem 3.18 to

$$a_1(\mu) = \mu^3 + \mu^2(-\tfrac{1}{2} + \tfrac{3}{2}i) + \mu(-\tfrac{9}{2} - \tfrac{5}{2}i) + (3 - i) \quad (3.123)$$

(a) The array defined by (3.69), (3.118), and (3.119) is

$$
\begin{array}{cccc}
c_{1j} & 1 & -\tfrac{1}{2}+\tfrac{3}{2}i & -\tfrac{9}{2}-\tfrac{5}{2}i & 3-i \\
d_{1j} & 3+i & -\tfrac{9}{2}+\tfrac{5}{2}i & -\tfrac{1}{2}-\tfrac{3}{2}i & 1 \\
c_{2j} & -\tfrac{3}{2}(1+i) & \tfrac{21}{2}(1+i) & -8 \\
d_{2j} & -8 & \tfrac{21}{2}(1-i) & \tfrac{3}{2}(-1+i) \\
c_{3j} & \tfrac{21}{5}(5+8i) & -\tfrac{119}{2} \\
d_{3j} & -\tfrac{119}{2} & \tfrac{21}{2}(5-8i) \\
c_{41}\ (=d_{41}) & 6272
\end{array}
$$

Thus the sequence (3.72) gives

$$P_1 = -d_{21} > 0, \quad P_2 = d_{21}d_{31} > 0, \quad P_3 = -d_{21}d_{31}d_{41} < 0$$

showing that $a_1(\mu)$ in (3.123) has one root inside and two outside the unit disk.

(b) It is left as an exercise for the reader to write down the 6×6 matrix in (3.120) and to compute the determinants required for application of (3.81).

(c) The matrix K in (3.95) is now replaced by

$$K = \begin{bmatrix} 1 - \alpha_3\bar{\alpha}_3 & \bar{\alpha}_1 - \alpha_2\bar{\alpha}_3 & \bar{\alpha}_2 - \alpha_1\bar{\alpha}_3 \\ \alpha_1 - \bar{\alpha}_2\alpha_3 & 1 + \alpha_1\bar{\alpha}_1 - \alpha_2\bar{\alpha}_2 - \alpha_3\bar{\alpha}_3 & \bar{\alpha}_1 - \alpha_2\bar{\alpha}_3 \\ \alpha_2 - \bar{\alpha}_1\alpha_3 & \alpha_1 - \bar{\alpha}_2\alpha_3 & 1 - \alpha_3\bar{\alpha}_3 \end{bmatrix} \qquad (3.124)$$

and substituting the values of the coefficients in (3.123) gives

$$K = \begin{bmatrix} -9 & \frac{21}{2}(1 + i) & \frac{3}{2}(-1 - i) \\ \frac{21}{2}(1 - i) & -33 & \frac{21}{2}(1 + i) \\ \frac{3}{2}(-1 + i) & \frac{21}{2}(1 - i) & -9 \end{bmatrix} \qquad (3.125)$$

The expression (3.94) shows that there are

$$k = 3 - V(1, -9, \frac{153}{2}, 1479)$$

$$= 1$$

roots inside the unit disk.

(d) Here

$$\hat{a}_1(\mu) = (3 + i)\mu^3 + (-\frac{9}{2} + \frac{5}{2}i)\mu^2 + (-\frac{1}{2} - \frac{3}{2}i)\mu + 1$$

and the companion matrix of $a_1(\mu)$ is

$$C = \begin{bmatrix} 0 & 1 & 0 \\ 0 & 0 & 1 \\ -3 + i & \frac{9}{2} + \frac{5}{2}i & \frac{1}{2} - \frac{3}{2}i \end{bmatrix}$$

To construct $\hat{a}_1(C)$ we can use (1.42), with the first row being given by (1.51) [in fact, equal to the first row in (3.124)]. We then obtain

$$\hat{a}_1(C) = \begin{bmatrix} -9 & \frac{21}{2}(1+i) & \frac{3}{2}(-1-i) \\ 3(2+i) & -12 - \frac{21}{2}i & \frac{15}{2} + 12i \\ -\frac{69}{2} - \frac{57}{2}i & \frac{39}{4} + \frac{303}{4}i & \frac{39}{4} - \frac{63}{4}i \end{bmatrix}$$

and it is routine to check that this has the same leading minors as K in (3.125).

Remarks on Theorem 3.18   (1) The details of critical cases are again complicated, and will not be gone into here.

   (2) As for the real case (see Remark 1 on Theorem 3.13), KJ is the matrix of a bezoutian expression associated with $a_1(\mu)$ and $\mu^n \bar{a}_1(1/\mu)$.

   (3) To obtain an analogue of Theorem 3.15, which provided a real Routh array for a complex polynomial, one can use the fact that the polynomial

$$a_1(\mu)\bar{a}_1(\mu) \tag{3.126}$$

has the same distribution of roots relative to the unit circle as does $a_1(\mu)$. This is because the roots of $\bar{a}_1(\mu) = \bar{\alpha}_0\mu^n + \bar{\alpha}_1\mu^{n-1} + \cdots + \bar{\alpha}_n$ are the complex conjugates of those of $a_1(\mu)$. Since (3.126) is a real polynomial of degree 2n, we can apply to it the real Jury-Marden array of Theorem 3.10.

3.2.4   Recapitulation

It is interesting to point out the four main strands which run through the results we have presented in Sections 3.2.1 to 3.2.3. These are (i) the tabular algorithms of Routh and Jury-Marden; (ii) the Sylvester-type matrices in which the elements consist of the coefficients of the polynomial, the Hurwitz and Schur-Cohn matrices being the prime examples; (iii) the symmetric (or hermitian) matrices — for example, those of Hermite and of Schur-Cohn-Fujiwara — in which the elements are computed from the underlying bezoutian definition; and (iv) the companion matrix

formulations, which provide a link with the controllability/
observability matrices of Chapter 2.

The reader will recognize these four themes from their
appearance in Chapter 1 for the g.c.d. problem.

Problems

3.6 Test whether the following polynomials are asymptotically
     stable.

     (i) $\lambda^3 + 13\lambda^2 + 5\lambda + 1$

     (ii) $\lambda^4 + \lambda^3 + 4\lambda^2 + 4\lambda + 3$

     (iii) $4\lambda^5 + 5\lambda^4 + 26\lambda^3 + 30\lambda^2 + 7\lambda + 2$

3.7 Use all the methods of Section 3.2.1 to determine the numbers
     of roots of the following polynomials to the left and right of
     the imaginary axis.

     (i) $\lambda^3 + \lambda^2 - 2\lambda - 1$

     (ii) $\lambda^4 + 2\lambda^3 + 3\lambda^2 + 4\lambda + 5$

     (iii) $\lambda^6 - 5\lambda^5 + 2\lambda^4 - \lambda^2 + \lambda + 3$

3.8 Determine in each of the following cases the range of values
     of the real parameter k such that the polynomial is
     asymptotically stable.

     (i) $\lambda^4 + 2\lambda^3 + 3\lambda^2 + 4\lambda + k$

     (ii) $(3 - k)\lambda^3 + 2\lambda^2 + (5 - 2k)\lambda + 2$

3.9 A linear system (3.1) has characteristic polynomial

     $$\lambda^4 + 10\lambda^3 + 35\lambda^2 + 50\lambda + 24 + k$$

     Determine the range of values of the real parameter k such
     that $x(t) \to 0$ faster than e^{-t}, as $t \to \infty$.

3.10 Investigate the root distribution relative to the imaginary
     axis of the following polynomials, using the critical-case
     analysis of the Routh array.

   (i) $\lambda^4 + 2\lambda^3 + 3\lambda^2 + 6\lambda + 1$

   (ii) $\lambda^6 + 2\lambda^5 + 3\lambda^4 + 4\lambda^3 + 3\lambda^2 + 2\lambda + 1$

   (iii) $\lambda^5 + 2\lambda^4 + 3\lambda^3 + 4\lambda^2 + 3\lambda + 2$

3.11 Use Orlando's formula (3.34) to deduce that the last Hurwitz
     determinant H_n is zero if and only if $a(\lambda)$ has at least one
     pair of roots λ_0, $-\lambda_0$, or a zero root.

3.12 Show that necessary and sufficient conditions for a real
     polynomial $a(\lambda)$ to be asymptotically stable are:

   (i)   When $n = 2$, $a_1 > 0$, $a_2 > 0$ (compare with Problem 3.1).

   (ii)   When $n = 3$, $a_1 > 0$, $a_3 > 0$, $a_1 a_2 - a_0 a_3 > 0$.

3.13 A simple sufficient test for instability of the real
     polynomial $a(\lambda)$ in (3.13) with $a_i > 0$, for all i, states that
     $a(\lambda)$ is unstable if $\eta_i > 1$ for some $i = 1, 2, \ldots, n - 2$
     $(n \geqslant 3)$, where $\eta_i = a_{i-1} a_{i+2} / a_i a_{i+1}$. Use this to show that

$$10^{-6}\lambda^7 + 2.10^{-5}\lambda^6 + 6.10^{-4}\lambda^5 + 5.10^{-3}\lambda^4 + 10^{-2}\lambda^3 + 0.8\lambda^2$$
$$+ k_1 \lambda + k_2$$

   is unstable for any positive values of k_1 and k_2.

3.14 Some simple sufficient conditions for stability of the real
     polynomial $a(\lambda)$ in (3.13) with $a_i > 0$, for all i, state that
     $a(\lambda)$ is asymptotically stable if *either*

$$\eta_i < \eta^{(1)}, \quad i = 1, 2, \ldots, n - 2 \quad (n \geqslant 5),$$

   where the η_i are defined in Problem 3.13, and $\eta^{(1)}$ is the real
   root of $\lambda(\lambda + 1)^2 = 1$ (i.e., $\eta^{(1)} \doteq 0.465$), *or*

$$\eta_i + \eta_{i+1} < \eta^{(2)}, \qquad i = 1, 2, \ldots, n - 3 \quad (n \geqslant 5)$$

where $\eta^{(2)} = 3(4)^{-1/3} - 1 \simeq 0.890$. Use these results to show that

$$a(\lambda) = 10^{-5}\lambda^6 + 4.10^{-4}\lambda^5 + (1.5)10^{-2}\lambda^4 + 0.12\lambda^3 + 0.5\lambda^2$$
$$+ 2\lambda + 1.7$$

is asymptotically stable.

3.15 Verify that the following polynomials satisfy the necessary conditions (3.88) and (3.89) for convergence. Then apply the Jury-Marden test to determine whether they are indeed convergent.

(i) $288\mu^3 - 336\mu^2 + 146\mu - 25$

(ii) $2\mu^4 - 3\mu^3 + 2\mu^2 - \mu + 1$

3.16 Use all the methods of Section 3.2.2 to determine the numbers of roots of the following polynomials inside and outside the unit disk.

(i) $2\mu^3 - 3\mu^2 - 4\mu + 6$

(ii) $\mu^4 + 2\mu^3 + \mu^2 + 3\mu + 2$

3.17 Use the Jury-Marden test to determine the range of values of the real parameter k such that the polynomial $8\mu^2 + (2k - 4)\mu - k$ is convergent.

3.18 Test whether all the roots of the polynomial in Problem 3.15(i) have moduli less than 3/4.

3.19 Investigate the root distribution of the following polynomials using the critical-case analysis of the Jury-Marden array.

(i) $\mu^3 + \mu^2 + \mu + 1$

(ii) $\mu^3 + 3.3\mu^2 + 3\mu + 0.8$

3.20 Apply (3.85) to $a_1(\mu)$ in (3.73), and hence verify (3.86) in this case.

3.21 Show that necessary and sufficient conditions for the second-degree real polynomial $a_1(\mu)$ to be convergent are $a_1(1) > 0$, $a_1(-1) > 0$, $\alpha_0 > \alpha_2$ (compare with Problem 3.1).

3.22 Determine the root distribution of

$$\lambda^3 + (1 + i)\lambda^2 + (2 - 3i)\lambda + 7$$

using Theorems 3.15 and 3.16.

3.23 The complex characteristic polynomial

$$\lambda^2 + (2k\omega + 2\Omega i)\lambda + \omega^2 - \Omega^2$$

arises in the theory of whirling shafts, where Ω is the angular velocity, ω is the frequency of undamped oscillations, and k is a damping coefficient. Show that the system is asymptotically stable if and only if $\Omega < \omega$.

3.24 Determine the number of roots of the complex polynomial

$$8\mu^3 + \mu^2(34 + 16i) + \mu(2 + 67i) + (-24 + 12i)$$

inside and outside the unit disk.

3.25 Apply the suggestion in Remark 3 on Theorem 3.18 to $a_1(\mu)$ in (3.123), thereby testing its root location relative to the unit circle using a real Jury-Marden array.

3.26 Prove that in general the Schur-Cohn-Fujiwara matrix K in Theorem 3.18c satisfies the relationship $K = M\hat{a}_1(C)$, where $\hat{a}_1(\mu)$ is defined in (3.122), $M = [m_{ij}]$ is a lower triangular matrix having $m_{ii} = 1$, $m_{ij} = \alpha_{i-j}$, $i > j$, and C is the companion matrix of $a_1(\mu)$ in the form (1.26), with $\alpha_0 = 1$. Notice that JM has the same form as T in (1.35).

3.27 If $a(\lambda)$ is asymptotically stable, then the negative number

$$\sigma \overset{\Delta}{=} \max_i \mathrm{Re}(\lambda_i)$$

is called the *abscissa of stability* of $a(\lambda)$. Two upper bounds for σ are given by

$$\sigma \leqslant -2^{-\epsilon(n)} R^{-(n-1)(n+2)/2} r_{11} r_{21} \cdots r_{n1}$$

$$\sigma < -2^{-n} R^{-2n+2} r_{n1} r_{n-1,1}$$

where $R = 2 \max_i |a_i/a_0|^{1/i}$, $\epsilon(n)$ is defined in (1.22), and r_{i1} are the first-column elements in the Routh array (3.15). Apply these results to the relevant polynomials in Problem 3.6.

3.28 Another theorem due to Schur states that the complex polynomial $a_1(\mu)$ in (3.68), with $|\alpha_0| > |\alpha_n|$, is convergent if and only if the polynomial $[\bar{\alpha}_0 a_1(\mu) - \alpha_n \hat{a}_1(\mu)]/\mu$ is also convergent, where $\hat{a}_1(\mu)$ is defined in (3.122). Assume in the following that all the α_i are real.

(i)   Use the result above, together with that in Problem 3.21, to show that if $n = 3$, and

$$\alpha_0 > \alpha_1 > \alpha_2 > \alpha_3 > 0$$

then $a_1(\mu)$ is convergent.

(ii)   It can be shown that the result in (i) holds for any value of n, namely that a *sufficient* condition for $a_1(\mu)$ to be convergent is

$$\alpha_0 > \alpha_1 > \alpha_2 > \cdots > \alpha_n > 0$$

Construct an example to show that this is not a necessary condition.

(iii)   Deduce the implication of the condition

$$\alpha_n > \alpha_{n-1} > \cdots > \alpha_1 > \alpha_0 > 0$$

on the location of the roots of $a_1(\mu)$.

3.29 Let $\alpha_0 = 1$ in (3.68). A result of Cauchy states that if

$$\bar{\alpha} \stackrel{\Delta}{=} \max_i |\alpha_i|$$

then the polynomial (3.68) has all its roots inside the disk
$|\mu| \leqslant 1 + \bar{\alpha}$. A recent improvement states that the roots are
contained within the annulus

$$\frac{|\alpha_n|}{2(1 + \bar{\alpha})^{n-1}(n\bar{\alpha} + 1)} \leqslant |\mu| \leqslant 1 + \left[1 - \frac{1}{(1 + \bar{\alpha})^n}\right]\bar{\alpha}$$

(in both cases the α_i are allowed to be complex). Apply these
result to the polynomials in Problem 3.16.

3.30 Use the formula (3.15) to show that if $n + 1$ nonzero real
numbers are specified as the first column of a Routh array
defined in Theorem 3.3, then the real polynomial to which it
corresponds is unique (see also Problem 1.41).

3.3 SOME KEY PROOFS

It would almost take a book of its own to give individual proofs of
the results in Section 3.2. Fortunately, this is unnecessary; only
a few key theorems have to be proved, and those remaining can then
be deduced from the various interrelationships between the
different criteria. In keeping with the objectives of this book
we devote this section entirely to relatively recent methods
involving matrices; a brief account will be given in Section 3.4 of
one of the important classical techniques, but these are readily
available in many existing works.

3.3.1 Matrix Equation Approach: Continuous-Time Case

We begin with theorems on location of the eigenvalues of an
arbitrary square $n \times n$ complex matrix A, relative to the imaginary
axis. These are then applied to appropriate companion matrices to
produce proofs of some of the theorems in Section 3.2.1 and 3.2.3.
The discrete-time case will be dealt with similarly in Section
3.3.2.

Define the *inertia* of a real or complex n × n matrix A,
denoted by In(A), to be the ordered triple of integers (π, ν, δ),
where $\pi = \pi(A)$, $\nu = \nu(A)$, $\delta = \delta(A)$ are the numbers of eigenvalues
of A having positive, negative, and zero real parts, respectively,
so that

$$\pi + \nu + \delta = n$$

In particular, A is the matrix of an asymptotically stable
continuous-time linear system if and only if In(A) = (0, n, 0).
If H is an arbitrary hermitian matrix, then a well-known result of
Sylvester (often expressed in terms of hermitian forms, and proved
in most elementary texts on matrix algebra) states that for any
nonsingular matrix T

$$\text{In}(T^*HT) = \text{In}(H) \tag{3.127}$$

Furthermore, another classical theorem, due to *Jacobi*, states
that if H is nonsingular [i.e., $\delta(H) = 0$], then

$$\nu(H) = V(1, H_1, H_2, \ldots, H_n) \tag{3.128}$$

The basic idea behind the matrix equation approach is to express
the inertia of an arbitrary complex (or real) matrix in terms of a
hermitian (or real symmetric) matrix. In the case of stability
matrices, the result is as follows:

Theorem 3.19 (Liapunov)   There exists a negative definite hermitian
matrix P [i.e., In(P) = (0, n, 0)] such that

$$A^*P + PA = Q \tag{3.129}$$

where Q is an arbitrary positive definite hermitian matrix, if and
only if A is a stability matrix, i.e., In(A) = (0, n, 0).

Proof. Suppose that there exists a negative definite P satisfying
(3.129), and let λ_k, w_k be corresponding eigenvalue and right
eigenvector of A. We have

$$Aw_k = \lambda_k w_k, \qquad w_k^*A^* = \bar{\lambda}_k w_k^*$$

the second part being obtained by taking the conjugate transpose of
the first. Equation (3.129) gives

$$w_k^* Q w_k = w_k^* (A^* P + PA) w_k$$

$$= (\bar{\lambda}_k w_k^*) P w_k + w_k^* P (\lambda_k w_k)$$

$$= (\lambda_k + \bar{\lambda}_k)(w_k^* P w_k)$$

Since Q is positive definite and P is negative definite, we have

$$\text{Re}(\lambda_k) = \frac{1}{2}(\lambda_k + \bar{\lambda}_k)$$

$$= \frac{1}{2} \frac{w_k^* Q w_k}{w_k^* P w_k} < 0$$

for all k, showing that A is a stability matrix.

Conversely, suppose that all the eigenvalues of A have
negative real parts. We wish to show that there exists a negative
definite hermitian matrix P satisfying (3.129). We can in fact
exhibit an explicit expression for a solution of (3.129). Consider
a matrix defined by

$$X(t) = -e^{A^* t} Q e^{At} \tag{3.130}$$

Using the properties of the matrix exponential, it follows that

$$\frac{dX(t)}{dt} = -A^* e^{A^* t} Q e^{At} - e^{A^* t} Q e^{At} A$$

$$= A^* X(t) + X(t) A \tag{3.131}$$

Now integrate both sides of (3.131) with respect to time from zero
to some positive value T, to obtain

$$X(T) - X(0) = A^* \left[\int_0^T X(t) \, dt \right] + \left[\int_0^T X(t) \, dt \right] A \tag{3.132}$$

Since by assumption A is a stability matrix, we know from Remark 4
on Theorem 3.1 that $X(T) \to 0$ as $T \to \infty$, and clearly from (3.130)
$X(0) = -Q$. Thus allowing $T \to \infty$ in (3.132) reduces this equation to
(3.129), where

$$P = \int_0^\infty X(t)\ dt$$

$$= -\int_0^\infty e^{A^*t} Q e^{At}\ dt \tag{3.133}$$

The integral (3.133) converges because A is a stability matrix (to verify this, consider norms). Furthermore, it is obvious that P given by (3.133) is hermitian. Finally, if $v \neq 0$ is an arbitrary, constant column n-vector, then

$$v^*Pv = -\int_0^\infty v^* e^{A^*t} Q e^{At} v\ dt$$

$$= -\int_0^\infty (e^{At}v)^* Q (e^{At}v)\ dt \tag{3.134}$$

Since Q is positive definite, and exp(At) is nonsingular for all $t \geq 0$, the integrand in (3.134) is positive for all t, showing that $v^*Pv < 0$, i.e., P is negative definite. ∎

Remarks on Theorem 3.19   (1) Equation (3.129) is quite often stated in the form

$$A^*P + PA = -Q \tag{3.135}$$

which is equivalent to replacing P in the original form by (-P), so the solution of (3.135) is to be positive definite. This is important when (3.129) is related to Liapunov's theory of stability of differential equations. In this case

$$V(x) = x^*(t)Px(t) > 0$$

is a *Liapunov function* for the system of differential equations (3.1). The time derivative of V with respect to the system equations is

$$\frac{dV}{dt} = \dot{x}^*Px + x^*P\dot{x}$$

$$= x^*A^*Px + x^*PAx, \quad \text{using (3.1)}$$

$$= -x^*Qx < 0, \quad \text{using (3.135)}$$

A stability theorem of Liapunov then establishes that the origin of (3.1) is asymptotically stable. In particular, when A is real, then P is also real, and symmetric.

(2) The solution P of (3.129) is unique if and only if there are no eigenvalues λ_i, λ_j of A such that $\bar{\lambda}_i + \lambda_j = 0$ (see Problem 3.31). In particular, this condition will be satisfied if A is a stability matrix.

(3) Notice that there is no loss of generality in replacing Q in (3.129) or (3.135) by the unit matrix. For since Q is positive definite we can write it in a factorized form $Q_1^*Q_1$, where Q_1 is nonsingular. Multiply (3.129) on the left and right by $(Q_1^*)^{-1}$ and Q_1, respectively, to obtain

$$\hat{A}^*P_1 + P_1\hat{A} = I_n \tag{3.136}$$

where

$$\hat{A} = Q_1AQ_1^{-1}, \quad P_1 = (Q_1^{-1})^*PQ_1^{-1}$$

Clearly, A has the same eigenvalues as $\hat{A}$, and by (3.127) P_1 and P have the same inertia, so (3.136) is equivalent to (3.129).

Thus, to test whether A is a stability matrix, solve (3.129) with $Q \equiv I$. If P is negative definite, then A is a stability matrix. If P is positive definite or indefinite, then A has at least one eigenvalue whose real part is positive or zero. Since P is hermitian, (3.129) represents $\frac{1}{2}n(n + 1)$ linear equations to be solved for the $\frac{1}{2}n(n + 1)$ unknown elements of P. This represents a reasonably practical test for applying to A, and there is in consequence a large literature on the numerical solution of (3.129).

(4) By replacing A by (-A) in (3.129) we see that A has all
its eigenvalues with *positive* real parts if and only if the
solution of (3.129) is positive definite. By a mild abuse of
nomenclature, A is called in this case a *positive stability* matrix.

(5) A converse results is interesting: If A is a stability
matrix, and P ranges over the set of all negative definite
hermitian matrices, then Q defined by (3.129) ranges over the set
of hermitian matrices having at least one positive eigenvalue.

Example 24   Consider the stability matrix

$$A = \begin{bmatrix} 1 & -3 \\ 2 & -4 \end{bmatrix}$$

which has eigenvalues -1, -2. It is easily checked that (3.129) is
satisfied by a particular pair

$$P = \begin{bmatrix} -3 & 4 \\ 4 & -20 \end{bmatrix}, \quad Q = \begin{bmatrix} 10 & -43 \\ -43 & 136 \end{bmatrix}$$

and here P is negative definite, Q is indefinite. This illustrates
Remark 5 above, and also reinforces the point made in Remark 3:
Selecting a negative definite P and then obtaining Q from (3.129)
will in general provide no information, via Theorem 3.19, as to
whether or not A is a stability matrix. We must take Q positive
definite, and then the resulting solution P of (3.129) will
completely resolve the question.

Theorem 3.19 can be extended to deal with eigenvalue location
in the left and right halves of the complex plane, but a
preliminary result is needed first.

Theorem 3.20   If an n × n hermitian matrix H can be partitioned as

$$H = \begin{bmatrix} H_1 & H_2 \\ H_2^* & H_3 \end{bmatrix} \begin{matrix} m \\ n-m \end{matrix} \qquad (3.137)$$

$$\quad\quad m \quad n-m$$

where H_1 is positive definite hermitian, and H_3 is negative definite hermitian, then

$$\text{In}(H) = (m, n - m, 0) \tag{3.138}$$

Proof. If we let

$$T = \begin{bmatrix} I_m & -H_1^{-1}H_2 \\ 0 & I_{n-m} \end{bmatrix}$$

(which is obviously nonsingular), then it is easy to verify (see Problem 1.27) that

$$T*HT = \begin{bmatrix} H_1 & 0 \\ 0 & H_4 \end{bmatrix}$$

where

$$H_4 = H_3 - H_2^*H_1^{-1}H_2 \tag{3.139}$$

Since T*HT is block diagonal, its eigenvalues are those of H_1 (all positive), together with those of H_4. However, H_1^{-1} is positive definite, and hence $-H_2^*H_1^{-1}H_2$ is negative definite or semidefinite. Thus H_4 is negative definite, i.e., all its eigenvalues are negative, so we have shown that

$$\text{In}(T^*HT) = (m, n - m, 0)$$

The desired result (3.138) now follows at once from (3.127). ∎

It is of some interest to recall that H_4 in (3.139) is the Schur complement (H/H_1), defined in equation (1.90).

We are now ready to prove a major theorem on location of eigenvalues.

Theorem 3.21

(i) The condition $\delta(A) = 0$ is necessary and sufficient for the existence of a hermitian matrix P such that the hermitian matrix Q defined by

$$A^*P + PA = Q \tag{3.140}$$

is positive definite.

(ii) In this case

$$In(P) = In(A) \tag{3.141}$$

Proof. Consider first part (i). Suppose that there exists some
hermitian matrix P such that Q in (3.140) is positive definite.
We wish to show that $\delta(A) = 0$. As in the proof of Theorem 3.19,
let λ_k, w_k ($\neq 0$) be corresponding eigenvalue and right eigenvector
of A. Then (3.140) gives

$$(\lambda_k + \bar{\lambda}_k)(w_k^* P w_k) = w_k^* Q w_k > 0 \tag{3.142}$$

since Q is positive definite, by assumption. Thus (3.142) shows
that $\lambda_k + \bar{\lambda}_k \neq 0$, for all k, i.e., A has no purely imaginary
eigenvalue.

Conversely, assuming that $\delta(A) = 0$ we wish to show that a
hermitian matrix P exists satisfying (3.140) with Q positive
definite. If $In(A) = (r, n - r, 0)$, say, then we can choose a
transformation matrix M such that Jordan form J of A is given by

$$A = MJM^{-1}, \quad A^* = (M^{-1})^* J^* M^* \tag{3.143}$$

where

$$J = \begin{bmatrix} J_1 & 0 \\ 0 & J_2 \end{bmatrix} \begin{matrix} r \\ n-r \end{matrix} \tag{3.144}$$
$$\quad\quad r \quad n-r$$

and J_1, J_2 contain the eigenvalues of A with positive and negative
real parts, respectively. Substitute (3.143) into (3.140) to
obtain

$$J^* \tilde{P} + \tilde{P} J = \tilde{Q} \tag{3.145}$$

where

$$P = (M^*)^{-1} \tilde{P} M^{-1}, \quad Q = (M^*)^{-1} \tilde{Q} M^{-1} \tag{3.146}$$

Since $(-J_1)$ and J_2 are both, by construction, stability matrices, Theorem 3.19 shows that there exist hermitian matrices $\tilde{P}_1$ (positive definite) and $\tilde{P}_2$ (negative definite) such that

$$J_1^*\tilde{P}_1 + \tilde{P}_1 J_1 = \tilde{Q}_1 \qquad (3.147)$$

$$J_2^*\tilde{P}_2 + \tilde{P}_2 J_2 = \tilde{Q}_2 \qquad (3.148)$$

with $\tilde{Q}_1$ and $\tilde{Q}_2$ positive definite hermitian [notice that (3.147) arises from (3.129) with $A = -J_1$, $P = -\tilde{P}_1$, $Q = \tilde{Q}_1$]. The equations (3.147) and (3.148) can be combined in the form

$$J^*\tilde{P} + \tilde{P}J = \tilde{Q} \qquad (3.149)$$

where now

$$\tilde{P} = \begin{array}{cc} \begin{array}{cc} r & n-r \end{array} & \\ \left[\begin{array}{cc} \tilde{P}_1 & 0 \\ 0 & \tilde{P}_2 \end{array} \right] & \begin{array}{c} r \\ n-r \end{array} \end{array}, \qquad \tilde{Q} = \begin{array}{cc} \begin{array}{cc} r & n-r \end{array} & \\ \left[\begin{array}{cc} \tilde{Q}_1 & 0 \\ 0 & \tilde{Q}_2 \end{array} \right] & \begin{array}{c} r \\ n-r \end{array} \end{array} \qquad (3.150)$$

We have therefore shown that the hermitian matrix P given by (3.146) and (3.150) makes the corresponding Q [also given in (3.146) and (3.150)] positive definite.

The inertia condition (3.141) in part (ii) is now readily established. From (3.127) and (3.146) we deduce that $In(P) = In(\tilde{P})$. Moreover, application of Theorem 3.20 to (3.150) shows that

$$In(\tilde{P}) = (r, n - r, 0) = In(A)$$

as required. ■

Remarks on Theorem 3.21   (1) Notice that it is not true that (3.140) has a solution P for an *arbitrary* positive definite Q, whenever $\delta(A) = 0$. This can be seen from the following simple example. If

$$A = \begin{bmatrix} -1 & 0 \\ 0 & 1 \end{bmatrix}, \qquad P = \begin{bmatrix} p_1 & p_2 \\ p_2 & p_3 \end{bmatrix}$$

then

$$A^*P + PA = \begin{bmatrix} -2p_1 & 0 \\ 0 & 2p_3 \end{bmatrix}$$

so Q must be diagonal.

(2) For the applications which we shall make to polynomials later in this section, we need the following generalization of Theorem 3.21. Surprisingly, this involves a controllability condition of the type encountered in Section 2.2.

Theorem 3.22 If in equation (3.140) there exists a hermitian matrix P such that Q is only positive semidefinite, but in addition the pair (A^*, Q) is c.c., then $\delta(A) = 0$, and $In(P) = In(A)$.

To develop the proof of this result we advance through a series of preliminary lemmas.

Lemma 3.22.1 If in equation (3.140) $\delta(A) = 0$, and there exists a hermitian P such that Q is positive semidefinite, then $\pi(P) \leq \pi(A)$, $\nu(P) \leq \nu(A)$.

Proof. By virtue of Theorem 3.21, the condition $\delta(A) = 0$ ensures that there exists a hermitian matrix P' such that $A^*P' + P'A$ is positive definite, and furthermore that $In(P') = In(A)$. Define a matrix

$$P_\varepsilon = P + \varepsilon P', \quad \varepsilon > 0 \tag{3.151}$$

where P is the matrix in the statement of the lemma. Clearly,

$$A^*P_\varepsilon + P_\varepsilon A = (A^*P + PA) + \varepsilon(A^*P' + P'A)$$

and the right-hand side is positive definite for all $\varepsilon > 0$, so by again appealing to Theorem 3.21 we deduce that $In(P_\varepsilon) = In(A)$, for $\varepsilon > 0$. We now use an argument of some subtlety: First note a standard result that the eigenvalues of matrix P_ε as expressed in (3.151) vary *continuously* with the parameter ε (of course, the eigenvalues are all real since P_ε is hermitian). Since A is nonsingular [because by assumption $\delta(A) = 0$], P_ε is also nonsingular for *all* $\varepsilon > 0$. This continuity therefore implies that as $\varepsilon \to 0$ none of the eigenvalues of P_ε passes *through* the origin. However, it may be that in the limit one or more of the eigenvalues of P_ε actually reduces to the origin, thus giving a possible reduction in the number of positive and negative eigenvalues of P_0 (= P) as compared with those of P_ε. Translated into symbolic terms, this means that

$$\pi(P) \leqslant \pi(P_\varepsilon) = \pi(A), \quad \nu(P) \leqslant \nu(P_\varepsilon) = \nu(A) \quad \blacksquare$$

Lemma 3.22.2   If in equation (3.140), there exists a nonsingular, hermitian P such that Q is positive semidefinite, then $\pi(P) \geqslant \pi(A)$, $\nu(P) \geqslant \nu(A)$.

Proof. We set $A_\varepsilon = A + \varepsilon P^{-1}$, $\varepsilon > 0$, and again invoke continuity of eigenvalues, here of A_ε. We have

$$A_\varepsilon^* P + PA_\varepsilon = A^* P + PA + 2\varepsilon I$$

$$= Q + 2\varepsilon I$$

and the right-hand side of this equation is positive definite for all $\varepsilon > 0$, so by Theorem 3.21, $In(A_\varepsilon) = In(P)$. Since P is a *fixed* matrix, and $\delta(P) = 0$, it follows that as $\varepsilon \to 0$ the numbers of eigenvalues of A_ε having positive and negative real parts cannot increase. Thus in the limit, $\pi(A_0) = \pi(A) \leqslant \pi(P)$, $\nu(A_0) = \nu(A) \leqslant \nu(P)$. $\blacksquare$

Combining the two preceding lemmas immediately produces:

Lemma 3.22.3   If in equation (3.140) $\delta(A) = 0$, and there exists a nonsingular hermitian P such that Q is positive semidefinite, then $In(P) = In(A)$.

We can now give:

Proof of Theorem 3.22. Suppose first that $\delta(A) \neq 0$, i.e., A does have an eigenvalue $i\alpha$ with zero real part, so that A^* has an eigenvalue $-i\alpha$. Let w^* be a corresponding left eigenvector of A^*, so that

$$w^*A^* = -i\alpha w^*, \qquad Aw = i\alpha w \qquad\qquad (3.152)$$

and (3.140) becomes

$$w^*(A^*P + PA)w = w^*w(-i\alpha + i\alpha)$$

$$= 0 = w^*Qw \qquad\qquad (3.153)$$

Since Q is positive semidefinite, (3.153) implies that $w^*Q = 0$ (to prove this, write Q in factorized form $Q_1^*Q_1$). Combining this fact with (3.152) shows that the nonzero vector w satisfies the condition

$$w^*[(-i\alpha)I - A^*, \ Q] = 0$$

and implies that rank $[(-i\alpha)I - A^*, \ Q] < n$. By Theorem 2.2, this contradicts the assumption that the pair (A^*, Q) is c.c. Thus we conclude that A has no such purely imaginary eigenvalue, i.e., $\delta(A) = 0$.

We next show that P in the statement of the theorem is nonsingular. For suppose that the contrary holds. Then there exists a row n-vector $v^* \neq 0$ such that $v^*P = 0$, $Pv = 0$. Hence

$$v^*Qv = v^*(A^*P + PA)v = 0$$

from which, as before, since Q is positive semidefinite, we deduce that $v^*Q = 0$. This in turn implies that

$$v^*(A^*P + PA) = 0$$

showing that $v^*A^*P = 0$. Similarly,

$$v^*(A^*QA)v = v^*[(A^*)^2PA + A^*PA^2]v$$

$$= v^*(A^*)^2(v^*A^*P)^* + (v^*A^*P)A^2v$$

$$= 0$$

whence $v^* A^* Q = 0$, since $A^* QA$ is positive semidefinite. This procedure can be repeated, leading to

$$v^*[Q, \ A^*Q, \ (A^*)^2 Q, \ \ldots, \ (A^*)^{n-1}Q] = 0$$

showing that rank $C(A^*, Q) < n$, a contradiction, so P must be nonsingular.

We have thus established that the assumptions of the theorem imply that the conditions of Lemma 3.22.3 hold, so $In(P) = In(A)$. ∎

There are a number of other results available on the determination of matrix inertia (relative to the imaginary axis) by means of appropriate matrix equations, but we have deliberately restricted ourselves to those which are needed for proving root location theorems for polynomials. We now present some of these proofs.

Proof of Routh and Hurwitz Theorems 3.3, 3.4. An obvious approach would be to set A in the matrix equation (3.140) equal to a companion form, so that the results on inertia of matrices would become results on the location of the roots of the polynomial $a(\lambda)$ in (3.13) [remember that we are considering only real $a(\lambda)$; also, we can assume without loss of generality that $a_0 = 1$]. This procedure can in fact be developed, but we shall instead give a method which involves the *Schwarz form* matrix, defined by

$$
B = \begin{bmatrix}
0 & 1 & 0 & 0 & & 0 \\
-b_n & 0 & 1 & 0 & & 0 \\
0 & -b_{n-1} & 0 & 1 & & \cdot \\
\cdot & \cdot & \cdot & \cdot & \cdot & \cdot \\
& & & \cdot & \cdot & \cdot \\
\cdot & & & \cdot & 0 & 1 \\
0 & \cdot & \cdot & 0 & -b_2 & -b_1
\end{bmatrix} \qquad (3.154)
$$

The elements b_i along the subdiagonal, and in the (n, n) position, are all real.

Step 1. The crucial property of B is that it satisfies the
equation

$$B^T P + PB = Q \qquad (3.155)$$

with

$$P = \text{diag}[b_1 b_2 \cdots b_n, \; b_1 b_2 \cdots b_{n-1}, \; \ldots, \; b_1 b_2, \; b_1] \qquad (3.156)$$

$$Q = -2b_1^2 \, \text{diag}[0, \, 0, \, \ldots, \, 0, \, 1] \qquad (3.157)$$

as can be easily be verified by direct substitution into (3.155).
Since Q in (3.157) is only semidefinite (assuming that $b_1 \neq 0$) we
wish to appeal to Theorem 3.22, which requires $C(B^T, Q)$ to have
rank n. In view of the special form of Q, it is sufficient to
consider $C(B^T, e_n)$, where e_n is the last column of I_n. It is
straightforward to compute that

$$
C(B^T, e_n) =
\begin{bmatrix}
0 & 0 & 0 & . & . & (-1)^{n-1} b_2 b_3 \cdots b_n \\
0 & 0 & 0 & & & . \\
. & . & . & . & & . \\
. & . & . & . & . & . \\
 & & 0 & b_2 b_3 & . & . \\
0 & -b_2 & x & & & . \\
1 & x & x & . & . & . & x
\end{bmatrix}
$$

where the x's denote elements of no significance. Clearly,
$C(B^T, e_n)$ has rank n provided that $b_i \neq 0$, for all i, so the
conditions of Theorem 3.22 are then satisfied. In other words, if
$b_i \neq 0$, for all i, then B has no purely imaginary eigenvalues, and
$\text{In}(B) = \text{In}(-P)$ [the negative sign arises because Q in (3.157) is
negative semidefinite]. Since P is diagonal, its inertia is equal
to (n - k, k, 0), where k is the number of negative terms in the
sequence

$$b_1, \; b_1 b_2, \; b_1 b_2 b_3, \; \ldots, \; b_1 b_2 b_3 \cdots b_n \qquad (3.158)$$

so B has k eigenvalues with positive real parts.

Step 2. The inertia of B is now determined in a different way by constructing the Routh array for its characteristic polynomial. To clarify the subsequent argument, it is helpful to first consider a specific case, say n = 5. It is easy to establish by direct expansion that

$$\det(\lambda I_5 - B) = \lambda^5 + b_1\lambda^4 + (b_2 + b_3 + b_4 + b_5)\lambda^3$$

$$+ b_1(b_3 + b_4 + b_5)\lambda^2$$

$$+ (b_2 b_4 + b_2 b_5 + b_3 b_5)\lambda + b_1 b_3 b_5$$

Using (3.14) and (3.15), the Routh array for this polynomial is easily computed, and has first column 1, b_1, b_2, $b_1 b_3$, $b_2 b_4$, $b_1 b_3 b_5$.

By Theorem 3.3, the inertia of B is determined from the number of sign changes in this first column. We now show that this pattern holds in general: the first column of the Routh array for $\det(\lambda I_n - B)$ is

$$1, \quad b_1, \quad b_2, \quad b_1 b_3, \quad b_2 b_4, \quad b_1 b_3 b_5, \quad b_2 b_4 b_6, \quad b_1 b_3 b_5 b_7, \quad \ldots$$

$$(3.159)$$

where the final term in (3.159) ends in b_n. To prove (3.159), let M_r, r = 1, ..., n denote the rth leading principal minor of $\det(\lambda I_n - B)$. Expansion of M_r by its last column leads to

$$M_n = (\lambda + b_1)M_{n-1} + b_2 M_{n-2} \qquad (3.160)$$

$$M_r = \lambda M_{r-1} + b_{n-r+2}M_{r-2}, \qquad r = 3, \ldots, n - 1 \qquad (3.161)$$

Write

$$M_{n-i} = m_{i1}\lambda^{n-i} + m_{i2}\lambda^{n-i-2} + \cdots + m_{i,j+1}\lambda^{n-i-2j} + \cdots,$$

$$i = 1, \ldots, n - 1 \qquad (3.162)$$

where we note that the powers of λ decrease by two, and $m_{i1} = 1$, for all i. Substitute (3.162) into (3.161) with r = n - i, and equate coefficients of λ^{n-i-2j}, to obtain

$$m_{i,j+1} = m_{i+1,j+1} + b_{i+2}m_{i+2,j} \tag{3.163}$$

Using the forms of M_{n-1} and M_{n-2} provided by (3.162), it follows from (3.160) that the first two rows of the Routh array $\{r_{ij}\}$ for M_n $[= \det(\lambda I_n - B)]$ are the coefficients of $\lambda M_{n-1} + b_2 M_{n-2}$ and $b_1 M_{n-1}$, i.e.,

$$\begin{aligned}
r_{0j}&: \quad m_{11} \quad m_{12} + b_2 m_{21} \quad m_{13} + b_2 m_{22} \quad \cdots \\
r_{1j}&: \quad b_1 m_{11} \quad\quad b_1 m_{12} \quad\quad\quad b_1 m_{13} \quad\quad \cdots
\end{aligned} \tag{3.164}$$

The standard Routh formula (3.15) gives the third row to be

$$r_{2j} = -\frac{1}{b_1 m_{11}} \begin{vmatrix} m_{11} & m_{1,j+1} + b_2 m_{2j} \\ b_1 m_{11} & b_1 m_{1,j+1} \end{vmatrix}$$

$$= b_2 m_{2j}, \quad j = 1, 2, \ldots \tag{3.165}$$

We can now establish by induction that in general

$$r_{i1} = b_i r_{i-2,1}, \quad r_{ij} = r_{i1} m_{ij}, \quad i \geqslant 1 \tag{3.166}$$

It is obvious from (3.164) and (3.165) that (3.166) is true for $i = 2$. Consider row $(i + 3)$ of the array:

$$r_{i+2,j} = -\frac{1}{r_{i+1,1}} \begin{vmatrix} r_{i1} & r_{i,j+1} \\ r_{i+1,1} & r_{i+1,j+1} \end{vmatrix}$$

$$= -\frac{1}{r_{i+1,1}} \begin{vmatrix} r_{i1} & r_{i1} m_{i,j+1} \\ r_{i+1,1} & r_{i+1,1} m_{i+1,j+1} \end{vmatrix}, \quad \text{by (3.166)}$$

$$= r_{i1}(m_{i,j+1} - m_{i+1,j+1})$$

$$= r_{i1} b_{i+2} m_{i+2,j}, \quad \text{by (3.163)}$$

$$= r_{i+2,1} m_{i+2,j}, \quad \text{by (3.166)}$$

which shows that the induction hypothesis is verified. In particular, (3.166) implies that the elements r_{i1} take the required form (3.159).

The reader can verify that the number k of negative terms in (3.158) is equal to the number of variations in sign in the sequence (3.159), i.e., in the first column of the Routh array for $\det(\lambda I - B)$.

Step 3. It remains to relate the preceding to $a(\lambda)$ itself. Let H_i (assumed nonzero) be the Hurwitz determinants, defined in Theorem 3.4, for $a(\lambda)$. If we make the following special choice for the elements in the Schwarz matrix:

$$b_1 = H_1, \qquad b_2 = \frac{H_2}{H_1}, \qquad b_3 = \frac{H_3}{H_2 H_1}$$

$$b_r = \frac{H_{r-3} H_r}{H_{r-2} H_{r-1}}, \qquad r = 4, \ldots, n \tag{3.167}$$

then (3.159) becomes

$$1, \quad H_1, \quad \frac{H_2}{H_1}, \quad \frac{H_3}{H_2}, \quad \ldots, \quad \frac{H_n}{H_{n-1}} \tag{3.168}$$

which is exactly the same as the sequence of first-column elements in the Routh array of $a(\lambda)$, as given by (3.32). It therefore follows that with the choice above for the b's, the polynomials $a(\lambda)$ and $\det(\lambda I - B)$ are identical, since their Routh arrays have identical first columns (see Problem 3.30).

Summary. The matrix B in (3.154) has the given polynomial $a(\lambda)$ as its characteristic polynomial when the b's are determined via (3.167). In Step 1 of the proof the inertia of B was determined using Theorem 3.22, showing that $\pi(B)$ is the number k of negative terms in (3.158). In Step 2 we saw that k is equal to the number of variations in sign in (3.159), which is the first column of the Routh array for $\det(\lambda I - B)$. Finally, in Step 3 the sequence (3.159) was seen to coincide with the Routh-Hurwitz sequence (3.168) for $a(\lambda)$. ∎

Example 25 We illustrate the proof for the polynomial $a(\lambda)$ in
(3.27), whose Hurwitz determinants are evaluated in (3.29).
Substituting these values into (3.167) gives

$$b_1 = 1, \quad b_2 = 16, \quad b_3 = -27, \quad b_4 = -1$$

so the Schwarz form (3.154) is

$$B = \begin{bmatrix} 0 & 1 & 0 & 0 \\ 1 & 0 & 1 & 0 \\ 0 & 27 & 0 & 1 \\ 0 & 0 & -16 & -1 \end{bmatrix}$$

and the reader can check that the characteristic polynomial of B is
indeed (3.27). From (3.156) the solution of

$$B^T P + PB = -2 \, \text{diag}[0, 0, 0, 1]$$

is

$$P = \text{diag}[432, -432, 16, 1]$$

Since $\text{In}(-P) = (1, 3, 0) = \text{In}(B)$, we deduce that $a(\lambda)$ has,
respectively, one and three roots with positive and negative real
parts, agreeing with what was found in Example 7.

Remarks   (1) The procedure we have followed is attractive for two
reasons: first, the controllability condition needed for the
application of Theorem 3.22 is easily established; and second, it
involves the Schwarz form matrix (3.154), which is of interest in
its own right (see Problems 3.34 to 3.36).

(2) Continuing from the last point, it can be shown that if A
is the usual companion matrix (1.26) associated with $a(\lambda)$, then A
is similar to B in (3.154), with $b_1 = r_{11}$, $b_2 = r_{21}$,
$b_i = r_{i1}/r_{i-2,1}$ $(i > 2)$, where $\{r_{ij}\}$ is the Routh array of
Theorem 3.3. Thus the Schwarz matrix is a *real* canonical form to
which any nonderogatory matrix is similar.

(3) As we noted earlier, the various results in Section 3.2.1
can all be interconnected, so once the Routh-Hurwitz theorem has

been verified they can, in principle, then be deduced from it. However, it is of some interest to derive these alternative forms directly from the inertia theorem. We indicate briefly how this can be done for the Hermite results.

<u>Theorem 3.8</u> (Real Polynomial) Equation (3.140) is satisfied with A equal to the usual companion matrix (1.26),

$$Q = -q^T q, \quad \text{where } q = 2[\ldots, 0, a_3, 0, a_1]$$

and P equal to twice the modified Hermite matrix $H^e_{(1)}$ defined in (3.61). It is somewhat complicated to check that the controllability condition in Theorem 3.22 holds [namely, that rank $C(A^T, q) = n$]. It then follows that the location of the roots of $a(\lambda)$ is determined from the inertia of $H^e_{(1)}$. Since the latter is a real, symmetric nonsingular matrix, application of Jacobi's formula (3.128) produces the desired result (3.58).

<u>Example 26</u> We illustrate the procedure just described by again using the polynomial $a(\lambda)$ in (3.27), whose companion matrix A has last row [16, 28, 12, -1]. The associated Hermite matrix H^e was obtained in Example 12, and reversing its rows and columns according to (3.61) gives

$$H^e_{(1)} = \begin{bmatrix} 448 & 0 & -16 & 0 \\ 0 & 352 & 0 & -28 \\ -16 & 0 & 16 & 0 \\ 0 & -28 & 0 & 1 \end{bmatrix}$$

It is then easy to verify that the desired form of (3.140) is

$$A^T(2H^e_{(1)}) + (2H^e_{(1)})A = -4 \begin{bmatrix} 0 \\ -28 \\ 0 \\ 1 \end{bmatrix} [0, -28, 0, 1]$$

Theorem 3.16 (Complex Polynomial) The form of the theorem used
here is that involving the real symmetric matrix H^r defined in
(3.105); denote its leading principal minors by $H_1^r, \ldots, H_n^r$
(assumed nonzero). The equation (3.140) is satisfied with A equal
to a certain tridiagonal matrix B whose characteristic polynomial
is $a(\lambda)$, and

$$P = \text{diag}\left[\frac{H_n^r}{H_{n-1}^r}, \frac{H_{n-1}^r}{H_{n-2}^r}, \ldots, \frac{H_2^r}{H_1^r}, H_1^r\right]$$

$$Q = -\text{diag}[0, 0, \ldots, 0, (H_1^r)^2]$$

The matrix B is in fact the complex form of Schwarz's matrix
(3.154), and has rows

first row: $[-iq_n, 1, 0, \ldots, 0]$

kth row: $[0, \ldots, 0, -p_{n-k+2}, -iq_{n-k+1}, 1, 0, \ldots, 0]$,

$\leftarrow$ (k-2) $\rightarrow$ $k = 2, \ldots, n-1$

nth row: $[0, \ldots, 0, -p_2, -p_1 - iq_1]$

$$(3.169)$$

The numbers q_k are irrelevant, but a rather lengthy argument shows
that

$$p_1 = \frac{1}{2}H_1^r, \quad p_k = \frac{H_{k-2}^r H_k^r}{(H_{k-1}^r)^2}, \quad k = 2, \ldots, n \quad (H_0^r = 1)$$

After establishing (again nontrivially) that rank $C(B^T, Q) = n$,
the required information (3.103) on the root location of $a(\lambda)$ then
follows from the inertia of P.

3.3.2 Matrix Equation Approach: Discrete-Time Case

We now parallel the treatment of Section 3.3.1 for a complex $n \times n$ matrix A_1 whose eigenvalues are to be located in relation to the unit circle $|\mu| = 1$. We define the *c-inertia* of A_1 to be $In_c(A_1) = (\pi_c, \nu_c, \delta_c)$, where π_c, ν_c, δ_c are the respective numbers of eigenvalues of A_1 outside, inside, and on the boundary of the unit disk; in particular, recall that A_1 is convergent if and only if $In_c(A_1) = (0, n, 0)$. Let A_1 be related to A via the transformation (3.11), in which it is assumed that 1 is not an eigenvalue of A_1, so that $A_1 - I$ is nonsingular. In view of the property (3.9) of the bilinear mapping (3.10), it follows that $In(A) = In_c(A_1)$. Substituting the expression (3.11) for A into the matrix equation (3.140), which for ease of reference we reproduce:

$$A^*P + PA = Q \qquad\qquad\qquad (3.170)$$

transforms it into

$$A_1^*PA_1 - P = Q_1 \qquad\qquad\qquad (3.171)$$

where

$$Q_1 = \frac{1}{2}(A_1^* - I)Q(A_1 - I) \qquad\qquad\qquad (3.172)$$

In view of the assumption that $A_1 - I$ is nonsingular, (3.172) shows that Q_1 and Q have the same sign property. Thus (3.171) is the discrete-time equivalent of (3.170), and Theorem 3.19 becomes the following:

<u>Theorem 3.23 (Stein)</u> There exists a negative definite hermitian matrix P satisfying (3.171), where Q_1 is arbitrary positive definite hermitian, if and only if A_1 is a convergent matrix, i.e., $In_c(A_1) = (0, n, 0)$.

Direct Proof. It is of interest to provide an alternative proof which does not rely on the continuous-time result.

The first part closely follows the corresponding part of the proof of Theorem 3.19. Suppose a negative definite P satisfying

(3.171) exists, and let μ, w be an eigenvalue and right eigenvector of A_1. Then

$$A_1 w = \mu w, \quad w^* A_1^* = \bar{\mu} w^*$$

and substituting these expressions into (3.171) produces

$$w^* Q_1 w^* = \bar{\mu} w^* P \mu w - w^* P w$$

$$= (|\mu|^2 - 1) w^* P w$$

Since P is negative definite and Q_1 positive definite, it follows that $|\mu|^2 < 1$ for each eigenvalue μ of A_1.

To establish the converse, we assume that A_1 is convergent and consider the matrix

$$P_N = -Q_1 - A_1^* Q_1 A_1 - (A_1^*)^2 Q_1 A_1^2 - \cdots - (A_1^*)^N Q_1 A_1^N \qquad (3.173)$$

Since $-Q_1$ is negative definite, each additional term on the right in (3.173) is negative definite (or semidefinite if A_1 is singular), so P_N is negative definite for all N. Furthermore, trivial calculations show that

$$A_1^* P_N A_1 - P_N = Q_1 - (A_1^*)^{N+1} Q_1 A_1^{N+1} \qquad (3.174)$$

and

$$P_{N+1} - P_N = -(A_1^*)^{N+1} Q_1 A_1^{N+1} \qquad (3.175)$$

Since A_1 is convergent, we have $\|A_1\|^{N+1} \to 0$ as $N \to \infty$, and

$$\| (A_1^*)^{N+1} Q_1 A_1^{N+1} \| \leqslant \|A_1^*\|^{N+1} \|Q_1\| \, \|A_1\|^{N+1} \qquad (3.176)$$

using a property of matrix norms. It therefore follows from (3.176) that

$$(A_1^*)^{N+1} Q_1 A_1^{N+1} \to 0 \quad \text{as } N \to \infty$$

so from (3.174) and (3.175) we can deduce that the matrix $\lim_{N \to \infty} P_N$ exists and is the desired negative definite solution of (3.171). ∎

Remarks on Theorem 3.23   These also closely mirror the remarks on Theorem 3.19 and will therefore be kept brief.

(1) If P is replaced by (-P) in (3.171), then $V(x) = x^*(k)Px(k) > 0$ is a Liapunov function for the system of difference equations (3.2), and

$$V[x(k + 1)] - V[x(k)] = x^*(k + 1)Px(k + 1) - x^*(k)Px(k)$$

$$= x^*(k)A_1^*PA_1x(k) - x^*(k)Px(k)$$

$$= -x^*(k)Q_1x(k) < 0$$

(2) To test whether A_1 is a convergent matrix, (3.171) is solved for P with $A_1 = I$. The series expression in (3.173) can in fact be developed into a useful procedure for determining P.

(3) The solution P of (3.171) is unique if and only if there are no eigenvalues μ_i, μ_j of A_1 such that $\bar{\mu}_i\mu_j = 1$ (see Problem 3.31).

(4) As in Remark 5 on Theorem 3.19, a converse result states that if A_1 is a convergent matrix, and P ranges over the set of all negative definite hermitian matrices, then Q_1 defined by (3.171) ranges over the set of hermitian matrices having at least one positive eigenvalue.

It would be tedious to reproduce the development of the inertia theorems of the preceding section for the present case, so we merely state without proof the analogue of Theorems 3.21 and 3.22:

Theorem 3.24   Consider the equation (3.171).

(i)   A necessary and sufficient condition for the existence of a hermitian solution P such that Q_1 is positive definite is that $\delta_c(A_1) = 0$.

(ii) If a hermitian P exists with Q_1 positive semidefinite, then $\delta_c(A_1) = 0$ provided that the pair (A_1^*, Q_1) is c.r.

In either case

$$In(P) = In_c(A_1) \tag{3.177}$$

Notice that as we are dealing with the discrete-time system (3.2), we have preferred to state the condition on the rank of $C(A_1^*, Q_1)$ as being one for complete reachability (see Section 2.2). We now use Theorem 3.24 to prove one of the basic discrete-time results of Section 3.2.2.

Proof of Schur-Cohn-Fujiwara Theorem 3.13. Let A_1 in (3.171) be the companion form associated with the real polynomial $a_1(\mu)$ in (3.68), having last row $[-\alpha_n, \ldots, -\alpha_1]$ (we again assume for convenience that $\alpha_0 = 1$). Then equation (3.171) is satisfied with P equal to the Schur-Cohn matrix K defined in (3.93), and $Q_1 = -qq^T$ where

$$q = [1 - \alpha_n^2, \; \alpha_1 - \alpha_n \alpha_{n-1}, \; \alpha_2 - \alpha_n \alpha_{n-2}, \; \ldots, \; \alpha_{n-1} - \alpha_n \alpha_1]^T \tag{3.178}$$

In order to apply Theorem 3.24 it remains to verify that $C(A_1^T, Q_1)$ has rank n, for which it suffices to test $C(A_1^T, q)$. The root location of $a_1(\mu)$ is thus given by the inertia of K, which by the Jacobi formula (3.128) leads to the desired result (3.94).

An interesting alternative proof shows that with the same companion matrix A_1, but with $Q_1 = \text{diag}[0, 0, \ldots, 0, 1]$, then the solution of (3.171) is equal to the *inverse* of the Schur-Cohn matrix.

Similar developments are possible for the complex case, Theorem 3.18c.

Example 27   We illustrate the matrix equation proof of Theorem 3.13 for the polynomial $a_1(\mu)$ in (3.73), converted to monic form so that

$$\hat{a}_1(\mu) = \mu^3 + 2\mu^2 - \frac{5}{2}\mu + \frac{3}{2}, \quad A_1 = \begin{bmatrix} 0 & 1 & 0 \\ 0 & 0 & 1 \\ -\frac{3}{2} & \frac{5}{2} & -2 \end{bmatrix}$$

The corresponding Schur-Cohn matrix is $\hat{K} = (1/\alpha_0^2)K = \frac{1}{4}K$, where K is the matrix found in Example 20 for $a_1(\mu)$. The reader is left to confirm that the desired form of (3.171) is

$$A_1^T\hat{K}A_1 - \hat{K} = -\begin{bmatrix} -\frac{5}{4} \\ \frac{23}{4} \\ -\frac{11}{2} \end{bmatrix}\begin{bmatrix} -\frac{5}{4}, & \frac{23}{4}, & -\frac{11}{2} \end{bmatrix}$$

where the right side has been obtained using (3.178).

3.3.3 Alternative Proof of the Complex Schur-Cohn Theorem

We present in this section a recent, novel proof of the complex Schur-Cohn Theorem 3.18(iii). This is included because the argument used is very straightforward, and can be applied to obtain a generalization of the result, although we will not describe this here. A similar approach to proving the complex Hermite Theorem 3.16 has also been developed.

Consider the complex polynomial (3.68), with $\alpha_0 > 0$, together with the "complex reverse" polynomial $\hat{a}_1(\mu)$ defined as in (3.122):

$$\hat{a}_1(\mu) = \mu^n\bar{a}_1(1/\mu) = \bar{\alpha}_n\mu^n + \bar{\alpha}_{n-1}\mu^{n-1} + \cdots + \bar{\alpha}_0 \tag{3.179}$$

Let S denote the standard $n \times n$ companion matrix in (1.26) when all a_i are set equal to zero, i.e.,

$$S = \begin{bmatrix} 0 & & \\ \cdot & & I_{n-1} \\ \cdot & & \\ 0 & \cdots & 0 \end{bmatrix} \tag{3.180}$$

(This is sometimes called a *shift* matrix, since premultiplying any
n × n matrix by S has the effect of shifting each row one level
upwards.)   It is then trivial to show that

$$a_1(S) = \alpha_0 S^n + \alpha_1 S^{n-1} + \cdots + \alpha_n I$$

$$= \begin{bmatrix} \alpha_n & \alpha_{n-1} & \cdots & \alpha_1 \\ 0 & \alpha_n & \cdots & \alpha_2 \\ . & . & \cdots & . \\ 0 & 0 & \cdots & \alpha_n \end{bmatrix} \qquad (3.181)$$

with a similar expression for $\hat{a}_1(S)$.   Routine algebraic
manipulations then establish that the elements of the Schur-Cohn
matrix K as defined in (3.121) can be obtained from the identity

$$K \equiv [\hat{a}_1(S)]^* \hat{a}_1(S) - [a_1(S)]^* a_1(S) \qquad (3.182)$$

Let the roots of $a_1(\mu)$ be $\mu_1, \ldots, \mu_n$ (assuming again that none has
unit modulus), so that

$$a_1(\mu) = \alpha_0(\mu - \mu_1)(\mu - \mu_2) \cdots (\mu - \mu_n) \qquad (3.183)$$

and hence

$$\hat{a}_1(\mu) = \bar{\alpha}_0(1 - \bar{\mu}_1\mu)(1 - \bar{\mu}_2\mu) \cdots (1 - \bar{\mu}_n\mu) \qquad (3.184)$$

If we define matrices

$$B_i = S - \mu_i I_n, \quad C_i = I_n - \bar{\mu}_i S, \quad i = 1, 2, \ldots, n \qquad (3.185)$$

then (3.183) and (3.184) imply that

$$a_1(S) = \alpha_0 B_1 B_2 \cdots B_n, \quad \hat{a}_1(S) = \bar{\alpha}_0 C_1 C_2 \cdots C_n$$

and these expressions reduce (3.182) to

$$K = |\alpha_0|^2 (C_n^* \cdots C_1^* C_1 \cdots C_n - B_n^* \cdots B_1^* B_1 \cdots B_n) \qquad (3.186)$$

Using the fact that B_i commutes with B_j, C_j, for all i, the key
step of the proof is to realize that (3.186) can be written as

$$K = |\alpha_0|^2 \sum_{j=1}^{n} C_n^* \cdots C_{j+1}^* B_{j-1}^* \cdots B_1^* (C_j^* C_j - B_j^* B_j) B_1 \cdots$$

$$B_{j-1} C_{j+1} \cdots C_n \quad (3.187)$$

For example, when $n = 3$, (3.187) becomes

$$K = |\alpha_0|^2 \Big[C_3^* C_2^* (C_1^* C_1 - B_1^* B_1) C_2 C_3 + C_3^* B_1^* (C_2^* C_2 - B_2^* B_2) B_1 C_3$$

$$+ B_2^* B_1^* (C_3^* C_3 - B_3^* B_3) B_1 B_2 \Big]$$

$$= |\alpha_0|^2 (C_3^* C_2^* C_1^* C_1 C_2 C_3 - B_3^* B_2^* B_1^* B_1 B_2 B_3)$$

The expression (3.187) can be reduced further, since by (3.185) we have

$$C_j^* C_j - B_j^* B_j = (1 - |\mu_j|^2)(I - S^T S)$$

and it is trivial to show that

$$I - S^T S = \text{diag}[1, 0, \ldots, 0] \qquad (3.188)$$

$$= e_1^T e_1$$

where e_1 denotes the first row of I_n. Therefore, (3.188) reduces (3.187) to

$$K = |\alpha_0|^2 \sum_{j=1}^{n} (1 - |\mu_j|^2) d_j d_j^* \qquad (3.189)$$

where $d_j = C_n^* \cdots C_{j+1}^* B_{j-1}^* \cdots B_1^* e_1^T$. In matrix terms, (3.189) is equivalent to

$$K = |\alpha_0|^2 D \text{ diag}[(1 - |\mu_1|^2), \ldots, (1 - |\mu_n|^2)] D^*$$

$$= |\alpha_0|^2 D M D^*, \quad \text{say} \qquad (3.190)$$

where $D = [d_1, d_2, \ldots, d_n]$. If K is nonsingular, then taking determinants of both sides of (3.190) shows that D is also nonsingular, and that no $|\mu_i|$ is equal to unity. In this case Sylvester's result (3.127) shows that K has the same inertia as the real diagonal matrix M. Clearly, from the expression in (3.190), M

has k positive and n - k negative eigenvalues, where k is the
number of roots of $a_1(\mu)$ inside the unit disk. Hence, by Jacobi's
formula (3.128)

$$n - k = \nu(K) = V(1, K_1, \ldots, K_n)$$

and this establishes Theorem 3.18(iii) [see (3.94)]. ∎

It can be shown that (3.190) leads to the expression

$$\det K = |\alpha_0|^{2n} \prod_{i,j=1}^{n} (1 - \mu_i \bar{\mu}_j) \qquad\qquad (3.191)$$

which can be regarded as the analogue of Orlando's formula (3.34)
and (3.104).

Problems

3.31 Let $v(\cdot)$ denote the stacking operator, defined in (A3) in
    Appendix A.
    (i) Show that equation (3.129) can be written in the form

$$(A^* \otimes I_n + I_n \otimes A^T) v(P) = v(Q)$$

        Hence deduce using (A2) in Appendix A, that (3.129) has a
        unique solution P if and only if A has no eigenvalues λ_i,
        λ_j such that $\bar{\lambda}_i + \lambda_j = 0$.
    (ii) Express (3.171) similarly, and deduce that the condition
        for uniqueness in this case is $\bar{\mu}_i \mu_j \neq 1$, for any
        eigenvalues μ_i, μ_j of A_1.

3.32 Prove the following stronger version of Theorem 3.20:
    Provided that H_1 in (3.137) is nonsingular, then

$$\text{In}(H) = \text{In}(H_1) + \text{In}(H_4)$$

    where H_4 is the Schur complement (H/H_1) in (3.139). Hence
    deduce that if H is positive definite, then so are H_1 and H_3.

3.33 Use the method in Remark 1 on Theorem 3.19 to show that if A
is a stability matrix, and $x(t)$ is the corresponding solution
of the system of differential equations (3.1), then

$$\int_0^\infty x^*(t)Qx(t) \ dt = x^*(0)Px(0)$$

where P and Q satisfy (3.135). By differentiating $tx^*(t)Px(t)$
with respect to t, show further that

$$\int_0^\infty tx^*(t)Qx(t) \ dt = x^*(0)P_1x(0)$$

where $A^*P_1 + P_1A = -P$.

3.34 Use the result of the preceding problem to show that if
$\dot{x}(t) = Bx(t)$, where B is the Schwarz form (3.154) with $b_i > 0$,
for all i, so that B is asymptotically stable, then

$$\int_0^\infty x_n^2(t) \ dt = \frac{1}{2b_1^2} \sum_{i=1}^n b_1 b_2 \cdots b_{n-i+1} x_i^2(0)$$

3.35 Let the Schwarz matrix B in (3.154) be asymptotically stable,
i.e., $b_i > 0$, for all i, and define

$$Y = J_n \text{diag}[(-1)^{n+1}(b_1 b_2 \cdots b_n)^{1/2}, \ (-1)^n(b_1 \cdots b_{n-1})^{1/2},$$

$$\ldots, \ -(b_1 b_2)^{1/2}, \ b_1^{1/2}]$$

Show that $YBY^{-1} = R$, where R is the tridiagonal matrix

$$\begin{bmatrix} -b_1 & b_2^{1/2} & 0 & 0 & & & \\ -b_2^{1/2} & 0 & b_3^{1/2} & 0 & & 0 & \\ 0 & -b_3^{1/2} & 0 & b_4^{1/2} & \cdot & \cdot & \cdot \\ \cdot & \cdot & \cdot & \cdot & \cdot & 0 & b_n^{1/2} \\ & 0 & & & \cdot & -b_n^{1/2} & 0 \end{bmatrix}$$

termed the *Routh canonical form*.

3.36 Show that R in Problem 3.35 satisfies

$$R^T + R = -2b_1 \, \text{diag}[1, 0, 0, \ldots, 0]$$

Hence deduce, using the result of Problem 3.33, that if
$\dot{x}(t) = Rx(t)$, then

$$\int_0^\infty x_1^2(t) \, dt = \frac{1}{2b_1} \sum_{i=1}^n x_i^2(0)$$

3.37 In the Schwarz matrix B in (3.154) set

$$b_1 = 1/k_1, \quad b_i = 1/k_{i-1}k_i, \quad i = 2, \ldots, n$$

and show that $(k_1 k_2 k_3 \cdots k_n) \det(\lambda I_n - B)$ is equal to a
tridiagonal determinant having $k_n\lambda$, $k_{n-1}\lambda$, $\ldots$, $k_2\lambda$, $(1 + k_1\lambda)$
along the principal diagonal, all elements on the super-
diagonal equal to -1, and all elements on the subdiagonal
equal to 1. It then follows that any Hurwitz polynomial can
be expressed as this tridiagonal determinant with $k_i > 0$, for
all i, and conversely.

3.38 Let A and P in (3.129) be real, with A assumed nonsingular.
Show that the equation can be reduced to

$$A^T S + SA = (A^T Q - QA)$$

where $P = \frac{1}{2}(Q - S)A^{-1}$, and S is a real skew symmetric matrix
to be determined. (Notice that the number of equations and
unknowns is thereby reduced by n.)

3.39 Let $A = [a_{ij}]$ be a real tridiagonal matrix, i.e., $a_{ij} = 0$ if
$|i - j| \geq 2$. By taking

$$P = \text{diag}\left[-1, \frac{a_{12}}{a_{21}}, \frac{-a_{12}a_{23}}{a_{21}a_{32}}, \ldots, (-1)^n \frac{a_{12}a_{23} \cdots a_{n-1,n}}{a_{21}a_{32} \cdots a_{n,n-1}}\right]$$

in (3.129), show that sufficient conditions for A to be a
stability matrix are $a_{ii} < 0$, $i = 1, \ldots, n$, and

$$a_{i,i+1}a_{i+1,i} < 0, \quad i = 1, \ldots, n - 1.$$

3.40 Let $B = [b_{ij}]$ be a tridiagonal matrix, with its only nonzero elements defined by

$$b_{11} = \alpha_1 + i\beta_1, \quad \alpha_1 \neq 0$$

$$b_{kk} = i\beta_k, \quad b_{k-1,k} = \alpha_k, \quad b_{k,k-1} = -\gamma_k, \quad k = 2, 3, \ldots, n$$

where α_k, β_k, γ_k are real, and $\alpha_k \neq 0$, $\gamma_k \neq 0$, for all k. Use Theorem 3.22 to prove that $In(B) = In(P)$, where

$$P = diag\left[\alpha_1, \frac{\alpha_1\alpha_2}{\gamma_2}, \frac{\alpha_1\alpha_2\alpha_3}{\gamma_2\gamma_3}, \ldots, \frac{\alpha_1\alpha_2 \cdots \alpha_n}{\gamma_2\gamma_3 \cdots \gamma_n}\right]$$

When $\gamma_k = 1$, for all k, then B is a complex version of the Schwarz form in (3.154) [compare with (3.169)].

3.41 Let A be the companion matrix in the usual form (1.26) for the polynomial in Problem 3.6(i). Hence test whether the polynomial is asymptotically stable by using Theorem 3.19.

3.42 Let

$$P = (1 - \varepsilon)P_0 + \varepsilon I, \quad 0 \leq \varepsilon \leq 1$$

where P_0 is a real symmetric nonsingular matrix. Use a "continuity of eigenvalues" argument to show that P is nonsingular if and only if P_0 has no negative eigenvalue.

3.43 Let A be a real n × n matrix, with symmetric and skew symmetric parts

$$A_s = \frac{1}{2}(A + A^T), \quad A_k = \frac{1}{2}(A - A^T)$$

Assume that A_s is negative semidefinite. Let λ be an eigenvalue of A, with w a corresponding left eigenvector, and assume that it satisfies the additional condition $wA_s = 0$.

(i) Show that λ is an eigenvalue of A_k, and hence deduce that A is not a stability matrix.

(ii) Use the result stated in Problem 2.18 to show that rank $C(A_k, A_s) < n$.

In fact, the following theorem can be proved, and is noteworthy because of the unexpected appearance of a
controllability condition: If A_s is negative semidefinite,
then A is a stability matrix if and only if the pair (A_k, A_s)
is c.c.

3.44 Show that if A_1 is a convergent matrix, and x(k) is the
solution of the difference equations (3.2), then

$$\sum_{k=0}^{\infty} x^*(k)Q_1 x(k) = -x^*(0)Px(0)$$

where P and Q_1 satisfy (3.171). This is the discrete-time
analogue of the first result in Problem 3.33.

3.45 Let $A_c = A + iB$, where A and B are real n × n matrices, and
consider the matrix equation (3.129), i.e.,

$$A_c^* P + PA_c = -Q$$

where P $(= P_1 + iP_2)$ and Q $(= Q_1 + iQ_2)$ are hermitian. Show
that this equation is equivalent to the *real* form

$$A_r^T P_r + P_r A_r = -Q_r$$

where A_r is the matrix defined in (3.12), and P_r, Q_r are
defined in the same way, i.e.,

$$P_r = \begin{bmatrix} P_1 & P_2 \\ -P_2 & P_1 \end{bmatrix}, \qquad Q_r = \begin{bmatrix} Q_1 & Q_2 \\ -Q_2 & Q_1 \end{bmatrix}$$

Verify that P_r and Q_r are symmetric.

3.4 CAUCHY INDEX METHOD

We restrict attention in this section to the continuous-time
problem. The reader will have noticed that in the proofs in
Section 3.3.2 we ignored any critical cases of roots lying on the
imaginary axis. This deficiency of the matrix equation approach
does not apply to the classical methods based on complex variable

theory. We will indicate only the salient features here, as most of the material is well covered in standard textbooks.

The *Cauchy index* of a real rational function $f(\lambda)$ between the limits $\lambda = \alpha$ and $\lambda = \beta$ (where α, β are real numbers, or $\pm\infty$) is

$$I_\alpha^\beta f(\lambda) = \text{(number of jumps of } f(\lambda) \text{ from } -\infty \text{ to } +\infty)$$
$$- \text{(number of jumps of } f(\lambda) \text{ from } +\infty \text{ to } -\infty) \qquad (3.192)$$

as λ moves along the real axis from α to β (ignoring any discontinuity at these end points).

Example 28 Consider the simple function

$$f(\lambda) = \frac{2\lambda - 1}{\lambda^2 - \lambda - 6} \equiv \frac{1}{\lambda + 2} + \frac{1}{\lambda - 3}$$

Clearly, $f(\lambda)$ has discontinuities at $\lambda = -2$ and $\lambda = +3$; as λ moves along the real axis from left to right, $f(\lambda)$ jumps from $-\infty$ to $+\infty$ at $\lambda = -2$, and from $-\infty$ to $+\infty$ at $\lambda = 3$. For example, the definition (3.192) gives

$$I_{-4}^{-1} f(\lambda) = 1, \qquad I_{-1}^{+1} f(\lambda) = 0, \qquad I_{-\infty}^{+\infty} f(\lambda) = 2$$

An interesting special case is obtained when

$$f(\lambda) = \frac{g'(\lambda)}{g(\lambda)} \qquad (3.193)$$

where $g(\lambda)$ is a real polynomial in the factorized form (1.8), i.e.,

$$g(\lambda) = g_0(\lambda - \lambda_1)^{r_1}(\lambda - \lambda_2)^{r_2} \cdots (\lambda - \lambda_p)^{r_p}$$

Suppose that the roots of $g(\lambda)$ have been ordered so that only the first s ($\leqslant p$) are real. Then it is easy to see that (3.193) gives

$$f(\lambda) = \sum_{j=1}^{p} \frac{r_j}{\lambda - \lambda_j}$$

$$= \sum_{j=1}^{s} \frac{r_j}{\lambda - \lambda_j} + f_1(\lambda) \qquad (3.194)$$

where $f_1(\lambda)$ is a real rational function having *no* real poles. Each time λ passes through one of the real poles, the Cauchy index increases by one. We therefore have:

Theorem 3.25   The index

$$I_\alpha^\beta \frac{g'(\lambda)}{g(\lambda)}$$

is equal to the number of distinct real roots of $g(\lambda)$ in the interval $\alpha < \lambda < \beta$.

The reader should check the application of the theorem to $f(\lambda)$ in Example 28, where

$$g(\lambda) = \lambda^2 - \lambda - 6 \equiv (\lambda + 2)(\lambda - 3)$$

By using the encirclement theorem from complex variable theory, it is possible to obtain the following fundamental result on root location for a real polynomial $a(\lambda)$, in the usual form (3.13).

Theorem 3.26   The number of roots of $a(\lambda)$ with positive real parts is given by

$$k = \frac{1}{2}\left[n - s - I_{-\infty}^\infty \frac{a_2(\lambda)}{a_1(\lambda)}\right] \qquad (3.195)$$

where s is the number of purely imaginary roots of $a(\lambda)$, and

$$a_1(\lambda) = a_0\lambda^n - a_2\lambda^{n-2} + a_4\lambda^{n-4} - \cdots \qquad (3.196)$$

$$a_2(\lambda) = a_1\lambda^{n-1} - a_3\lambda^{n-3} + a_5\lambda^{n-5} - \cdots \qquad (3.197)$$

Different procedures for evaluating the Cauchy index in (3.195) lead to some of the classical theorems of Section 3.2.1. In view of this common origin, it is therefore not so surprising that we were able to exhibit connections between the various results. Our primary aim is to show how the Routh array arises, and for this we need the concept of a *Sturm sequence:*

$$k_0 f_0(\lambda), \quad k_1 f_1(\lambda), \quad k_2 f_2(\lambda), \quad k_3 f_3(\lambda), \quad \ldots \qquad (3.198)$$

associated with *any* two real polynomials $f_0(\lambda)$ and $f_1(\lambda)$ with $\delta f_1 < \delta f_0$, the k's being arbitrary positive constants. This sequence (3.198) is defined as the set of *remainders* obtained by applying Euclid's algorithm (Section 1.6.1) to $f_0(\lambda)$ and $f_1(\lambda)$, but with the *signs* of the remainders *reversed*, i.e., equation (1.133) is replaced by

$$f_i(\lambda) = f_{i+1}(\lambda)q_{i+1} - f_{i+2}(\lambda), \quad i = 0, 1, 2, \ldots \qquad (3.199)$$

If $f_m(\lambda)$ is the last nonvanishing remainder, then since changing the signs does not affect the g.c.d. procedure, $f_m(\lambda)$ is a g.c.d. of $f_0(\lambda)$ and $f_1(\lambda)$. The importance of the Sturm sequence is that it enables the Cauchy index to be calculated as follows:

Theorem 3.27 (Sturm)

$$I_\alpha^\beta \frac{f_1(\lambda)}{f_0(\lambda)} = V(\alpha) - V(\beta) \qquad (3.200)$$

where $V(\theta)$ denotes the number of variations in sign in the sequence (3.198) for a fixed real value $\lambda = \theta$ [any term $f_i(\theta) \equiv 0$ in the sequence is omitted].

Clearly, application of (3.200) to (3.195) with $f_0(\lambda) = a_1(\lambda)$, $f_1(\lambda) = a_2(\lambda)$, will enable k to be determined, provided that s is known; we will discuss evaluation of the latter shortly. Routh's contribution was to devise the tabular scheme of Theorem 3.3, which eliminates the need to determine the polynomials in (3.198) by actual division. Specifically, it can be shown that if we obtain $f_2(\lambda)$ from (3.199) with $f_0(\lambda) = a_1(\lambda)$, $f_1(\lambda) = a_2(\lambda)$, where these are defined in (3.196) and (3.197), then

$$f_2(\lambda) = r_{21}\lambda^{n-2} - r_{22}\lambda^{n-4} + r_{23}\lambda^{n-6} - \cdots$$

where $\{r_{21}, r_{22}, r_{23}, \ldots\}$ is the third row of the Routh array generated by (3.15), with initial two rows as in (3.14), i.e.,

$$\{a_0, a_2, a_4, \ldots\}, \quad \{a_1, a_3, a_5, \ldots\}$$

The argument can be continued, showing that the Sturm sequence
polynomials generated by $a_1(\lambda)$ and $a_2(\lambda)$ are given by the rows of
the Routh array, i.e.,

$$f_i(\lambda) = r_{i1}\lambda^{n-i} - r_{i2}\lambda^{n-i-2} + r_{i3}\lambda^{n-i-4} - \cdots, \quad i \geqslant 2 \quad (3.201)$$

The regular case occurs when $r_{i1} \neq 0$, for all i, when each
polynomial in (3.201) has degree one less than its predecessor.
The last term is then $f_n(\lambda) = r_{n1}$, implying that $a_1(\lambda)$ and $a_2(\lambda)$
are relatively prime. This in turn implies that s = 0; see Remark
3 on Theorem 3.3. Combining (3.195) and (3.200) then shows that

$$k = \frac{1}{2}[n - V(-\infty) + V(\infty)] \qquad\qquad (3.202)$$

where $V(\cdot)$ refers to the number of variations in sign in the
sequence (3.201), with $f_0(\lambda) = a_1(\lambda)$, $f_1(\lambda) = a_2(\lambda)$. We see from
(3.201) that the sign of $f_i(\infty)$ is the same as that of r_{i1}, so that

$$V(\infty) = V(r_{01}, r_{11}, r_{21}, \ldots)$$

and similarly,

$$V(-\infty) = V[(-1)^n r_{01}, (-1)^{n-1} r_{11}, (-1)^{n-2} r_{21}, \ldots]$$

Since in this regular case the Sturm sequence contains n + 1 terms,
it follows that $V(-\infty) = n - V(\infty)$, which on substitution into
(3.202) gives $k = V(r_{01}, r_{11}, r_{21}, \ldots)$, which is the result stated
in Theorem 3.3.

When the array is not regular, then for singularities of type
(ii), as described in Remark 3 on Theorem 3.3, the number s of
purely imaginary roots of $a(\lambda)$ is nonzero, and is equal to the
number of real roots of the g.c.d. $g(\lambda)$ of $a_1(\lambda)$ and $a_2(\lambda)$ [this is
the same as the number of real roots of $\delta(i\omega)$, where $\delta(\lambda)$ is the
g.c.d. as defined in (3.20)]. As we pointed out in our earlier
discussion, the g.c.d. is obtained from the last nonvanishing row

in the array, and an illustration was given in Example 4. There
the number s was found by inspection, but in general we can use
Theorem 3.25, in which the Cauchy index can be computed from the
Sturm sequence given by Theorem 3.27. We again achieve this by
using the Routh algorithm, but it now turns out that the polynomials in the sequence are given by *alternate* rows in the array,
i.e.,

$$f_0(\lambda) = g(\lambda), \quad f_1(\lambda) = g'(\lambda) \tag{3.203}$$

$$f_i(\lambda) = s_i(r_{2i-1,1}\lambda^{N-i} + r_{2i-1,2}\lambda^{N-i-1} + \cdots), \quad i \geqslant 2 \tag{3.204}$$

where $\delta g = N$,

$$s_i = \text{sgn}(r_{11}r_{21}r_{31} \cdots r_{2i-2,1}) \tag{3.205}$$

and the array begins with the coefficients in (3.203).

Example 29   Suppose that we wish to determine the number of
distinct real roots of

$$g(\lambda) = \lambda^4 - 3\lambda^3 - \lambda^2 + 8\lambda - 4 \tag{3.206}$$

in the interval $-1 < \lambda < 1$. The Routh array (3.15) is constructed,
using as initial two rows the coefficients of $g(\lambda)$ and $g'(\lambda)$:

r_{0j}	1	-3	-1	8	-4
r_{1j}	4	-9	-2	8	
r_{2j}	-3	-2	24	-16	
r_{3j}	-7	18	-8		
r_{4j}	-17	48	-28		
r_{5j}	-1	2			
r_{6j}	1	-2			
r_{71}	0				

Notice that at each step we have deleted suitable positive common
factors from each row, so as to remove fractions and simplify the

arithmetic. From (3.204) and (3.205) the Sturm sequence polynomial
polynomials are

$$f_2(\lambda) = \text{sgn}(4.-3)(-7\lambda^2 + 17\lambda - 8)$$

$$= 7\lambda^2 - 18\lambda + 8$$

$$f_3(\lambda) = \text{sgn}(4.-3.-7.-17)(-\lambda + 2)$$

$$= \lambda - 2$$

In order to use Theorems 3.25 and 3.27 we need to compute

$$V(-1) = V[g(-1), g'(-1), f_2(-1), f_3(-1)]$$

$$= V(-9, -3, 33, -3) = 2$$

and

$$V(1) = V[g(1), g'(1), f_2(1), f_3(1)]$$

$$= V(4, 1, -3, -1) = 1$$

Hence the number of distinct real roots is $V(-1) - V(1) = 1$.

If $g(\lambda)$ has repeated real roots, then construct $d_1(\lambda) = (g(\lambda),$
$g'(\lambda))$, $d_2(\lambda) = (d_1(\lambda), d_1'(\lambda))$, and so on. If in $\alpha < \lambda < \beta$, the
respective numbers of distinct real roots of $g(\lambda)$, $d_1(\lambda)$, $d_2(\lambda)$,
..., are k_0, k_1, k_2, ..., then $g(\lambda)$ has $k_0 - k_1$ simple real roots,
$k_1 - k_2$ real roots of multiplicity two, and so on. Thus for $g(\lambda)$
in (3.206), since $r_{71} = 0$, we obtain from the preceding row
$d_1(\lambda) = \lambda - 2$, so that by inspection $k_1 = 1$. Hence in the interval
$(-1, 1)$, $g(\lambda)$ has no simple real roots and one real root of
multiplicity two.

We now link the Cauchy index and Sturm sequence to some of the
other results in Section 3.2.1. The companion matrix result is
attractively simple: Let G be the companion matrix in the standard
form (1.26) for an arbitrary real Nth-degree polynomial $g(\lambda)$, and
write $L = g'(G)J_N$. Then a Sturm sequence associated with $f_0 = g(\lambda)$
and $f_1 = g'(\lambda)$ is

$$f_i(\lambda) = L_{ii}\lambda^{N-i} + L_{i,i+1}\lambda^{N-i-1} + \cdots + L_{iN}, \quad i = 2, 3, \ldots$$
$$(3.207)$$

where L_{ij} stands for the minor found from rows 1, 2, ..., i - 1, i and columns 1, 2, ..., i - 1, j of L, with j ⩾ i.

Example 30 If $g(\lambda)$ is the polynomial in (3.206), then g'(G) is constructed in the usual way from Theorem 1.2 to give, after reversal of columns,

$$L = \begin{bmatrix} 4 & -9 & -2 & 8 \\ 3 & 2 & -24 & 16 \\ 11 & -21 & -8 & 12 \\ 12 & 3 & -76 & 44 \end{bmatrix}$$

Equation (3.207) gives

$$f_2(\lambda) = \begin{vmatrix} 4 & -9 \\ 3 & 2 \end{vmatrix}\lambda^2 + \begin{vmatrix} 4 & -2 \\ 3 & -24 \end{vmatrix}\lambda + \begin{vmatrix} 4 & 8 \\ 3 & 16 \end{vmatrix}$$

$$= 5(7\lambda^2 - 18\lambda + 8)$$

$$f_3(\lambda) = \begin{vmatrix} 4 & -9 & -2 \\ 3 & 2 & -24 \\ 11 & -21 & -8 \end{vmatrix}\lambda + \begin{vmatrix} 4 & -9 & 8 \\ 3 & 2 & 16 \\ 11 & -21 & 12 \end{vmatrix}$$

$$= 250(\lambda - 2)$$

which agree with the expressions found in Example 29.

Precisely the same correspondence holds in the general case, when the Sturm sequence is generated by $a_1(\lambda)$ and $a_2(\lambda)$ in (3.196) and (3.197). This provides the basis for the derivation of Theorem 3.6, whereby the minors in (3.48) and (3.49) are related to the leading coefficients in the members of the Sturm sequence.

Two alternative ways of calculating the Cauchy index are now shown to lead to Theorem 3.7, involving Markov parameters, and Hermite's symmetric matrix result in Theorem 3.8. Let $f_0(\lambda)$, $f_1(\lambda)$ be arbitrary real polynomials as before with $\delta f_1 < \delta f_0 = M$. Following (3.51), define the associated Markov parameters s_i by

$$\frac{f_1(\lambda)}{f_0(\lambda)} = \frac{s_0}{\lambda} + \frac{s_1}{\lambda^2} + \frac{s_2}{\lambda^3} + \cdots \tag{3.208}$$

and let $S = [s_{ij}]$ be defined by $s_{ij} = s_{i+j-2}$. Also, let Z denote the bezoutian matrix associated with $f_0(\lambda)$ and $f_1(\lambda)$ defined as in (1.111).

Theorem 3.28

$$I_{-\infty}^{\infty} \frac{f_1(\lambda)}{f_0(\lambda)} = \text{signature } (S) \tag{3.209}$$

$$= \text{signature } (Z) \tag{3.210}$$

Taking $f_0(\lambda)$ and $f_1(\lambda)$ in (3.209) to be $a_1(\lambda)$ and $a_2(\lambda)$, as required for (3.195) in Theorem 3.26, produces the stability criterion of Theorem 3.7. Taking $f_0(\lambda)$ and $f_1(\lambda)$ in (3.210) to be $a(\lambda)$ and $a(-\lambda)$ [see (3.62)] leads to Theorem 3.8, in terms of the modified Hermite matrix in (3.61). Note that the signature of an arbitrary nonsingular hermitian (or real symmetric) matrix H is equal to $n - 2\nu(H)$, where $\nu(H)$ is determined from (3.128) in terms of the leading principal minors H_i.

Further insight can be gained into the role of the bezoutian matrix by employing the concept of realization, introduced in Section 2.5. Assume that in (3.209) $f_0(\lambda)$ is monic and relatively prime to $f_1(\lambda)$. Let $R = \{F, g, h\}$ be any *real* minimal realization of f_1/f_0, and suppose there exists a matrix P such that

$$PF = F^T P, \quad Pg = h^T \tag{3.211}$$

Then (3.21) gives

$$PFg = F^T Pg = F^T h^T$$

$$PF^2 g = (F^T)^2 Pg = (F^T)^2 h$$

and so on, leading to

$$P(g, Fg, F^2 g, \ldots, F^{M-1} g) = (h^T, F^T h^T, (F^T)^2 h^T, \ldots, (F^T)^{M-1} h^T)$$

Using the definition of controllability matrix in (2.23), we can
write

$$PC(F, g) = C(F^T, h^T)$$

$$= [O(F, h)]^T, \quad \text{by (2.49)}$$

Since the realization R is minimal, it follows from Theorem 2.14
that $C(F, g)$ and $O(F, h)$ are both nonsingular, so P is uniquely
determined as

$$P = [O(F, h)]^T [C(F, g)]^{-1} \tag{3.212}$$

It is easy to show that P in (3.212) is also symmetric (see
Problem 3.48). The importance of this matrix in our present
context is as follows:

Theorem 3.29

$$I_{-\infty}^{\infty} \frac{f_1(\lambda)}{f_0(\lambda)} = \text{signature (P)} \tag{3.213}$$

where P is the real symmetric matrix in (3.212).

This result is particularly interesting, because it relates
the ideas of minimal realization, and controllability/observability
matrices [through (3.212)] to evaluation of a Cauchy index, and
hence to the determination of root location, or number of real
roots, of a polynomial. A further surprise is in store: If we
choose R to be the controllable canonical form realization (2.108)
of f_1/f_0, i.e., F is the companion matrix of $f_0(\lambda)$ (with
coefficients along its last row) and

$$g = [0, 0, \ldots, 0, 1]^T, \quad h = [\phi_m, \phi_{m-1}, \ldots]$$

where $f_1(\lambda) = \phi_m + \phi_{m-1}\lambda + \cdots$, then comparison of (3.212) and
(2.114) reveals that in this case P is precisely the bezoutian
matrix Z associated with $f_0(\lambda)$ and $f_1(\lambda)$. We can now see that
(3.210) is merely a special case of (3.213), so in theory at least,
the bezoutian/Hermite matrix for evaluating a Cauchy index is just

one out of an infinite number of possible choices.

Let us conclude this discussion by briefly indicating how the problem of root location when $a(\lambda)$ has *complex* coefficients can be dealt with. For simplicity, suppose that $a(\lambda)$ has no purely imaginary roots, and write

$$a(i\omega) = \gamma_0\omega^n + \gamma_1\omega^{n-1} + \cdots + \gamma_n + i(\epsilon_0\omega^n + \epsilon_1\omega^{n-1} + \cdots + \epsilon_n)$$

$$= \gamma(\omega) + i\epsilon(\omega)$$

where $\gamma(\omega)$ and $\epsilon(\omega)$ are real polynomials. There is no loss of generality in assuming that $\epsilon_0 \neq 0$. Then the generalization of Theorem 3.26 is that the number of roots of $a(\lambda)$ with positive real parts is given by

$$k = \frac{1}{2}\left[n - I_{-\infty}^{\infty}\frac{\gamma(\omega)}{\epsilon(\omega)}\right] \qquad (3.214)$$

The Cauchy index in (3.214) can be evaluated by various methods, as before. If the Routh scheme is used, then Theorem 3.15 is obtained; if the bezoutian expression (3.210) is applied, then Theorem 3.16 is derived in terms of the real Hermite matrix in (3.105).

Problems

3.46 Use the Routh-Sturm algorithm to compute the total number of distinct real roots of $\lambda^4 - 6\lambda^3 + 13\lambda^2 - 12\lambda + 4$, together with their multiplicities.

3.47 Let C be the companion form (1.26), and define $K = a'(C)J_n$. It follows from the result in Problem 1.23 that the total number of distinct roots of a real polynomial $a(\lambda)$ is equal to the rank r of K. It can also be shown that the number of distinct *real* roots of $a(\lambda)$ is equal to

$$r - 2V(1, K_1, K_2, \ldots, K_r)$$

Apply these results to $a(\lambda) = \lambda^4 - \lambda^3 - \lambda - 1$.

3.48 Using the notation of (3.211), show that the matrix
$O(F, h)C(F, g)$ is symmetric, and hence show that P defined in
(3.212) is also symmetric.

3.5 ADDITIONAL TOPICS

3.5.1 Bilinear Transformation

The bilinear transformation

$$\lambda = \frac{\mu + 1}{\mu - 1}, \qquad \mu = \frac{\lambda + 1}{\lambda - 1} \tag{3.215}$$

was mentioned in passing in Section 3.1, equation (3.10). This is
a one-to-one mapping between the regions $\text{Re}(\lambda) < 0$ and $|\mu| < 1$.
Thus, if $a(\lambda)$ is the polynomial (3.13) for the continuous-time
problem, application of (3.215) produces

$$a\left(\frac{\mu + 1}{\mu - 1}\right) \triangleq \frac{b(\mu)}{(\mu - 1)^n} \tag{3.216}$$

where

$$b(\mu) = b_0 \mu^n + b_1 \mu^{n-1} + \cdots + b_n \tag{3.217}$$

The polynomial $b(\mu)$ has the same distribution of roots relative to
the unit circle as does $a(\lambda)$ relative to the imaginary axis.
Similarly, substituting for μ from (3.215) into the discrete-time
polynomial $a_1(\mu)$ in (3.68) gives

$$a_1\left(\frac{\lambda + 1}{\lambda - 1}\right) = \frac{b_1(\lambda)}{(\lambda - 1)^n} \tag{3.218}$$

where

$$b_1(\lambda) = \beta_0 \lambda^n + \beta_1 \lambda^{n-1} + \cdots + \beta_n \tag{3.219}$$

It therefore follows that the root location problems for $a(\lambda)$ and
$a_1(\mu)$ are equivalent to those for $b(\mu)$ and $b_1(\lambda)$, respectively.
For example $a_1(\mu)$ could be tested by applying any of the theorems
of Section 3.2.1 to $b_1(\lambda)$, so that only one set of algorithms would
be needed for both the continuous- and discrete-time problems.
This may be useful when the coefficients of the polynomial are

known numerically; when parameters are involved, it is still
advantageous to use a directly applicable theorem.

Our task is to obtain the coefficients b_i in (3.217) in terms
of the coefficients a_i of $a(\lambda)$ from the relationship (3.216). On
comparing the latter with (3.218) it can be seen that this involves
exactly the same operations as obtaining the β_i in terms of the α_i,
so we need not consider this problem separately. We present two
different methods, both being simple to apply. The first provides
a matrix representation:

Theorem 3.30   The coefficients b_i in (3.217) are determined by the
relationship

$$[b_n, b_{n-1}, \ldots, b_1, b_0] = [a_n, a_{n-1}, \ldots, a_1, a_0]\Gamma_{n+1} \quad (3.220)$$

where the $(n + 1) \times (n + 1)$ matrix $\Gamma_{n+1} = [\gamma_{ij}]$, i, j = 0, 1, $\ldots$,
n, is constructed from the formula

$$\gamma_{ij} = \gamma_{i,j+1} + \gamma_{i-1,j+1} + \gamma_{i-1,j}, \quad \begin{matrix} i = 1, 2, \ldots, n \\ j = n-1, n-2, \ldots, 0 \end{matrix} \quad (3.221)$$

subject to

$$\gamma_{in} = 1, \quad i = 0, 1, \ldots, n \quad (3.222)$$

The elements γ_{0j}, j = 0, $\ldots$, n, of the first row of Γ_{n+1} are
the binomial coefficients in the expansion of $(-1 + \mu)^n$.

Proof.   From (3.216) and (3.13) we obtain

$$b(\mu) = (\mu - 1)^n a\left(\frac{\mu + 1}{\mu - 1}\right)$$

$$= \sum_{i=0}^{n} a_i(\mu + 1)^{n-i}(\mu - 1)^i \quad (3.223)$$

In (3.220) we have

$$b_{n-k} = \sum_{i=0}^{n} a_{n-i}\gamma_{ik}$$

so equating terms containing a_{n-i} in (3.223) gives

$$\sum_{k=0}^{n} \gamma_{ik}\mu^k = (\mu + 1)^i(\mu - 1)^{n-i} \qquad (3.224)$$

Hence

$$(\mu + 1)\sum_{k=0}^{n} \gamma_{ik}\mu^k = (\mu + 1)^{i+1}(\mu - 1)^{n-i}$$

$$= (\mu - 1)\sum_{k=0}^{n} \gamma_{i+1,k}\mu^k \qquad (3.225)$$

on replacing i by (i + 1) in (3.224). Equating coefficients of powers of μ^{j+1} in (3.225), and replacing i by i - 1 then establishes (3.221). It is obvious from (3.217) and (3.223) that

$$b_0 = \sum_{i=0}^{n} a_i$$

so in (3.220) the elements of the last column of Γ_{n+1} are all unity. Similarly, since the term in a_n in (3.223) is $(\mu - 1)^n$, the stated expression for the first row of Γ_{n+1} holds. ∎

Example 31 The formula (3.221) means that γ_{ij} is the sum of the three adjacent elements situated as follows:

```
+-----------------------+
| γ_{i-1,j}   γ_{i-1,j+1}|
+--------+              |
         |              |
  γ_{ij} | γ_{i,j+1}    |
         +--------------+
```

Since the first row and last column of Γ_{n+1} can be written down immediately, the nth column of Γ_{n+1} can be constructed using (3.221), from the second row downward, then the (n - 1)th column, and so on. For example, if n = 3, it is easy to proceed in this way to determine

$$\Gamma_4 = \begin{bmatrix} -1 & 3 & -3 & 1 \\ 1 & -1 & -1 & 1 \\ -1 & -1 & 1 & 1 \\ 1 & 3 & 3 & 1 \end{bmatrix} \tag{3.226}$$

If say

$$a(\lambda) = \lambda^3 - \lambda^2 - 10\lambda - 9 \tag{3.227}$$

then (3.220) produces in this case

$$b(\mu) = -19\mu^3 + 39\mu^2 - 13\mu + 1 \tag{3.228}$$

Of course, Γ_{n+1} need only be calculated once and for all, for each value of n. In fact, a further reduction in computational effort can be achieved as follows. In (3.226) the reader will notice that the first and second columns are the same as the fourth and third columns, respectively, apart from a change of signs in the first and third rows. This simplification is easily shown to hold in general: When n is odd, compute columns n + 1, n, ..., $\frac{1}{2}$(n + 3) of Γ_{n+1} using Theorem 3.30; then column k is equal to column (n - k + 2), for k = 1, 2, ..., $\frac{1}{2}$(n + 1), with the elements in rows 1, 3, 5, ..., n being multiplied by -1. Similarly, when n is even, compute columns n + 1, n, ..., $\frac{1}{2}$n + 1 using Theorem 3.30, and for k = 1, 2, ..., $\frac{1}{2}$n, column k is equal to column (n - k + 2) with the signs of the elements reversed in rows 2, 4, 6, ..., n.

The second method for calculating the b_i utilizes Horner's method, introduced in Section 1.1. We begin by setting $\lambda = 1 + \sigma$, so that $a(\lambda)$ becomes

$$a(1 + \sigma) = a_0(1 + \sigma)^n + a_1(1 + \sigma)^{n-1} + \cdots + a_n$$

$$\overset{\Delta}{=} \theta_0 \sigma^n + \theta_1 \sigma^{n-1} + \cdots + \theta_n \tag{3.229}$$

The coefficients θ_i in (3.229) can be obtained using Horner's method as described in Problem 1.7. Next, substituting for λ in (3.215) produces $\sigma = 2/(\mu - 1)$. When this is itself substituted into (3.229), we obtain

$$\frac{1}{(\mu - 1)^n} \left[2^n \theta_0 + 2^{n-1} \theta_1 (\mu - 1) + 2^{n-2} \theta_2 (\mu - 1)^2 + \cdots \right.$$

$$\left. + \theta_n (\mu - 1)^n \right] \quad (3.230)$$

Comparison of (3.230) and (3.216) shows that $b(\mu)$ is the expression within the square brackets. Finally, the required b_i's can be obtained by a second application of Horner's scheme to the coefficients θ_n, $2\theta_{n-1}$, $2^2 \theta_{n-2}$, $\ldots$, $2^n \theta_0$ in (3.230).

Example 32   We apply the preceding to $a(\lambda)$ in (3.227). The first step is equivalent to replacing λ by $(\sigma + 1)$ in (3.227). Following Problem 1.7, Horner's scheme for determining the coefficients θ_i is as follows:

	1	-1	-10	-9
1	1	0	-10	-19 (= θ_3)
	1	1	-9 (= θ_2)	
	1	2 (= θ_1)		
	1 (= θ_0)			

The second step is to multiply these θ_i by the appropriate powers of 2, giving -19, 2×-9, $2^2 \times 2$, $2^3 \times 1$. Finally, Horner's method is applied to these coefficients, to obtain $b(\mu)$ from (3.230) [notice that this time we replace μ by $(\mu - 1)$]:

	-19	-18	8	8
-1	-19	1	7	1 (= b_3)
	-19	20	-13 (= b_2)	
	-19	39 (= b_1)		
	-19 (= b_0)			

and these values agree with (3.228).

A rather different way of looking at the transformation (3.215) is to use the companion matrix C in the standard form (1.26), having characteristic polynomial $a(\lambda)$ (assumed monic).

If we apply the matrix form (3.11) of the bilinear transformation
to C, we obtain a matrix

$$B = (C - I)^{-1}(C + I) \tag{3.231}$$

and the characteristic polynomial of B will be the monic form of
(3.217), i.e., $b(\mu)/b_0$. Of course, B as given by (3.231) will not
itself be in companion form. However, it is an interesting fact
that the matrix Γ of Theorem 3.30 can be used to put B into
companion form. It turns out to be more convenient to use $(-B)$,
which has the same distribution of eigenvalues as B relative to the
unit circle. If we set $T = \Gamma_n J_n$, where Γ_n is defined in Theorem
3.30, then it can be shown that the similarity transformation
$T(-B)T^{-1}$ puts $(-B)$ into companion form. The characteristic
polynomial $\tilde{b}(\mu)$ of $-B$ can, of course, be written down by
inspection. The desired polynomial $b(\mu)/b_0$ is obtained on noting
that

$$\tilde{b}(\mu) = \det[\mu I - (-B)] = (-1)^n \det[(-\mu)I - B]$$

$$= \frac{(-1)^n b(-\mu)}{b_0}$$

whence

$$\frac{b(\mu)}{b_0} = (-1)^n \tilde{b}(-\mu) \tag{3.232}$$

Furthermore, it is easy to prove that $T^{-1} = 2^{1-n} J_n \Gamma_n$ (see Problem
3.51). It is remarkable that T depends only upon n, and not the
coefficients of $a(\lambda)$.

Example 33   For $a(\lambda)$ in (3.227) we have

$$C = \begin{bmatrix} 0 & 1 & 0 \\ 0 & 0 & 1 \\ 9 & 10 & 1 \end{bmatrix}$$

and from (3.231) we obtain

$$B = \frac{1}{19} \begin{bmatrix} -1 & 0 & 2 \\ 18 & 19 & 2 \\ 18 & 38 & 21 \end{bmatrix}$$

It is easy to use Theorem 3.30 to obtain

$$\Gamma_3 = \begin{bmatrix} 1 & -2 & 1 \\ -1 & 0 & 1 \\ 1 & 2 & 1 \end{bmatrix}$$

and the reader can check that with $T = \Gamma_3 J_3$, $T^{-1} = \frac{1}{4} J_3 \Gamma_3$, then

$$T(-B)T^{-1} = \begin{bmatrix} 0 & 1 & 0 \\ 0 & 0 & 1 \\ -\dfrac{1}{19} & -\dfrac{13}{19} & -\dfrac{39}{19} \end{bmatrix}$$

The characteristic polynomial of $-B$ is thus

$$\tilde{b}(\mu) = \mu^3 + \frac{39}{19}\mu^2 + \frac{13}{19}\mu + \frac{1}{19}$$

which, on applying (3.232), is seen to agree with (3.228).

3.5.2 Other Regions of the Complex Plane

In some applications it is of interest to determine the numbers of
roots of a polynomial inside and outside regions of the complex
plane other than the half-plane $Re(\lambda) < 0$ or the unit disk
$D: |\mu| < 1$. For the continuous-time problem, the simple case of
$Re(\lambda) < -\sigma$ was considered in Remark 5 on Theorem 3.3; the
equivalent region $|\mu| < \tau$ for the discrete-time case was referred
to in Remark 4 on Theorem 3.10. Both these situations were
concerned with ensuring that the transient solution of the
appropriate linear system (3.1) or (3.2) tended to zero at least as
fast as some specified function. Another region of similar
significance is the sector indicated by the shaded region in Figure
3.1. If all the roots of $a(\lambda)$ lie within this region, then it
follows not only that the associated continuous-time system is

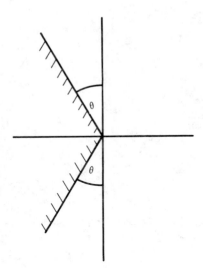

Figure 3.1 Stability sector.

asymptotically stable, but also that any oscillatory terms in the
solution will be at least damped by an amount proportional to
sin θ. If C is the standard companion matrix (1.26) associated
with real a(λ), consider the matrix

$$
C_a = \begin{bmatrix} \cos\theta & -\sin\theta \\ \sin\theta & \cos\theta \end{bmatrix} \otimes C
$$

$$
= \begin{bmatrix} (\cos\theta)C & -(\sin\theta)C \\ (\sin\theta)C & (\cos\theta)C \end{bmatrix} \tag{3.233}
$$

It is easy to check that the eigenvalues of the 2 × 2 matrix in
(3.233) are cos θ ± i sin θ = exp(±iθ). Using a result in (A2) in
Appendix A, it follows that the eigenvalues of the 2n × 2n matrix
C_a are

$$
\lambda_k^a = \lambda_k \exp(\pm i\theta), \quad k = 1, 2, \ldots, n
$$

where the λ_k are the eigenvalues of C, i.e., the roots of a(λ).
Hence

$$
\arg(\lambda_k) = \arg(\lambda_k^a) \pm \theta
$$

from which it is easily seen that the roots λ_k lie within the
shaded region in Figure 3.1 if and only if the eigenvalues λ_k^a lie
in the left half-plane. Thus the desired condition on $a(\lambda)$ will be
satisfied if and only if C_a is a stability matrix, and this can be
tested in the usual way, either by solving the Liapunov matrix
equation (3.129), or by finding its characteristic polynomial. If
$a(\lambda)$ is complex, we apply the preceding to the real polynomial
$a(\lambda)\bar{a}(\lambda)$, which has its roots inside the shaded sector in Figure
3.1 if and only if $a(\lambda)$ does.

When turning to arbitrary half-planes or circles, the ease
with which the special bilinear transformation (3.215) could be
handled naturally suggests using the general bilinear expression

$$\lambda = \frac{\alpha\mu + \beta}{\gamma\mu + \delta}, \qquad \alpha\delta - \beta\gamma \neq 0 \tag{3.234}$$

For example, (3.234) can represent a mapping between the unit disk
D in the μ-plane, and an arbitrary disk (or half-plane) $\hat{D}$ in the
λ-plane, by a suitable choice of the parameters α, β, γ, δ. If we
wish to determine the location of the roots of $a(\lambda)$ in (3.13)
relative to $\hat{D}$, we apply (3.234) to obtain

$$a\left(\frac{\alpha\mu + \beta}{\gamma\mu + \delta}\right) \triangleq \frac{b(\mu)}{(\gamma\mu + \delta)^n} \tag{3.235}$$

where $b(\mu)$ is expressed in (3.217). The problem is then solved by
finding the location of the roots of $b(\mu)$ relative to D. An
argument very similar to that used to establish Theorem 3.30 shows
that the coefficients of $b(\mu)$ in (3.235) are still given by an
expression of the form (3.221), but now the first row of
$\Gamma_{n+1} = [\gamma_{ij}]$ consists of the coefficients in the expansion of
$(\delta + \gamma\mu)^n$; the last column of Γ_{n+1} is

$$[\gamma^n, \alpha\gamma^{n-1}, \alpha^2\gamma^{n-2}, \ldots, \alpha^n]^T \tag{3.236}$$

and the recurrence formula (3.221) is replaced by

$$\gamma\gamma_{ij} = \alpha\gamma_{i-1,j} + \beta\gamma_{i-1,j+1} - \delta\gamma_{i,j+1}; \quad i \neq 0, \quad j \neq n \tag{3.237}$$

To deal with regions more general than circles, a rational transformation $\lambda = f(\mu)$ can be used. For example, if f has second degree, then it can be decomposed either into two or three successive transformations, each of which can be represented by a matrix multiplication on coefficients, as in Theorem 3.30.

It is worth pointing out here that we can generalize the matrix bilinear form in (3.231), i.e.,

$$B = (\gamma C + \delta I)^{-1}(\alpha C + \beta I) \tag{3.238}$$

where C is the companion matrix of $a(\lambda)$. The characteristic polynomial of B is the monic form of $b(\mu)$. As before, it can be shown that $T = \Gamma_n J$, where Γ_n is now associated with (3.234), transforms B in (3.238) via similarity into companion form, so again T depends only upon n and the parameters α, β, γ, δ.

A second way of tackling root location relative to general regions is suggested by our success in Section 3.3 with the application of appropriate linear matrix equations to polynomial problems. To begin with, consider the case of an ellipse in the λ-plane, having equation

$$(2k - 1)x^2 + y^2 - 1 = 0, \quad k \geqslant \tfrac{1}{2} \tag{3.239}$$

where $\lambda = x + iy$. For reasons which will become clear shortly, we associate with (3.239) the matrix equation

$$kA^*PA + \tfrac{1}{2}(k - 1)[(A^*)^2 P + PA^2] - P = Q \tag{3.240}$$

where Q is an arbitrary positive definite hermitian matrix. When k = 1 (3.239) represents the unit circle, and (3.240) reduces to the discrete-time Liapunov-type equation (3.171) (note that for convenience we have dropped here the suffix on A). We repeat the procedure used in the first part of the direct proof of Theorem 3.23: Let λ, v be a corresponding eigenvalue and right eigenvector of A. Using

$$Av = \lambda v, \quad v^*A^* = \bar{\lambda}v^*$$

in (3.240) produces

$$v^*Qv = [k|\lambda|^2 + \frac{1}{2}(k - 1)(\bar{\lambda}^2 + \lambda^2) - 1]v^*Pv$$

$$= [(2k - 1)x^2 + y^2 - 1]v^*Pv \qquad (3.241)$$

and the expression within square brackets in (3.241) is precisely
that in (3.239). This is the reason behind the form of (3.240).
Since Q is positive definite, it follows that if P is negative
definite, then the eigenvalue λ of A lies in the interior of the
ellipse whose boundary is described by (3.239). A converse
argument can also be developed, so we have:

Theorem 3.31 All the eigenvalues of A lie in the interior of the
ellipse (3.239) if and only if there is a unique negative definite
solution P of (3.240), for arbitrary positive definite hermitian Q.

To test a polynomial we apply Theorem 3.31 to its companion
matrix. However, there is no easy way known at present of
expressing this test in terms of the coefficients of the polynomial.
We close this chapter by briefly indicating two ways in which
extensions of Theorem 3.31 can be obtained.
The first approach is to define a rational transformation

$$\mu = \frac{\psi(\lambda)}{\phi(\lambda)} \qquad (3.242)$$

which sends D ($|\mu| < 1$) into the region

$$E: \quad |\psi(\lambda)|^2 - |\phi(\lambda)|^2 < 0 \qquad (3.243)$$

The corresponding matrix transformation is

$$A_1 = \psi(A)[\phi(A)]^{-1} \qquad (3.244)$$

assuming that the conditions for this to exist are satisfied.
Substitution of (3.244) into (3.171), viz., $A_1^*PA_1 - P = Q_1$,
produces

$$[\psi(A)]^*P\psi(A) - [\phi(A)]^*P\phi(A) = Q_2 \qquad (3.245)$$

and it follows from Theorem 3.24 that the inertia of A relative to
E is given by the inertia of the solution P of (3.245). Of course,
the bilinear transformation (3.215) is a special case of (3.242),
when

$$\psi(\lambda) = \lambda + 1, \quad \phi(\lambda) = \lambda - 1$$

E in (3.243) becomes the left half plane $x < 0$, and (3.245) becomes
the continuous-time Liapunov equation (3.129). It should be noted
that the region E defined by (3.243) does *not* contain ellipses or
parabolas, but does contain certain strips, sectors, and hyperbolas.

The second approach uses a region F of the λ-plane defined by

$$F: \quad f_{00} + f_{10}x + f_{01}y + f_{11}xy + f_{20}x^2 + f_{02}y^2 < 0 \qquad (3.246)$$

where $f_{20} + f_{02} \geqslant 0$. Then the following extension of Theorem 3.31
can be proved:

Theorem 3.32   All the eigenvalues of A lie within F defined by
(3.246) if and only if there is a unique negative definite
hermitian solution P of

$$f_{00}P + c_1 PA^* + \bar{c}_1 AP + c_2 APA^* + c_3 P(A^*)^2 + \bar{c}_3 A^2 P = Q \qquad (3.247)$$

where Q is arbitrary positive definite hermitian and

$$c_1 = \frac{1}{2}(f_{10} + if_{01}), \quad c_2 = \frac{1}{2}(f_{20} + f_{02}),$$

$$c_3 = \frac{1}{4}(f_{20} - f_{02} + if_{11}) \qquad (3.248)$$

Remark   When $f_{01} = f_{11} = 0$, then F is symmetric with respect to the
real axis, and c_1 and c_2 in (3.248) are real. The region F contains the interiors of all ellipses and parabolas, in addition to
certain hyperbolas, strips, sectors, and others. In general,
neither of the regions E in (3.248) or F in (3.246) contains the
other.

Problems

3.49 Compute Γ_6 in Theorem 3.30. Hence obtain the polynomial $b(\mu)$
     in (3.217) for the case when $a(\lambda)$ is the polynomial in Problem
     3.6(iii). Repeat the determination of $b(\mu)$ using Horner's
     method.

3.50 Let $a(\lambda)$ be the polynomial in Problem 3.6(ii), with companion
     matrix C. Determine the corresponding matrix B in (3.231).
     Compute Γ_4 in Theorem 3.30, and hence use the method of
     Example 33 to obtain the monic form of $b(\mu)$ in (3.217).

3.51 Prove that the matrix Γ_{n+1} defined in Theorem 3.30 satisfies
     the condition $(\Gamma_{n+1})^2 = 2^n I_{n+1}$.

3.52 Show that the transformation $\lambda = \mu/(\mu - 1)$ maps the line
     segment $0 \leqslant Re(\mu) \leqslant 1$ into the negative real axis in the
     λ-plane. When this transformation is applied to $a(\lambda)$, let the
     coefficients of the resulting polynomial $b(\mu)$ be expressed as
     in (3.220), i.e.,

     $$[b_n, \ldots, b_1, b_0] = [a_n, \ldots, a_1, a_0]\Gamma_{n+1}$$

     Show directly, without using (3.237), that in this case the
     first row and last column of Γ_{n+1} are the same as in Theorem
     3.30, but that the relation (3.221) is replaced by

     $$\gamma_{ij} = \gamma_{i,j+1} + \gamma_{i-1,j}$$

     Hence write down Γ_5. Deduce also that Γ_{n+1} is upper
     triangular and idempotent (i.e., $\Gamma_{n+1}^2 = I_{n+1}$).

3.53 Prove that when $\delta = -\alpha$ in (3.234), then the corresponding
     matrix Γ_{n+1} satisfies the condition

     $$(\Gamma_{n+1})^2 = (\alpha^2 + \beta\gamma)^n I_{n+1} \qquad (3.249)$$

     The results in Problem 3.51, and the last part of Problem
     3.52, are special cases of (3.249).

3.54 Consider the matrix equation

$$\sum_{i,j=0}^{N} \alpha_{ij}(A^T)^i P A^j = Q \qquad (3.250)$$

where A is a real n × n matrix, and P, Q are n × n real symmetric matrices. Define real skew symmetric matrices

$$R = A^T Q - QA, \qquad S = A^T P - PA$$

and hence show that (3.250) reduces to

$$\sum_{i,j=0}^{N} \alpha_{ij}(A^T)^i S A^j = R \qquad (3.251)$$

Show also that

$$P = (Q - F)E^{-1}$$

where

$$E = \sum_{i,j=0}^{N} \alpha_{ij} A^{i+j}, \qquad F = \sum_{j=0}^{N} \sum_{i=1}^{N} \alpha_{ij} \sum_{k=1}^{i} (A^T)^{i-k} S A^{j+k-1}$$

[it can be shown that if (3.249) has a unique solution, then E is nonsingular] If A and Q are given, then (3.251) represents only $\frac{1}{2}n(n-1)$ equations and unknowns, a saving of n compared with (3.249). The result in Problem 3.38 is a special case of the above.

3.55 Deduce that the eigenvalues of A all lie in the sector in Figure 3.1 with θ = 45°, if and only if there exist positive definite matrices P_1, P_2 such that matrices Q_1, Q_2 defined by

$$Q_1 = -(A^* P_1 + P_1 A), \qquad Q_2 = (A^*)^2 P_2 + P_2 A^2$$

are both positive definite.

3.56 Use Theorem 3.32 to show that the eigenvalues of A all lie in the parabolic region $y^2 < x$ if and only if there exists a unique negative definite hermitian solution P of the equation

$$2APA^* - 2PA^* - 2AP - P(A^*)^2 - A^2P = Q$$

for arbitrary positive definite hermitian Q.

BIBLIOGRAPHICAL NOTES

Section 3.0. Textbooks which present a good coverage of stability and root location are those by Jury (1974), Lehnigk (1966), Marden (1966), Obreschkoff (1963), Parodi (1959), and Porter (1967). Only those by Marden and by Jury give discrete-time as well as continuous-time results, and only the latter refers to the recent method of obtaining proofs of the classical results using linear matrix equations. This idea originated with Parks (1962), whose survey paper (1963a) is also useful. Some directly relevant material is contained in Chapters 2 and 4 of Barnett's book (1971a) and survey (1974c).

Section 3.1. A full discussion of stability concepts, including the work of Liapunov can be found in a number of textbooks, of which those by Hahn (1963) and Willems (1970) can be particularly recommended. The book by Barnett and Storey (1970) concentrates on the application of Liapunov theory to linear systems in state space form. A full proof of Theorem 3.1 was given by Lehnigk (1966). A standard work on numerical linear algebra is by Wilkinson and Reinsch (1971). The result of Problem 3.5 was obtained in a different fashion by Brenner (1961), and also applies for eigenvalue location in a sector (Anderson et al., 1975). An interesting iterative scheme for polynomial root location in both the continuous- and discrete-time cases has been recently proposed by Howland (1978); see also Miller (1971).

Section 3.2.1. Routh's original paper of 1877, together with some background notes and related papers, has been reprinted in a work edited by Fuller (1975). This author (Fuller, 1977) contributed an account of Routh's life to the Routh Centenary Issue of the *International Journal of Control* (August 1977). The statement of

Theorems 3.3 and 3.4 can be found in most standard texts on control
systems. For additional properties of the Hurwitz matrix in
(3.25), see Asner (1970), Kemperman (1982), and Lehnigk (1970).
A full treatment of the singular cases is given by Gantmacher
(1959); for discussion of the first method in Remark 4 on Theorem
3.3, see Jury (1975b). Precise results on the continuity of the
roots as a function of the polynomial coefficients can be found in
Chapter 2 of Marden (1966). Over the past few years there has been
a continuing correspondence on dealing with the singular cases
(e.g., Shamash, 1980), much of it spurious and indeed irrelevant
(White, 1979). A proof of Orlando's formula (3.34), the relationships (3.36), and Theorems 3.5 and 3.7 can also be found in
Gantmacher (1959). Theorem 3.6 is taken from Barnett (1971d) where
details of Remark 3 can be found. The formulae (3.56) were quoted
by Parks (1963a). A good account of Theorem 3.8 was provided by
Parks (1977a), who in an associated paper (Parks, 1977b) gave a
translation of Hermite's original work of 1854. The reduced
Hermite matrices of Remark 2 on Theorem 3.8, and Theorem 3.9, are
due to Anderson (1972a). The Hermite-Hurwitz link in (3.66) can be
nicely established by matrix multiplication (Liénard and Chipart,
1914; Parks, 1969), an alternative proof being available in
Lehnigk (1966). The formula (3.67) is taken from Jury and
Anderson (1972). A review of some of the interrelationships
between results on real polynomials was given by Anderson (1967).

Section 3.2.2. A standard source for Theorem 3.10 is Jury's text
(1964). Our treatment of the singularities in Theorem 3.10 follows
Kuo (1977), but Jury (1964) gave a different, and more detailed,
discussion. A comprehensive treatment of inners and their
applications has been given by Jury (1974, 1975a), and it is his
representation of Theorem 3.11 which we have quoted. The "left
triangle of zeros" displayed in (3.82) occurs frequently in inners
representations, and can be used to computational advantage when
evaluating the determinants in Theorems 3.11 and 3.12. For the
related notion of "outers," see Barnett (1976b). The work of Schur

and Cohn dates back to 1917 and 1922, respectively, so that there
was a surprisingly long delay in extending the results of Hermite,
Routh, and Hurwitz to the discrete-time case. A proof of Theorem
3.11 can be conveniently found in Marden (1966), where a derivation
of (3.83) is also given. The expression (3.86), involving the
array with divisions in (3.85), was given by Barnett (1973b,
1974b), but see also Jury (1971), where the stability criterion of
Theorem 3.12 can be found. A direct analogue of the Liénard-
Chipart criterion, which involves n constraints instead of the two
in (3.88) and (3.89), has been obtained by Anderson and Jury
(1973). Theorem 3.13 was obtained by Fujiwara (1926) as an
extension of earlier work of Schur and Cohn, and can be linked to
the Jury-Marden array (Sarma and Pai, 1968; Power, 1970). Theorem
3.14 was proved by Barnett (1970b), and the reduction outlined in
Remark 2 was obtained by Anderson and Jury (1973), who also demonstrated the relationship between this and the reduced Hermite
expression in (3.63) (Anderson and Jury, 1974) [for a further
simplification, see Jury and Anderson (1981)]. Numerical
application of the Schur-Cohn criterion was discussed by
Gargantini (1971).

Section 3.2.3. Complex polynomials arise, among other places, in
the stability study of multidimensional polynomials (Jury, 1978).
The references for the complex Hermite Theorem 3.16 are the same as
those for the real case (Theorem 3.8). The Bilharz and complex
Routh results were set out by Parks (1963a), and the same author
(Parks, 1977a) derived the complex Orlando formula (3.104). For
the continued fraction approach in Remark 2 on Theorem 3.15, see
Cutteridge (1959), Talbot (1960), Henrici (1977), and Jones and
Thron (1980). Lehnigk (1966) gave a full treatment of the real
symmetric Hermite matrix in Remark 5 on Theorem 3.16. The
relationship (3.105) between the two Hermite matrices was pointed
out by Maroulas and Barnett (1979a), together with some further
results, including that mentioned in Remark 2 on Theorem 3.17; the
latter theorem was proved by Barnett (1971e). The complex results

in Theorem 3.18 are taken from the same sources as their real
counterparts. It can be noted that Schur and Cohn actually used
the complex conjugate of K in (3.121). A survey of the Hermite and
Schur-Cohn results is recently available in a new translation
(Krein and Naimark, 1981). The results in Problems 3.13 and 3.14
are samples from an interesting collection by Lipatov and Sokolov
(1979). The results quoted in Problem 3.27 are from Henrici
(1970), the Schur theorem of Problem 3.28 is in the book by Marden
(1966, p. 198), and Problem 3.29 is based on Datt and Govil (1978).
Formulae for obtaining all the elements in a Routh array from those
in the first column (Problem 3.30) were set out by Tsoi (1979).

The total number of publications in the field of root location
and stability is vast, and continues to attract interest worldwide
(Hamada, 1979). We have tried to quote only references of primary
importance, and many others can be found in the books listed for
Section 3.0.

Section 3.3.1. Liapunov's theorem dates back to 1892. The proof
given here, involving differential equations, is based on Bellman
(1970), but a purely algebraic proof has been given by Hahn (1956).
The result of Remark 5 on Theorem 3.19 was proved by Stein (1965).
Theorem 3.20 is a special case of a more general result of
Haynsworth (1968) (see Problem 3.32). Theorem 3.21 was originally
derived by Ostrowski and Schneider (1962), but the shorter proof we
have given is due to Wimmer (1973) (see also Taussky, 1961a).
Wimmer (1974a) is also responsible for Theorem 3.22; for the lemmas
needed to prove this result, see Ostrowski and Schneider (1962) and
Carlson and Schneider (1963). For further results on control-
ability and inertia, see (Carlson and Hill, 1976, 1977) and
(Carlson et al., 1982). The proof of the Routh and Hurwitz
theorems closely follows Parks (1962) [see also Mansour (1965)].
The matrix in (3.154) was given by Schwarz (1956) for both the real
and complex cases [for the latter, see (3.169) and Problem 3.40].
A further study of the Schwarz matrix, together with an extension
to the discrete-time case, was made by Anderson et al. (1976b) [see

also Power (1967a, 1969)]. The similarity transformation between
companion and Schwarz matrices was given by Chen (1974). For the
real Hermite theorem case, see Parks (1963b) and Anderson (1967),
whereas the rather different development outlined for the complex
Hermite case follows Lehnigk (1967). The paper by Anderson gave a
connection with the Markov parameters of Theorem 3.7 [see also
Datta (1978)].

Section 3.3.2. The first direct statement and proof of the
discrete-time version (Theorem 3.23) of Liapunov's linear stability
theorem of 1892 seems to have been by Stein (1952), a surprisingly
long delay. Our proof of the necessity part of the theorem is
similar to the approach of Cullen (1965) and Redheffer (1965).
Remark 4 on Theorem 3.23 was given by Stein (1965). The link
between the continuous- and discrete-time versions of the matrix
stability and inertia theorems was described by Taussky (1964).
One of the few textbooks to include a treatment of Theorem 3.23 is
by Anderson and Moore (1979). The application of discrete-time
Liapunov theory to convergence of iterative processes was discussed
by Ortega (1973). Wimmer's paper (1973), referred to earlier in
relation to Theorem 3.21, also contained a similar proof of part
(i) of Theorem 3.24. Some further aspects of stability were
discussed by Taussky-Todd (1968), and of inertia by Cain (1980).
The indicated proof of Theorem 3.13 is due to Parks (1964) [see
also Mansour (1965)]. The alternative scheme, involving the
inverse of the Schur-Cohn matrix, was developed by Bitmead and
Weiss (1979) using bezoutian matrices [see also Dickinson (1980)].
Other alternative proofs of Theorem 3.13 are by Kalman (1965a),
Datta (1978), and Vieira and Kailath (1977). It is interesting
that the latter paper uses Theorem 1.14, on the inverse of the
bezoutian matrix. For further results involving Toeplitz matrices,
see Bitmead (1981). A Liapunov proof of a version of the Jury-
Marden array (Theorem 3.10) has recently been devised (Harn and
Chen, 1981; Mansour, 1982).

Section 3.3.3. The proof given of Theorem 3.18(c) is a special
case of that of Pták and Young (1980), who showed that there is an
infinite class of hermitian matrices having the same inertia as K.
This extends an earlier paper by Young (1979a), in which the
formula (3.191) was given. Pták and Young (1982) have also covered
the singular case, continuing studies by Young (1979b, c, 1980a) of
polynomials and linear matrix equations including a proof of the
Hermite theorem [Young (1983)]. Pták (1981) has used (3.182) to
obtain an alternative proof of the result of Bitmead and Weiss
(1979), including the complex case. The method of Problem 3.31
follows Bellman (1970). The result of Problem 3.33 is due to
MacFarlane (1963a), and can be extended to integrals containing
t^r ($r = 2, 3, \ldots$). The expression in Problem 3.34 can be put in
terms of the Hurwitz determinants, using (3.167) [see Parks
(1963c)]. The connection between the Routh and Schwarz matrices in
Problem 3.35 originated with Power (1969), but the Routh form
itself, and the application to evaluation of the integral in
Problem 3.36, are due to Puri and Weygandt (1964), who also
extended this to include time-weighted integrals. The canonical
form for Hurwitz polynomials in Problem 3.37 is due to Bückner
(1952), and a network theory interpretation has been given by
Bashkow and Desoer (1957). The reduction using a skew symmetric
matrix in Problem 3.38 is from Barnett and Storey (1966), but see
also (Taussky, 1961b). The result in Problem 3.39 is due to
Barnett and Storey (1968), and the extension of Schwarz's form to
the complex case in Problem 3.40 was given by Wimmer (1974b). The
theorem in Problem 3.43 was derived by Hahn (1971), but an
interesting control theory proof has been presented by Heymann and
Feuer (1974). The method of Problem 3.44 can also be extended to
time-weighted sums (Barnett, 1974d), although the development is a
little more complicated than for the continuous-time case.

The numerical solution of linear matrix equations is itself a
much-studied topic, which however lies outside the scope of this
book. Some references worth consulting are Bartels and Stewart

(1972), Belanger and McGillivray (1976), Barroud (1977), Howland
and Senez (1970), and Golub et al. (1979).

Section 3.4. The treatment given of the Cauchy index method has
been kept deliberately sketchy, as a more comprehensive treatment
can be found in many standard texts, including Gantmacher (1959),
Marden (1966), Mishina and Proskuryakov (1965), and Porter (1967).
Theorems 3.28 and 3.29 are taken from (Anderson, 1972b). The
expression (3.207), and the result in Problem 3.47, are from two
papers of Barnett (1971d, 1970c, respectively). The theorems of
Sturm and Cauchy were used in Routh's original monograph, and
Fuller (1975) in his introduction to the reprint of this work gave
a useful account of these results, including proofs. The Cauchy
index method can also be applied to the discrete-time problem
(Anderson et al. 1976a).

Section 3.5.1. Theorem 3.30 was obtained by Duffin (1969), and the
simplification in the construction of Γ_{n+1} was pointed out by Jury
(1973), who also gave several further references, including an
alternative method of Power (1967b). Some other relevant results
are contained in Jury and Chan (1973). The idea of using Horner's
method to perform the bilinear transformation is due to Davis
(1974), and Policastro (1979). The similarity transformation,
involving Γ, of the matrix bilinear expression (3.231) was
discovered by Shane and Barnett (1974), and Young (1980b) has
given a more elegant proof, which leads to certain extensions.

Section 3.5.2. The Kronecker product expression in (3.233) follows
Anderson et al. (1974); see also Problem 3.5 and the bibliographical note thereon. Properties of the bilinear mapping
(3.234) can be found in standard texts such as (Nehari, 1952). The
formula (3.237) was derived by Barnett (1973d), who also gave
further detail on the decomposition of second-degree transformations. For the similarity transformation of (3.238), see Shane
and Barnett (1974) and Young (1979c, 1980b). For Theorem 3.31,

see papers by Jury and Ahn (1972, 1974) and especially Ahn (1979).
A transformation for polynomials relative to the ellipse has been
given by Walach and Zeheb (1982). The expression (3.245) was
obtained by Jury and Ahn (1972). Theorem 3.32 is due to Gutman
(1979), being an improvement on a similar rdsult of Jury and Ahn
(1974). The most comprehensive treatment to date of the problem of
eigenvalue location of a matrix in regions of the complex plane has
been provided by Gutman and Jury (1981), including detailed
discussion of the regions defined by (3.243) and (3.246) [see also
Kharitonov (1981), and Gutman (1982)]. Barnett (1975b) commented
on matrix inertia relative to an ellipse, and Barnett and Scraton
(1982) used a certain block companion matrix. Hill (1977) and
Mazko (1980) obtained some inertia theorems for certain regions,
and Howland (1983) gave an interesting procedure for root location
relative to rectangles. A transformation method for polynomials
relative to other conics has been given by Zeheb and Hertz (1982).

Due to lack of space, we can merely quote some references on
further properties of the Routh and Jury-Marden arrays. The extent
to which these algorithms are interchangeable has been demonstrated
by Barnett (1973a,b, 1974b). For the relationship between the
Routh array and bezoutian matrices, see Barnett (1977), Barnett and
Maroulas (1978), and Maroulas (1982).

The result in Problem 3.51 was obtained by Duffin (1969), and
its extension in (3.249) by Barnett (1973d). The transformation in
Problem 3.52 was discussed, among others, by Jury and Chan (1973).
The reduction in Problem 3.54 was derived by Barnett (1976a) as a
generalization of earlier results for Liapunov-type equations
(Barnett and Storey, 1966; Barnett, 1974e). Jury and Ahn (1974)
gave the result in Problem 3.56.

4

Feedback, Realization, and
Polynomial Matrices

4.0   INTRODUCTION

We return to the linear system representations of Chapter 2,

$$\dot{x}(t) = Ax(t) + Bu(t) \qquad\qquad (4.1)$$

$$x(k + 1) = Ax(k) + Bu(k) \qquad\qquad (4.2)$$

where in each case x is the n-dimensional state vector, and u is
the m-dimensional vector of control (or input) variables.   As
before, t ($\geqslant$ 0) denotes a continuous-time variable, and k (= 0, 1,
2, 3, ...) represents discrete intervals of time.   The
r-dimensional output vectors are also as given in Chapter 2:

$$y(t) = Hx(t) \qquad\qquad (4.3)$$

$$y(k) = Hx(k) \qquad\qquad (4.4)$$

We shall assume, except where stated, that A, B, and H are real.
The use of the term "multivariable" in the control literature is
rather misleading, since it is often used to mean that m > 1, but
we shall understand by this that either m, or r, or both, are
greater than unity.   Our aims in this chapter are twofold: first,
to extend to the multivariable case a number of results which were
obtained in Chapter 2 for systems with a single input and output.
This leads to an involvement with polynomial matrices, so the

second main thrust is to study relevant properties of polynomial
(and rational) matrices.

Recall that we wish to determine a control program which will
make the state variables behave in a satisfactory manner, according
to some design criteria, based on a knowledge of the output
variables. For example, if y(t) is to be kept as close as possible
(in some defined sense) to a given reference vector R(t), the
system is called a *servomechanism*, and if R is a constant, a
regulator. The water tank level, and room temperature models,
mentioned in Section 2.1, are simple examples of regulators. If we
choose coordinates so that R = 0, then we usually require x(t) to
tend to zero, as t tends to infinity, i.e., the closed-loop system
is to be asymptotically stable.

Section 4.1 is devoted to an investigation of the principle of
linear feedback, which was considered for single-input systems in
Section 2.6. We begin in Section 4.1.1 with the case when the
state is itself measurable, so that state feedback is possible, and
defer the general situation of output feedback until Section 4.1.4.
The intermediate sections develop some interesting material on
canonical forms (4.1.2) and feedback obtained by minimizing an
integral of quadratic forms (4.1.3).

In Section 4.2 we consider transfer function matrices, first
defined in Section 2.1. Generalizations of results in Section 2.5
are obtained. Our point of view is to examine different representations, either in state space terms (Section 4.2.2) or as a matrix
fraction (Section 4.2.3). In the course of this development we
need to define the concepts of relative primeness and greatest
common divisor for polynomial matrices. In view of their
importance, we devote Section 4.3 to an investigation of these
topics. Various extensions are presented of some of the results
for scalar polynomials, given in Chapter 1. In particular,
interesting generalizations arise of our four recurrent themes:
companion, Sylvester and bezoutian matrices, and the Routh array.

4.1 LINEAR FEEDBACK

4.1.1 Eigenvalue Assignment by State Feedback

We assume in this section that the state is itself measurable, and generalize the treatment given in Section 2.6 for the single-input case. Thus, each of the m control variables is now a linear combination of the states, i.e.,

$$u_i = k_{i1}x_1 + k_{i2}x_2 + \cdots + k_{in}x_n, \quad i = 1, \ldots, m \tag{4.5}$$

or, in vector-matrix notation,

$$u = Kx \tag{4.6}$$

where K is the *feedback matrix*. Substitution of (4.6) into (4.1) and (4.2) produces, respectively, the closed-loop systems

$$\dot{x}(t) = (A + BK)x(t) \tag{4.7}$$

$$x(k + 1) = (A + BK)x(k) \tag{4.8}$$

so that in either case the system *closed-loop matrix* is

$$A_c = A + BK \tag{4.9}$$

As we remarked for the single-input case in Section 2.6, the behavior of the solution x(t) of (4.7) [or x(k) of (4.8)] as a function of time will depend upon the state transition matrix associated with A_c, as defined in equations (2.9) and (2.10), respectively. To determine this time behavior *completely* requires a knowledge of the Jordan form of A_c. However, we know from Theorems 3.1 and 3.2 that if K can be chosen so that all the eigenvalues of A_c have negative real parts (respectively, moduli less than unity), then (4.7) [respectively, (4.8)] is asymptotically stable. What can in fact be achieved is stated in the following crucial result, which is a generalization of Theorem 2.16.

<u>Theorem 4.1</u> Let the characteristic polynomial of $A_c = A + BK$ be

$$\phi(\lambda) = \det(\lambda I_n - A_c)$$

$$\triangleq \lambda^n + \phi_1 \lambda^{n-1} + \cdots + \phi_n \qquad\qquad (4.10)$$

where $\phi_1, \ldots, \phi_n$ are real numbers. There exists a real matrix K such that the coefficients $\phi_1, \ldots, \phi_n$ are equal to any set of n real numbers, if and only if

$$\text{rank } C(A, B) = n \qquad\qquad (4.11)$$

Proof.

(i) <u>Sufficiency</u>. We prove that if the characteristic polynomial of A_c can be arbitrarily assigned, then (4.11) holds. For suppose that (4.11) is not true, i.e., rank $C(A, B) < n$. By Theorem 2.2 this implies that there is at least one eigenvalue $(\lambda_k$, say) of A such that

$$\text{rank}[\lambda_k I - A, B] < n \qquad\qquad (4.12)$$

It was seen in the first part of the proof of Theorem 2.2 that (4.12) implies the existence of a row n-vector $p \neq 0$ such that

$$pA = \lambda_k p, \quad pB = 0$$

Hence

$$p(A + BK) = \lambda_k p$$

showing that λ_k is an eigenvalue of A_c for *all* m × n matrices K. In other words, λ_k is always a root of the characteristic polynomial of A_c. This contradicts the assumption that this polynomial can be made arbitrary, so we can conclude that no eigenvalue λ_k exists satisfying (4.12). Hence, by Theorem 2.2, the pair (A, B) is c.c., i.e., (4.11) holds.

(ii) <u>Necessity</u>. We prove that if (4.11) holds, then K can be chosen so that the coefficients ϕ_i in (4.10) are arbitrary. To achieve this we need two auxiliary results:

Lemma 4.1.1 If (4.11) holds, then for any column $b \neq 0$ of B, there exist m-vectors, v_1, v_2, $\ldots$, v_{n-1} such that the set $\{z_i\}$ defined by

$$z_1 = b, \quad z_{k+1} = Az_k + Bv_k, \quad k = 1, 2, \ldots, n - 1 \qquad (4.13)$$

is linearly independent.

Proof. For convenience, and without loss of generality, suppose that b is the first column b_1 of B. Then from (4.13)

$$z_2 = Ab_1 + Bv_1 \qquad (4.14)$$

and it is always possible to choose v_1 such that z_2 is independent of b_1. For if not, (4.14) would then imply that $Ab_1 + Bv_1$ is dependent on b_1 for *all* v_1, so in particular, taking $v_1 = 0$ shows that Ab_1 is dependent on b_1. Thus by (4.14) Bv_1 must also be dependent on b_1 for *all* v_1. Taking v_1 to be successive columns of I_m shows that in this case the columns b_2, $\ldots$, b_m of B are dependent on b_1. It similarly follows that Ab_2, $\ldots$, Ab_m are also dependent on b_1, and so are A^2b_1, A^2b_2, $\ldots$, A^2b_m, A^3b_1, $\ldots$, $A^{n-1}b_m$. However, since

$$C(A, B) = [b_1, \ldots, b_m, Ab_1, \ldots, Ab_m, A^2b_1, \ldots, A^2b_m,$$

$$\ldots, A^{n-1}b_m] \qquad (4.15)$$

the preceding argument shows that $C(A, B)$ has rank one, which contradicts the assumption that (4.11) holds.

Having now established that it is possible to make z_2 and z_1 linearly independent, we advance by a similar step. Let L denote the space (of dimension 2) spanned by z_1 and z_2, and consider

$$z_3 = Az_2 + Bv_2 \qquad (4.16)$$

If $z_3 \in L$ for all possible v_2, then setting $v_2 = 0$ in (4.16) shows that $Az_2 \in L$. Hence $Bv_2 \in L$, for all v_2, so as before we can deduce that *all* the columns b_1, $\ldots$, b_m of B are contained in L.

However, from (4.14)

$$Ab_1 = z_2 - Bv_1$$

and since in particular $Bv_1 \in L$, we conclude that $Ab_1 \in L$. This in turn implies that $Ab_2 \in L$, ..., $Ab_m \in L$, and similarly, $A^2b_1 \in L$, $A^2b_2 \in L$, ..., $A^2b_m \in L$, and so on. We have therefore established that rank $C(A, B) = 2$. This contradicts (4.11), so it must be possible to choose v_2 in (4.16) so that z_3, z_2, z_1 are linearly independent. The lemma is then proved by repeating this argument until n linearly independent vectors z_1, z_2, ..., z_n are obtained. ∎

Lemma 4.1.2 If (4.11) holds, and b is the vector in Lemma 4.1.1, then there exists an m × n matrix M such that rank $C(A + BM, b) = n$.

Proof. Define a matrix M by

$$M[z_1, z_2, \ldots, z_n] = [v_1, v_2, \ldots, v_n] \qquad (4.17)$$

where z_1, ..., z_n and v_1, ..., v_{n-1} are the vectors in Lemma 4.1.1, and v_n is arbitrary. Notice that by construction, the matrix postmultiplying M in (4.17) is nonsingular, and that

$$v_k = Mz_k, \quad k = 1, 2, \ldots, n \qquad (4.18)$$

Hence, from (4.13) and (4.18)

$$
\begin{aligned}
z_2 &= Az_1 + BMz_1 = (A + BM)z_1 \\
z_3 &= Az_2 + BMz_2 = (A + BM)^2 z_1 \\
&\vdots \\
z_n &= (A + BM)^{n-1} z_1
\end{aligned}
\qquad (4.19)
$$

Since z_1 (= b), z_2, ..., z_n are linearly independent, (4.19) shows that

$$C(A + BM, b) = [b, (A + BM)b, (A + BM)^2 b, \ldots, (A + BM)^{n-1}b]$$

has rank n, thus establishing the result. ∎

We can now complete the proof of Theorem 4.1. Construct a matrix M as in Lemma 4.1.2, corresponding to any nonzero column b

(the rth, say) of B. Since the pair (A + BM, b) is c.c. by
construction, Theorem 2.16 states that a row n-vector f can be
found such that

$$A + BM + bf \tag{4.20}$$

has the desired characteristic polynomial (4.10). Writing (4.20)
as

$$A + B(M + F)$$

where F is an m × n matrix having all its rows zero except the rth,
which is equal to f, shows that the desired feedback matrix in
(4.9) is K = M + F. Since A, B and the coefficients ϕ_i are real,
so are M and f. ∎

Remarks on Theorem 4.1   (1) The comment on terminology, made in
Remark 1 on Theorem 2.16, applies equally here. Clearly, the
eigenvalues of A cannot be arbitrarily assigned, due to the fact
that A, B, and K, and hence the ϕ_i in (4.10) are real. Hence, any
complex roots of $\phi(\lambda)$ must occur in conjugate pairs.

   (2) The latter restriction is removed by noting that the proof
of the theorem still holds when A, B, K and the coefficients ϕ_i
belong to an arbitrary number field. In particular, if A and B are
real, but the ϕ_i are complex, then K will in general also be
complex.

   (3) Recall that the condition (4.11) is equivalent to the
system (4.1) being c.c., or the system (4.2) being c.r. (Theorems
2.1 and 2.3, respectively). Thus complete controllability
(respectively, reachability) is equivalent to arbitrary assignment
of the closed-loop polynomial. This is an interesting fact, since
it provides a mathematical link between the notions of
controllability/reachability (as defined in physical terms in
Section 2.2) and that of linear feedback.

Example 1   The constructive procedure of the second part of the
proof of Theorem 4.1 is now applied when

$$A = \begin{bmatrix} 0 & 1 & 0 \\ 0 & 0 & 1 \\ 0 & -1 & -2 \end{bmatrix}, \quad B = \begin{bmatrix} 1 & 2 \\ 0 & 1 \\ 1 & 1 \end{bmatrix} \qquad (4.21)$$

Suppose that the closed-loop matrix (4.9) is required to have eigenvalues -2, -3, -4, so that

$$\phi(\lambda) = \lambda^3 + 9\lambda^2 + 26\lambda + 24 \qquad (4.22)$$

The first column of B in (4.21) is nonzero, so we can set

$$z_1 = \begin{bmatrix} 1 \\ 0 \\ 1 \end{bmatrix}$$

and from (4.13)

$$z_2 = Az_1 + Bv_1$$

$$= \begin{bmatrix} 0 \\ 1 \\ -2 \end{bmatrix} + \begin{bmatrix} v_{11} + 2v_{12} \\ v_{12} \\ v_{11} + v_{12} \end{bmatrix}$$

where the elements v_{11} and v_{12} of v_1 can be any numbers such that z_2 is independent of z_1. The choice $v_{11} = 2$, $v_{12} = -1$ gives $z_2 = [0, 0, -1]^T$. Similarly, taking $v_2 = [1, 0]^T$ gives

$$z_3 = Az_2 + Bv_2 = [1, -1, 3]^T$$

Equation (4.17) becomes

$$M \begin{bmatrix} 1 & 0 & 1 \\ 0 & 0 & -1 \\ 1 & -1 & 3 \end{bmatrix} = \begin{bmatrix} 2 & 1 & v_{31} \\ -1 & 0 & v_{32} \end{bmatrix}$$

where v_{31}, v_{32} are arbitrary. Taking $v_{31} = 0$, $v_{32} = 1$ gives

$$M = \begin{bmatrix} 3 & 0 & -1 \\ -1 & -2 & 0 \end{bmatrix}, \quad A + BM = \begin{bmatrix} 1 & -3 & -1 \\ -1 & -2 & 1 \\ 2 & -3 & -3 \end{bmatrix} \qquad (4.23)$$

We can now apply the method of Theorem 2.16 to the pair $(A + BM, b)$ to obtain a vector f such that $(A + BM) + bf$ has the characteristic polynomial (4.22). It is left as an exercise for the reader to obtain $f = [-8, 3, 3]$, whence the desired feedback matrix is

$$K = M + \begin{bmatrix} -8 & 3 & 3 \\ 0 & 0 & 0 \end{bmatrix} = \begin{bmatrix} -5 & 3 & 2 \\ -1 & -2 & 0 \end{bmatrix}$$

where M is given by (4.23). The reader can also check that the closed-loop matrix $A + BK$ does have the required characteristic polynomial (4.22).

(4) Unlike the single-input problem, the solution for K is *not* in general unique — for example, a different choice for v_{31} and v_{32} in Example 1 would produce a different feedback matrix (see Problem 4.1). This freedom can be exploited to give some choice in the selection of the eigenvectors of $A + BK$.

(5) An ingenious idea for testing whether the pair (A, B) is c.c. is to proceed as follows: Compute the eigenvalues of A, and of $A + BK$, where K is an arbitrary n × m matrix (with elements obtained by using a pseudo-random number generator, say). Then it follows from Theorem 4.1 that the pair (A, B) is c.c. if and only if A and $A + BK$ have no eigenvalues in common, for such eigenvalues are unaffected by linear feedback. If there is any doubt, due to limitations of accuracy, as to whether two eigenvalues are indeed coincident, then a different K can be selected and the procedure repeated. Advantage can be taken of powerful standard programs for computing the eigenvalues, and the method has been found efficient (as compared with the rank tests of Theorems 2.1 and 2.2) for systems having n as large as 100.

There are various other methods of computing suitable matrices K. We now quote one result which relies on a knowledge of the eigenvalues $\lambda_1, \ldots, \lambda_n$ of A, assumed to be distinct, together with associated eigenvectors $w_1, \ldots, w_n$.

Theorem 4.2 Suppose that the eigenvalues of A are ordered so that those of A_c are to be $\mu_1, \ldots, \mu_p, \lambda_{p+1}, \ldots, \lambda_n$ $(p \leqslant n)$. Then K in Theorem 4.1 is given by

$$K = -hg\hat{W} \tag{4.24}$$

where $W = [w_1, w_2, \ldots, w_n]$, $\hat{W}$ consists of the first p rows of W^{-1},

$$g \triangleq \left[\frac{\alpha_1}{\beta_1}, \frac{\alpha_2}{\beta_2}, \ldots, \frac{\alpha_p}{\beta_p} \right] \tag{4.25}$$

$$\alpha_i = \frac{\displaystyle\prod_{j=1}^{p} (\lambda_i - \mu_j)}{\displaystyle\prod_{\substack{j=1 \\ j \neq i}}^{p} (\lambda_i - \lambda_j)}, \quad p > 1 \tag{4.26}$$

$$\alpha_1 = \lambda_1 - \mu_1, \quad p = 1 \tag{4.27}$$

$$\beta = [\beta_1, \beta_2, \ldots, \beta_p]^T \triangleq \hat{W}Bh \tag{4.28}$$

and h is any column m-vector such that all $\beta_i \neq 0$.

Proof. This consists of direct verification, and will be omitted. The condition (4.11) ensures that $\hat{W}B$ has no zero row, so that a suitable vector h in (4.28) always exists. It should be noted that (i) the rows of $\hat{W}$ are left eigenvectors of A corresponding to λ_1, ..., λ_p, and (ii) the relative simplicity of the expression for K is due to the fact that the formula (4.24) produces a matrix having unit rank. ∎

If the pair (A, B) is not c.c., then a more limited objective would be to apply linear feedback such that the resulting closed-loop matrix A_c is asymptotically stable. If this is possible, the system (4.1) (or (4.2)) is called *stabilizable*. The condition for stabilizability is made clear by the next result, for whose appreciation the reader must recall the following definition, given in (2.53): the system S described by (4.1) and (4.3) is *algebraically equivalent* to

$$\frac{d\hat{x}}{dt} = \hat{A}\hat{x} + \hat{B}u, \quad \hat{y} = \hat{H}\hat{x}$$

where

$$\hat{x} = Px \tag{4.29}$$

P is an arbitrary n × n nonsingular matrix, and

$$\hat{A} = PAP^{-1}, \quad \hat{B} = PB, \quad \hat{H} = HP^{-1} \tag{4.30}$$

Theorem 4.3   If rank $C(A, B) = \nu < n$, then there exists a system algebraically equivalent to S having the form

$$\frac{d}{dt}\begin{bmatrix} x^{(1)} \\ x^{(2)} \end{bmatrix} = \begin{bmatrix} A_1 & A_2 \\ 0 & A_3 \end{bmatrix}\begin{bmatrix} x^{(1)} \\ x^{(2)} \end{bmatrix} + \begin{bmatrix} B_1 \\ 0 \end{bmatrix}u \tag{4.31}$$

$$y = [H_1, \ H_2]\begin{bmatrix} x^{(1)} \\ x^{(2)} \end{bmatrix} \tag{4.32}$$

where $x^{(1)}$, $x^{(2)}$ have dimensions ν and $n - \nu$, respectively, and rank $C(A_1, B_1) = \nu$.

Proof. Let $c_1, c_2, \ldots, c_\nu$ be any set of ν linearly independent columns of $C(A, B)$, and consider (4.29) with P defined by

$$P^{-1} = [c_1, c_2, \ldots, c_\nu, c_{\nu+1}, \ldots, c_n] \tag{4.33}$$

where the remaining columns in (4.33) are any set of vectors which make P^{-1} nonsingular. Premultiplying both sides of (4.33) by P shows Pc_j is equal to the jth column e_j of I_n. Any column of $C(A, B)$ can be expressed as a linear combination of the basis c_1, $\ldots, c_\nu$, so in particular it follows that the vectors $Ac_1, \ldots, Ac_\nu$ can be expressed in terms of this basis, i.e.,

$$Ac_i = \sum_{j=1}^{\nu} \alpha_{ij}c_j, \quad i = 1, 2, \ldots, \nu \tag{4.34}$$

From (4.30) and (4.33) we have

$$\hat{A} = P[Ac_1, Ac_2, \ldots, Ac_n] \tag{4.35}$$

and by (4.34) we see that

$$PAc_i = \sum_{j=1}^{\nu} \alpha_{ij} Pc_j$$

$$= \sum_{j=1}^{\nu} \alpha_{ij} e_j$$

In other words, the first ν columns on the right in (4.35) take the form

$$[PAc_1, \ldots, PAc_\nu] = \begin{bmatrix} A_1 \\ 0 \end{bmatrix}$$

where A_1 is $\nu \times \nu$. Similarly, since $c_1, \ldots, c_\nu$ also form a basis for the columns of B, (4.30) gives

$$\hat{B} = PB = \begin{bmatrix} B_1 \\ 0 \end{bmatrix}$$

where B_1 is $\nu \times m$. We have thus established the form in (4.31). The matrices H_1 and H_2 in (4.32) have no special properties. To show that the pair (A_1, B_1) is c.c., we use the fact in (2.54) that

$$C(\hat{A}, \hat{B}) = PC(A, B)$$

which implies, by the nonsingularity of P, that $C(\hat{A}, \hat{B})$ has rank ν. However,

$$C(\hat{A}, \hat{B}) = \begin{bmatrix} B_1 & A_1 B_1 & A_1^2 B_1 & \cdots & A_1^{n-1} B_1 \\ 0 & 0 & 0 & & 0 \end{bmatrix}$$

$$= \begin{bmatrix} C(A_1, B_1) \\ 0 \end{bmatrix}$$

so we can deduce that rank $C(A_1, B_1) = \nu$. ∎

Remarks on Theorem 4.3 (1) Although stated in terms of the continuous-time system, the preceding result has exactly the same form for the discrete-time system described by (4.2) and (4.4).

(2) If we apply linear feedback

$$u = [K_1, K_2] \begin{bmatrix} x^{(1)} \\ x^{(2)} \end{bmatrix}$$

to (4.31), we obtain the closed-loop matrix

$$\hat{A}_c = \begin{bmatrix} A_1 & A_2 \\ 0 & A_3 \end{bmatrix} + \begin{bmatrix} B_1 \\ 0 \end{bmatrix} [K_1, K_2]$$

$$= \begin{bmatrix} A_1 + BK_1 & A_2 + B_1K_2 \\ 0 & A_3 \end{bmatrix} \tag{4.36}$$

Because the pair (A_1, B_1) is c.c., then by Theorem 4.1 we can choose K_1 so that $A_1 + B_1K_1$ has an arbitrary characteristic polynomial. Thus the eigenvalues of $\hat{A}_c$ will be those of $A_1 + B_1K_1$, together with those of A_3. The eigenvalues of the latter are unaffected by feedback. Indeed, from (4.31) we have

$$\frac{dx^{(2)}}{dt} = A_3 x^{(2)}$$

so the part of the state space described by $x^{(2)}$ cannot be influenced by the control variables, and hence is called *uncontrollable*. An immediate consequence is:

Corollary 4.3.1 The pair (A, B) is stabilizable if and only if the matrix A_3 in (4.31) is a stability matrix (for discrete-time systems, a convergent matrix).

Proof. We only need to point out that the pair $(\hat{A}, \hat{B})$ is stabilizable if and only if (A, B) is stabilizable, where P is *any* nonsingular matrix in (4.30). This is because (4.30) implies that

$$\hat{A} + \hat{B}\hat{K} = P(A + BK)P^{-1}$$

where $K = \hat{K}P$. Thus if there exists a $\hat{K}$ which stabilizes $(\hat{A}, \hat{B})$, then K stabilizes (A, B), and conversely. Taking the inverse of (4.33) as a particular choice for P establishes that we can decide stabilizability from $\hat{A}_c$ in (4.36). ∎

Example 2   We illustrate the constructive nature of the proof of Theorem 4.3 for the pair

$$A = \begin{bmatrix} 0 & 1 & 0 \\ -1 & 0 & 0 \\ 0 & 1 & 2 \end{bmatrix}, \quad B = \begin{bmatrix} 0 \\ -5 \\ 2 \end{bmatrix} \tag{4.37}$$

The controllability matrix is

$$C(A, B) = \begin{bmatrix} 0 & -5 & 0 \\ -5 & 0 & 5 \\ 2 & -1 & -2 \end{bmatrix}$$

which has rank 2. We take the first two columns of $C(A, B)$ to be c_1 and c_2, and write down from (4.33)

$$P^{-1} = \begin{bmatrix} 0 & -5 & 0 \\ -5 & 0 & 0 \\ 2 & -1 & 1 \end{bmatrix} \tag{4.38}$$

where a simple choice has been made for the third column in (4.38). It is easy to compute

$$P = \frac{1}{5} \begin{bmatrix} 0 & -1 & 0 \\ -1 & 0 & 0 \\ -1 & 2 & 5 \end{bmatrix}$$

whence from (4.30) we obtain

$$\hat{A} = \begin{bmatrix} 0 & -1 & 0 \\ 1 & 0 & 0 \\ 0 & 0 & 2 \end{bmatrix}, \quad \hat{B} = \begin{bmatrix} 1 \\ 0 \\ 0 \end{bmatrix}$$

Since $A_3 = 2$, we deduce that (4.37) is not a stabilizable pair, whether relating to a continuous- or discrete-time system.

It is worth noting that the test for controllability described
in Remark 5 in Theorem 4.1 is easily modified for stabilizability:
For a randomly selected matrix K, the pair (A, B) is stabilizable
if and only if any common eigenvalues of A and A + BK have negative
real parts (for discrete systems, modulus less than one).

4.1.2 Canonical Forms

We now conaider a generalization of the canonical forms of Section
2.4, to the multivariable case. This will enable us to present a
precise result concerning the structure which can be given to A_c in
(4.9) by linear feedback, in addition to assignment of eigenvalues.
Consider

$$C(A, B) = [B, AB, \ldots, A^{n-1}B] \qquad (4.39)$$

which by assumption has rank n. We can therefore select n linearly
independent columns in (4.39) to provide a transformation matrix P
in (4.29), which will express the system relative to a new set of
coordinates. As we shall see, there are several ways of doing this
in order that the resulting state matrix $\hat{A}$ in (4.30) has a
relatively simple form, but unlike the single-input case there is
no unique controllable canonical form.

It is somewhat more convenient to work with $T = P^{-1}$, so that
(4.30) is replaced by

$$\hat{A} = T^{-1}AT, \qquad \hat{B} = T^{-1}B \qquad (4.40)$$

One way of constructing T is to take

$$T = [b_1, Ab_1, A^2b_1, \ldots, A^{r_1-1}b_1, b_2, Ab_2, \ldots, A^{r_2-1}b_2, \ldots] \qquad (4.41)$$

where as before b_i denotes the ith column of B. In (4.41), r_i is
the smallest integer such that $A^{r_i}b_i$ is linearly dependent on all
the preceding columns, the process being continued until n linearly
independent columns of $C(A, B)$ have been selected. Notice that in
(4.41), it is merely for convenience that the suffices on the b's
have been assumed to be consecutive integers. It is then

straightforward to verify (see the subsequent discussion on
Example 3) that the matrices of the transformed system, as defined
by (4.40), take the following form. We write

$$\hat{A} = [A_{ij}], \quad \hat{B} = [\hat{b}_1, \hat{b}_2, \ldots, \hat{b}_m] \tag{4.42}$$

where A_{ii} is $r_i \times r_i$, A_{ij} is $r_i \times r_j$, for i, j = 1, 2, ..., μ, and
the last column in (4.41) involves b_μ. The diagonal blocks of $\hat{A}$
have companion form, and the off-diagonal blocks have at most only
their last column zero, i.e.,

$$A_{ii} = \begin{bmatrix} 0 & 0 & & x \\ 1 & 0 & 0 & x \\ & 1 \ddots & & \vdots \\ 0 & & 1 & x \end{bmatrix}, \quad A_{ij} = \begin{bmatrix} 0 & \cdots & 0 & x \\ & \cdots & 0 & x \\ & \cdots\cdots & & \\ 0 & \cdots & 0 & x \end{bmatrix}, \quad i < j \tag{4.43}$$

$$= 0, \quad i > j$$

where the x's denote (possibly) nonzero elements. Only the first μ
columns of $\hat{B}$ in (4.42) have a special form, given by

$$\hat{b}_i = [0, 0, \ldots\ldots\ldots\ldots 0, 1, 0, \ldots, 0]^T, \quad i = 1, \ldots, \mu$$
$$\overset{\longleftarrow}{}(r_1 + \cdots + r_{i-1})\overset{\longrightarrow}{} \tag{4.44}$$

where the arrows indicate the number of initial zeros, and $r_{-1} \overset{\Delta}{=} 0$,
so that b_1 has its first element unity. The expressions (4.42) to
(4.44) define a first type of controllable canonical form. It
should be noted that for a given ordering of the columns of B, the
integers r_i and the elements denoted by x's in (4.43), together
with the remaining elements of $\hat{B}$, are the same for all pairs
algebraically equivalent to (A, B). However, if the order of the
inputs is altered (equivalent to permuting the columns of B), then
in general the expressions for $\hat{A}$ and $\hat{B}$ (including the numbers r_i)
will also be changed. This disadvantage does not arise in our
second scheme for constructing a matrix T, now described.

Consider again C(A, B) in (4.39). Take linearly independent

columns from left to right in the order in which they occur, i.e.,
in the sequence

$$b_1, b_2, \ldots, b_m, Ab_1, Ab_2, \ldots, Ab_m, A^2b_1, \ldots \qquad (4.45)$$

At any stage in building up (4.45), reject A^ib_j ($i \geqslant 0$, $j \geqslant 1$) if
it is linearly dependent upon the previously selected members of
the basis. Notice that A^sb_j will also be deleted for $s > i$ (see
Problem 4.8). Also, we assume for convenience that the suffices on
the b's in (4.45) are consecutive integers. Indeed, if rank $B = m$,
then all the columns of B will be included in (4.45). *Reorder* the
vectors in (4.45) to produce

$$T = [b_1, Ab_1, \ldots, A^{k_1-1}b_1, b_2, Ab_2, \ldots, A^{k_2-1}b_2,$$

$$\ldots, b_r, \ldots, A^{k_s-1}b_s] \qquad (4.46)$$

It can be shown that the set of integers $k_1, \ldots, k_s$ is *uniquely*
determined by A and B, i.e., the k_i are invariant under transformations of the form (4.40), and are termed the *controllability
indices* of the pair (A, B). In particular, these indices are
unaltered by a reordering of the inputs. It is therefore always
possible to relabel the columns of B (if necessary) so that
$k_1 \geqslant k_2 \geqslant \ldots \geqslant k_s$.

Using T in (4.46), the canonical matrices $\hat{A}$, $\hat{B}$ defined by
(4.42) are such that A_{ii} has the same form as in (4.43), with
dimensions $k_i \times k_i$, A_{ij} also takes the same form as in (4.43) but
for *all* $i \neq j$, and the columns of $\hat{B}$ have the same general form as
(4.44), i.e.,

$$\hat{b}_i = [0, 0, \ldots\ldots\ldots\ldots\ldots, 0, 1, 0, \ldots, 0]^T,$$

$$\longleftarrow (k_1 + \cdots + k_{i-1}) \longrightarrow \quad i = 1, 2, \ldots, s \qquad (4.47)$$

Again, the unspecified elements of $\hat{A}$ and $\hat{B}$ are invariant under
algebraic equivalence.

Example 3   Suppose that $n = 5$, $m = 2$ and that we compute

$$
C(A, B) = \begin{bmatrix}
0 & 1 & 0 & 2 & 4 & 5 & 0 & . \\
0 & 0 & 0 & 0 & 1 & 0 & 1 & . \\
1 & 0 & 3 & -1 & 0 & -2 & -7 & . \\
0 & 1 & -2 & 0 & 3 & -3 & 7 & . \\
0 & 0 & 0 & 1 & 2 & 4 & 1 & .
\end{bmatrix} \qquad (4.48)
$$

$$
\quad b_1 \quad b_2 \quad Ab_1 \quad Ab_2 \quad A^2b_1 \quad A^2b_2 \quad A^3b_1
$$

Using the second scheme, we obtain as the columns of T the vectors

$$
b_1, \ Ab_1, \ A^2b_1, \ b_2, \ Ab_2 \qquad\qquad (4.49)
$$

showing that the controllability indices are $k_1 = 3$, $k_2 = 2$.
Furthermore, we can express subsequent columns of C in terms of the
basis (4.49) to give

$$
\begin{aligned}
A^2b_2 &= 2b_1 - 3b_2 + 4Ab_2 \\
A^3b_1 &= b_1 - 3Ab_1 + A^2b_1 - 2b_2 - Ab_2
\end{aligned} \qquad (4.50)
$$

There is no need to compute T^{-1} and use (4.40) in order to
determine the matrices $\hat{A}$ and $\hat{B}$ in the canonical form. To see this,
write

$$
\hat{A} = T^{-1}AT
$$

$$
= \begin{bmatrix}
\rho_1 \\
\rho_2 \\
\vdots \\
\rho_n
\end{bmatrix} [At_1, \ At_2, \ \ldots, \ At_n]
$$

$$
= \begin{bmatrix}
\rho_1(At_1) & \rho_1(At_2) & \cdots & \rho_1(At_n) \\
\vdots & \vdots & & \vdots \\
\rho_n(At_1) & \rho_n(At_2) & \cdots & \rho_n(At_n)
\end{bmatrix} \qquad (4.51)
$$

where T^{-1} has rows $\rho_1, \ \ldots, \ \rho_n$ and T has columns $t_1, \ \ldots, \ t_n$.

Express At_i in terms of the basis $t_1, \ldots, t_n$, i.e.,

$$At_i = \alpha_{i1}t_1 + \alpha_{i2}t_2 + \cdots + \alpha_{in}t_n \tag{4.52}$$

Then premultiplying (4.52) by ρ_j gives

$$\rho_j(At_i) = \alpha_{ij} \tag{4.53}$$

using the fact that $T^{-1}T \equiv I_n$, so $\rho_j t_i$ is equal to the Kronecker delta. The expressions (4.51) and (4.53) show that the ith *column* of $\hat{A}$ consists of the coefficients $\alpha_{i1}, \ldots, \alpha_{in}$ defined in (4.52), for $i = 1, \ldots, n$. In this example, the t's are given in (4.49), so we express Ab_1, A^2b_1, A^3b_1, Ab_2, A^2b_2 in terms of this basis. Using (4.50), we can write down without further effort

$$\hat{A} = \left[\begin{array}{ccc:cc} 0 & 0 & 1 & 0 & 2 \\ 1 & 0 & -3 & 0 & 0 \\ 0 & 1 & 1 & 0 & 0 \\ \hdashline 0 & 0 & -2 & 0 & -3 \\ 0 & 0 & -1 & 1 & 4 \end{array}\right] \tag{4.54}$$

Similarly, $\hat{B} = T^{-1}B$ implies that

$$\hat{b}_i = \begin{bmatrix} \rho_1 b_i \\ \vdots \\ \rho_n b_i \end{bmatrix}, \qquad i = 1, \ldots, m$$

so the ith column $\hat{b}_i$ of $\hat{B}$ is equal to the coefficients of b_i relative to the basis $t_1, \ldots, t_n$. In this example, since b_1 and b_2 are both in the basis, we obtain simply

$$\hat{B} = \left[\begin{array}{cc} 1 & 0 \\ 0 & 0 \\ 0 & 0 \\ \hline 0 & 1 \\ 0 & 0 \end{array}\right]$$

The relevance of the controllability indices for the linear state feedback problem lies in the following fundamental result, which we state without proof.

<u>Theorem 4.4</u> Let the pair (A, B) be c.c., with controllability indices $k_1 \geqslant k_2 \geqslant \cdots \geqslant k_s > 0$. Let $\psi_1(\lambda)$, $\psi_2(\lambda)$, ..., $\psi_q(\lambda)$ be any monic polynomials, over the same number field as A and B, satisfying the following conditions:

(i) $\psi_{i+1}(\lambda)$ divides $\psi_i(\lambda)$, $1 \leqslant i \leqslant q - 1 \leqslant m - 1$

(ii) $\sum\limits_{i=1}^{q} \delta\psi_i = n$

(iii) $\sum\limits_{i=1}^{p} \delta\psi_i \geqslant \sum\limits_{i=1}^{p} k_i$, for $1 \leqslant p \leqslant q$

Then there exists a matrix K such that $\lambda I - A_c = \lambda I - A - BK$ has $1, 1, \ldots, 1, \psi_q(\lambda), \ldots, \psi_1(\lambda)$ as its invariant factors.

<u>Remarks on Theorem 4.4</u> (1) Those unfamiliar with the notion of invariant factors of a polynomial matrix will find a definition in Theorem 4.17. Specifying the invariant polynomials of $\lambda I - A_c$ is equivalent to specifying the Jordan form of A_c. Thus only those Jordan forms can be achieved which correspond to invariant factors satisfying conditions (i) to (iii). Note that the characteristic polynomial of A_c is equal to the product $\psi_1(\lambda)\psi_2(\lambda) \cdots \psi_q(\lambda)$.

(2) In particular, if we take $\psi_1(\lambda)$ to have degree n, and $\psi_i(\lambda) = 1$, $i = 2, \ldots, q$, then conditions (i) to (iii) will always be satisfied, irrespective of the values of the controllability indices. A matrix having these invariant factors $1, 1, \ldots, 1, \psi_1(\lambda)$ is nonderogatory. In other words, for *any* c.c. system there always exists a feedback matrix K such that the closed-loop matrix is nonderogatory. This generalizes Remark 3 on Theorem 2.16 for the single-input problem.

(3) It can also be shown that the controllability indices are invariant under state feedback $u = Kx$.' That is, the indices for the pair $(A + BK, B)$ are the same as those for (A, B).

Example 4   Return to the problem in Example 3, for which n = 5,
m = 2, k_1 = 3, k_2 = 2.   Theorem 4.4 states that we can make the
invariant factors of $\lambda I - A_c$ equal to 1, 1, 1, $\psi_2(\lambda)$, $\psi_1(\lambda)$, where
$\psi_2(\lambda)$ divides $\psi_1(\lambda)$, $\delta\psi_1 + \delta\psi_2$ = 5, and $\delta\psi_1 \geqslant 3$.   Thus suppose, for
example, that we wish to make all the eigenvalues of A_c equal to
-2, i.e., the characteristic polynomial of A_c is to be $(\lambda + 2)^5$.
The only possible choices for $\psi_2(\lambda)$ and $\psi_1(\lambda)$ are

$$\psi_2(\lambda) = (\lambda + 2)^i, \quad \psi_1(\lambda) = (\lambda + 2)^{5-i}, \quad i = 0, 1, 2$$

Thus the Jordan form of A_c must be

$$\text{diag}[J_i, J_{5-i}], \quad i = 0, 1, 2$$

where J_i is an i × i Jordan block having diagonal elements equal to
-2.

4.1.3 Optimal Quadratic Regulator
As we mentioned in the introduction to this chapter, for regulator
problems we regard x(t) in (4.1) [or x(k) in (4.2)] as being
nominally zero. The system is disturbed by some amount

$$x(0) = x_0 \tag{4.55}$$

and we wish to apply control so that x(t) is returned to zero in a
"satisfactory" manner. Consider first the continuous-time case.
We can make x(t) decay to zero as quickly as we like by applying
linear feedback such that the real parts of the closed-loop
eigenvalues are sufficiently large and negative. However,
increasing the magnitudes of these negative real parts will require
a corresponding increase in the amount of control energy which must
be supplied to the system. The problem must therefore be expressed
so that there is a trade-off between the rate at which x(t) decays
to zero, and the amount of control energy involved. For the optimal
quadratic regulator formulation, we choose u(t) so as to minimize
the *quadratic performance index*

$$J = \int_0^\infty [x^T(t)Qx(t) + u^T(t)Ru(t)] \; dt \qquad (4.56)$$

The quadratic forms in the integrand in (4.56) can be regarded as
"generalized sums of squares" of the components of x and u, with Q
being real, symmetric, and positive semidefinite, and R real,
symmetric, and positive definite. The elements of Q and R are
weighting factors, whose values will depend upon physical
considerations. The infinite upper limit in (4.56) signifies that
we are interested in the steady-state control which makes $x(t) \to 0$
as $t \to \infty$.

Using the Euler-Lagrange equations in the calculus of
variations, it can be shown that the solution to the problem of
minimizing this *quadratic* performance index, subject to the *linear*
state equations (4.1) and (4.55), is precisely *linear* feedback of
the form (4.6):

Theorem 4.5   The control that minimizes (4.56) subject to (4.1) and
(4.55) is

$$u(t) = -R^{-1}B^TPx(t) \qquad (4.57)$$

where, provided that the pair (A, B) is c.c., and the pair (A, Q)
is c.o., then the real symmetric matrix P is the unique positive
definite solution of the quadratic matrix equation

$$PBR^{-1}B^TP - A^TP - PA - Q = 0 \qquad (4.58)$$

The importance of the expression (4.57) lies in the fact that
it provides a stabilizing control, even if the open-loop system is
unstable:

Theorem 4.6   The closed-loop system matrix obtained by applying
(4.57) to (4.1), namely

$$A_{co} = A - BR^{-1}B^TP \qquad (4.59)$$

is asymptotically stable.

Proof. It is easy to rearrange (4.58) into the equation

$$-(A - BR^{-1}B^TP)^TP - P(A - BR^{-1}B^TP) = Q + PBR^{-1}B^TP$$

that is,

$$A_{co}^T(-P) + (-P)A_{co} = Q + PBR^{-1}B^TP \qquad (4.60)$$

Since R is positive definite and B is m × n, with m < n, it follows that $(PB)R^{-1}(PB)^T$ is positive semidefinite, so the right-hand side of (4.60) is positive semidefinite. We wish to apply Theorem 3.22, and therefore need to show that the pair $(A_{co}^T, Q + PBR^{-1}B^TP)$ is c.c. By assumption, the pair (A^T, Q) is c.c., and

$$(A_{co}^T, Q + PBR^{-1}B^TP) \equiv (A^T - PC, Q + PCP)$$

where $C = BR^{-1}B^T$, so an immediate application of the result of Problem 2.18(ii) (with A, B in that result replaced by A^T, Q respectively) establishes the required fact. We can then deduce from Theorem 3.22 that A_{co} is a stability matrix, since (-P) is negative definite. ∎

Remarks on Theorems 4.5 and 4.6 (1) The fact that the conditions of Theorem 4.5 are sufficient to ensure the existence of a solution of (4.58) will be referred to again (Remark 1 on Theorem 4.7). However, if the existence of a positive definite symmetric solution of (4.58) is *assumed*, then it is easy to establish uniqueness. For suppose P_1, P_2 are two such solutions of (4.58). Substituting these into (4.58), and subtracting the resulting equations gives

$$(P_1 - P_2)A_1 + A_2^T(P_1 - P_2) = 0 \qquad (4.61)$$

where

$$A_1 = A - BR^{-1}B^TP_1, \qquad A_2 = A - BR^{-1}B^TP_2$$

By Theorem 4.6, both A_1 and A_2 are stability matrices. It therefore follows that there are no eigenvalues λ_{i1}, λ_{j2} of A_1 and

A_2, respectively, such that $\lambda_{i1} + \lambda_{j2} = 0$. It then follows (see Problem 3.31) that (4.61) has a unique solution $(P_1 - P_2) = 0$, i.e., $P_1 \equiv P_2$.

(2) The equation (4.58) is known in the literature as the *algebraic matrix Riccati equation*. This is because it is a special case of a matrix differential equation obtained when the upper limit on the integral in (4.55) is a fixed finite time (see Remark 6 below). This equation is itself a generalization of a scalar differential equation studied by Riccati in the early eighteenth century.

(3) Notice that if Q is positive definite, then so is the right-hand side of (4.60), and Theorem 4.6 then follows from the Liapunov Theorem 3.19. More generally, the proof of Theorem 4.6 shows that provided (A, Q) is c.o., then from Theorem 3.22 we can deduce that the solution P of (4.60) is such that P and A_{co} are nonsingular, and $In(A_{co}) = In(-P)$. In other words, provided that solutions P of (4.58) exist, then they are nonsingular, and satisfy this inertia condition (see also Problem 4.18).

(4) It is easy to incorporate output feedback u = Ky into this framework. We simply replace the quadratic term in the state in (4.56) by

$$y^T(t)Qy(t) = x^T(t)H^TQHx(t)$$

where H is the r × n matrix in (4.3). Note that when r < n, the matrix H^TQH will be only positive semidefinite even if Q is positive definite. The only change in Theorems 4.5 and 4.6 is that the pair (A, H) is to be c.o., instead of (A, Q) and (4.58) is replaced by

$$PBR^{-1}B^TP - A^TP - PA - H^TQH = 0 \qquad\qquad (4.62)$$

(5) Theorem 4.5 requires the pair (A, B) to be c.c., whereas we know from Section 4.1.2 that, by definition, a stabilizing feedback control exists provided that the pair (A, B) is stabilizable. The latter condition can indeed be shown to be sufficient for

Theorem 4.5 to still hold. In fact, a more general result can also be proved: Provided that the pairs (A, B) and (A^T, H^T) are each stabilizable [(A, H) is "detectable", see Section 4.1.4] then there is a unique positive semidefinite solution of (4.62) which makes A_{co} in (4.59) a stability matrix.

(6) It can be mentioned that if the upper limit on the performance index (4.56) is a fixed *finite* time τ, then the optimal control (4.57) remains linear, but P becomes the solution of the so-called matrix Riccati differential equation

$$\frac{dP(t)}{dt} = P(t)BR^{-1}B^T P(t) - A^T P(t) - P(t)A - Q, \quad P(\tau) = 0$$

Thus the feedback is time varying, and further discussion lies outside the scope of this book.

An important method for obtaining solutions of the quadratic matrix equations (4.58) and (4.62) is obtained by exploring the connection with the eigenvalue-eigenvector problem for the matrix

$$M = \begin{bmatrix} A & -BR^{-1}B^T \\ -H^T QH & -A^T \end{bmatrix} \tag{4.63}$$

Theorem 4.7 For the matrix M in (4.63) let

$$Z = \begin{bmatrix} \overset{n}{Z_1} & \overset{n}{Z_2} \\ Z_3 & Z_4 \end{bmatrix} \begin{matrix} n \\ n \end{matrix} \tag{4.64}$$

be any matrix which transforms it into (upper) Jordan form, i.e., $Z^{-1}MZ = J$. Then provided that Z_1 is nonsingular, $P = Z_3 Z_1^{-1}$ is a solution of (4.62).

Proof. Write

$$J = \begin{bmatrix} \beta_1 & \beta_2 \\ 0 & \beta_3 \end{bmatrix}$$

and equate blocks on both sides of the identity $MZ = ZJ$ to obtain

$$AZ_1 - BR^{-1}B^T Z_3 = Z_1 \beta_1 \qquad (4.65)$$

$$-H^T QHZ_1 - A^T Z_3 = Z_3 \beta_1 \qquad (4.66)$$

Since by assumption Z_1 is nonsingular, multiply both sides of
(4.65) on the left by $Z_3 Z_1^{-1}$, and on the right by Z_1^{-1}, and (4.66) on
the right by Z_1^{-1}. Subtracting the resulting expressions gives

$$Z_3 Z_1^{-1} A - Z_3 Z_1^{-1} BR^{-1}B^T Z_3 Z_1^{-1} + H^T QH + A^T Z_3 Z_1^{-1} = 0$$

which shows that $P = Z_3 Z_1^{-1}$ satisfies (4.62). ∎

Remarks on Theorem 4.7   (1) It can be shown that provided (A, B) is
c.c. and (A, H) is c.o., then M in (4.63) has no purely imaginary
eigenvalues. This ensures that there exist nonsingular matrices
Z_1, showing that under these conditions the equation (4.62) [and as
a special case (4.58)] does possess solutions.

(2) These solutions can be obtained from the first n columns
of any transformation matrix (4.64). For simplicity, we assume
from now on that all the eigenvalues of M in (4.63) are distinct,
so that the Jordan form of M reduces to

$$J = \mathrm{diag}[\theta_1, \theta_2, \ldots, \theta_{2n}] \qquad (4.67)$$

The matrix Z in (4.64) then consists of the eigenvectors of M,
which by assumption are linearly dependent. We wish to select the
positive definite solution of (4.62) [or (4.58)] which makes A_{co}
in (4.59) a stability matrix. The way in which this can be done is
made apparent by:

Theorem 4.8   If the first n columns of Z in (4.64) correspond to
eigenvalues $\theta_1, \ldots, \theta_n$ of M, then these are also eigenvalues of
the closed-loop matrix $A_{co} = A - BR^{-1}B^T Z_3 Z_1^{-1}$.

Proof. Multiplying (4.65) on the right by Z_1^{-1} gives

$$A - BR^{-1}B^T Z_3 Z_1^{-1} = Z_1 \beta_1 Z_1^{-1} \tag{4.68}$$

Thus A_{co} is similar to β_1, which has eigenvalues $\theta_1, \ldots, \theta_n$. ∎

Remarks on Theorem 4.8   (1) Equation (4.68) shows that the columns of Z_1 are eigenvectors of A_{co}.

(2) It can be shown (Problem 4.12) that if θ_i is an eigenvalue of M in (4.63), then so is $-\theta_i$. We have noted that M has no purely imaginary eigenvalues, so it therefore has precisely n eigenvalues with negative real parts. We can therefore obtain the optimal stabilizing feedback from $P = Z_3 Z_1^{-1}$ by taking the n eigenvectors of M which correspond to these left-half-plane eigenvalues. We note that it is not difficult to show that P is unaffected by the order in which the eigenvalues are selected.

(3) If the Jordan form of M is replaced by the upper triangular Schur canonical form, then a unitary transformation matrix Z exists. Theorems 4.7 and 4.8 still apply, and this modification has computational advantages.

Example 5   Consider the system described by

$$\dot{x}_1 = -x_2 + u_1, \qquad \dot{x}_2 = u_2$$

$$y_1 = 2x_1 + x_2$$

where we wish to determine u so as to minimize

$$\int_0^\infty (y_1^2 + u_1^2 + u_2^2) \, dt$$

Clearly,

$$A = \begin{bmatrix} 0 & -1 \\ 0 & 0 \end{bmatrix}, \qquad B = \begin{bmatrix} 1 & 0 \\ 0 & 1 \end{bmatrix}, \qquad H = [2,\ 1]$$

$$Q = [1], \qquad R = \begin{bmatrix} 1 & 0 \\ 0 & 1 \end{bmatrix}$$

and it is easy to check that the pair (A, B) is c.c. and the pair (A, H) is c.o. From (4.63)

$$M = \begin{bmatrix} 0 & -1 & -1 & 0 \\ 0 & 0 & 0 & -1 \\ -4 & -2 & 0 & 0 \\ -2 & -1 & 1 & 0 \end{bmatrix}$$

and the eigenvalues of M are -2, -1, +1, +2. Since the Jordan form of M is diagonal, the columns of Z are the eigenvectors of M. To determine Z_1 and Z_3 in (4.64) we compute eigenvectors corresponding to -2 and -1, respectively, $[1, 0, 2, 0]^T$, $[1, -1, 2, -1]^T$, and Z_1, Z_3 are equal to the upper and lower parts of these column vectors, i.e.,

$$Z_1 = \begin{bmatrix} 1 & 1 \\ 0 & -1 \end{bmatrix}, \quad Z_3 = \begin{bmatrix} 2 & 2 \\ 0 & -1 \end{bmatrix}$$

Hence from Theorems 4.7 and 4.8, the desired solution of the Riccati equation is

$$P = Z_3 Z_1^{-1} = \begin{bmatrix} 2 & 0 \\ 0 & 1 \end{bmatrix}$$

and from (4.57) the optimal feedback control is

$$u = -\begin{bmatrix} 1 & 0 \\ 0 & 1 \end{bmatrix}\begin{bmatrix} 1 & 0 \\ 0 & 1 \end{bmatrix}\begin{bmatrix} 2 & 0 \\ 0 & 1 \end{bmatrix} x$$

$$= -2x_1 - x_2$$

The closed-loop matrix in (4.59) is

$$A_{co} = A - \begin{bmatrix} 1 & 0 \\ 0 & 1 \end{bmatrix}\begin{bmatrix} 2 & 0 \\ 0 & 1 \end{bmatrix}$$

$$= \begin{bmatrix} -2 & -1 \\ 0 & -1 \end{bmatrix}$$

which also has eigenvalues -2, -1, confirming Theorem 4.8.

For completeness we now describe briefly the discrete-time
form of the optimal linear regulator, when the state equations are
(4.2) and (4.4), subject to a given initial condition (4.55). The
control vector is now to be chosen so as to minimize

$$\sum_{k=0}^{\infty} [y^T(k)Qy(k) + u^T(k)Ru(k)] \tag{4.69}$$

where the sign properties of Q and R are the same as for the
continuous-time case (4.56). The solution turns out to be

$$\begin{aligned} u(k) &= Kx(k) \\ &= -(R + B^TPB)^{-1}B^TPAx(k) \end{aligned} \tag{4.70}$$

where P satisfies a quadratic matrix equation of the form

$$P = A^T[P - PB(R + B^TPB)^{-1}B^TP]A + H^TQH \tag{4.71}$$

which corresponds to (4.62). If the pair (A, B) is completely
reachable, and the pair (A, H) is completely observable, then
(4.71) has a unique positive definite solution, and the closed-loop
system is asymptotically stable [i.e., all the eigenvalues of
A + BK have modulus less than unity, where K is given by (4.70)].
These statements are the analogues of Theorems 4.5 and 4.6; again,
as in Remark 5 on the latter, the conditions can be relaxed to
stabilizability and detectability to provide a unique positive
semidefinite stabilizing solution. An eigenvector solution
corresponding to that in Theorem 4.7, involving the matrix M in
(4.63), is now obtained by using

$$M_d = \begin{bmatrix} A + BR^{-1}B^TA^{-T}H^TQH & -BR^{-1}B^TA^{-T} \\ -A^{-T}H^TQH & A^{-T} \end{bmatrix}$$

The assumptions guarantee that M_d has no eigenvalues with unit
modulus, and the solution of (4.71) is $Z_3Z_1^{-1}$, where $\begin{bmatrix} Z_1 \\ Z_3 \end{bmatrix}$ is the

matrix of eigenvectors of M_d corresponding to eigenvalues inside the unit circle.

4.1.4 Output Feedback and Observers

In practice, the state variables are often not directly measurable, only a (smaller) number of output variables. We can apply linear *output feedback*

$$u = Ky$$

$$= KHx \qquad (4.72)$$

where K is m × r, H is r × n with r < n, m < n. For both the continuous- and discrete-time representations the closed-loop matrix is

$$A_{cy} = A + BKH \qquad (4.73)$$

which replaces A_c in (4.9) [see also equation (3.5)]. Unfortunately, Theorem 4.1 no longer holds for output feedback. That is, although the pair (A, B) is assumed c.c., it is not in general possible to choose K so that the characteristic polynomial of A_{cy} is arbitrary. Before discussing this point further, we show how a somewhat different approach *does* enable the closed-loop polynomial to be arbitrarily assigned. This relies on the principle that if the system is assumed to be c.o., then by definition the state is implicitly known. We again work only with the continuous-time case, but an analogous argument for discrete-time systems is easily constructed. We begin by applying the duality principle (Theorem 2.5) to Theorem 4.1, to obtain:

Theorem 4.9 There exists a real matrix L such that the characteristic polynomial of A + LH is the arbitrary real polynomial (4.10) if and only if rank $\mathcal{O}$(A, H) = n.

We wish to construct a model of a system, called a *state observer* (or *estimator*), whose state, denoted by $x_0(t)$, approaches the (unknown) state x(t) of the original system (4.1) as t → ∞.

This model is to have as its inputs the *known* input u(t) to the
original system, together with the *known* output y(t) of the
original system.

The basic step is to note that since the system is c.o.,
Theorem 4.9 implies that we can always find a matrix L such that
A + LH is asymptotically stable. If we set

$$e(t) = x(t) - x_0(t) \tag{4.74}$$

and write

$$\dot{e}(t) = (A + LH)e(t) \tag{4.75}$$

then, irrespective of e(0), e(t) has the desired behavior, i.e.,
e(t) → 0 as t → ∞. Substituting (4.74) into (4.75) gives

$$\dot{x}_0(t) = \dot{x}(t) - (A + LH)(x(t) - x_0(t))$$

$$= Ax(t) + Bu(t) - (A + LH)(x(t) - x_0(t))$$

$$= (A + LH)x_0(t) - Ly(t) + Bu(t) \tag{4.76}$$

on using the expression for the output in (4.3). The system
described by (4.76) achieves our objective: Its state $x_0(t)$
approximates to x(t) as t → ∞, and its inputs are u(t) and y(t).

The state observer (4.76) now allows us to arbitrarily assign
the closed-loop polynomial. Apply linear feedback in the *observer*
variables, i.e., u = Kx_0, which makes the original state equations
(4.1) become

$$\dot{x}(t) = Ax(t) + BKx_0(t) \tag{4.77}$$

Combining together (4.76) and (4.77) produces the overall system
equations

$$\frac{d}{dt}\begin{bmatrix} x(t) \\ x_0(t) \end{bmatrix} = \begin{bmatrix} A & BK \\ -LH & A + LH + BK \end{bmatrix}\begin{bmatrix} x(t) \\ x_0(t) \end{bmatrix} \tag{4.78}$$

Pre- and postmultiplication of the system matrix in (4.78) by

$$\begin{bmatrix} I_n & 0 \\ I_n & -I_n \end{bmatrix}$$

establishes that it is similar to

$$\begin{bmatrix} A + BK & -BK \\ 0 & A + LH \end{bmatrix} \tag{4.79}$$

The *overall* closed-loop characteristic polynomial is therefore
equal to that of (4.79), namely the product of the characteristic
polynomials of A + BK and A + LH. We have already discussed the
selection of L. Finally, since the original system (4.1) is c.c.,
we can determine K by the methods of Section 4.1 so that A + BK has
a preassigned characteristic polynomial.

It is of some interest to record at this point the dual of
Theorem 4.3, and its Corollary 4.3.1. We define the pair (A, H) to
be *detectable* if (A^T, H^T) is stabilizable.

Theorem 4.10   If rank $\mathcal{O}(A, H) = \mu < n$, then there exists a system
algebraically equivalent to S having the form

$$\frac{d}{dt}\begin{bmatrix} x^{(1)} \\ x^{(2)} \end{bmatrix} = \begin{bmatrix} A_1' & 0 \\ A_2' & A_3' \end{bmatrix}\begin{bmatrix} x^{(1)} \\ x^{(2)} \end{bmatrix} + \begin{bmatrix} B_1' \\ B_2' \end{bmatrix} u$$

$$y = [H_1' \quad 0]\begin{bmatrix} x^{(1)} \\ x^{(2)} \end{bmatrix} \tag{4.80}$$

where $x^{(1)}$, $x^{(2)}$ have dimensions μ and $n - \mu$, respectively, and
rank $\mathcal{O}(A_1', H_1') = \mu$. The pair (A, H) is detectable if and only if
A_3' in (4.80) is a stability (or convergent) matrix.

The part of the state space described by $x^{(2)}$ in (4.80) can be
called *unobservable*, since it does not influence the output. The
term "detectable" signifies that we can still determine that part
of the state space described by $x^{(1)}$, since the unobservable part
converges to zero as $t \to \infty$.

If we wish to determine feedback so that the closed-loop system is asymptotically stable, rather than have arbitrary eigenvalues, it is sufficient that the pair (A, B) be stabilizable and the pair (A, H) be detectable. Note that these are precisely the conditions mentioned in Remark 5 on Theorems 4.5 and 4.6, which allowed a state feedback control to be obtained via the Riccati equation (4.62).

We now return to the theme introduced at the beginning of this section, and demonstrate that even if a system is both c.c. and c.o., this does *not* necessarily mean that it is possible to choose K in (4.72) so as to arbitrarily assign the characteristic polynomial of A_{cy} in (4.73).

Example 6   It is easy to confirm that the single-input, single-output system

$$\dot{x} = \begin{bmatrix} 0 & 1 \\ -2 & 1 \end{bmatrix} x + \begin{bmatrix} 0 \\ 1 \end{bmatrix} u, \qquad y = [-2, \ 0]x$$

is both c.c. and c.o.   In this case K in (4.72) is a scalar, so (4.73) gives

$$A_{cy} = A + K \begin{bmatrix} 0 & 0 \\ -2 & 0 \end{bmatrix} = \begin{bmatrix} 0 & 1 \\ -(2 + 2K) & 1 \end{bmatrix}$$

and

$$\det(\lambda I - A_{cy}) = \lambda^2 - \lambda + 2 + 2K$$

showing that with feedback u = Ky the system is not even stabilizable for *any* value of K, because of the negative coefficient [see Problem 3.12(i)].   Indeed, this example illustrates the following result.   If HB = 0, then

$$tr(A + BKH) = tr(A) + tr(BKH)$$

$$= tr(A) + tr(HBK) = tr(A)$$

showing that in this case the system is not stabilizable if tr(A) is positive.

We shall continue to restrict ourselves to the case when A, B, H, and K are all real, and assume in addition that B and H have full rank. It can be shown that provided that the system is c.c. and c.o., then in general at least $q = \max(r, m)$ of the eigenvalues of A_{cy} can be assigned arbitrarily close to (but not necessarily coincident with) arbitrary prespecified values. We now give some special conditions under which it *is* possible to assign *all* the eigenvalues. It is necessary to assume that the inequality $m + r > n$ is satisfied.

Suppose that it is desired to make the eigenvalues of A_{cy} be an arbitrary distinct set of numbers $\eta_1, \eta_2, \ldots, \eta_n$ (by assumption, any complex eigenvalues will occur in conjugate pairs). Denote corresponding right and left eigenvectors of A_{cy} by $s_1, \ldots, s_n$ and $t_1^T, \ldots, t_n^T$. Thus we have

$$(A + BKH - \eta_i I)s_i = 0 \tag{4.81}$$

$$t_j^T(A + BKH - \eta_j I) = 0 \tag{4.82}$$

$$t_j^T s_i = \delta_{ij} \tag{4.83}$$

for $i, j = 1, \ldots, n$, where δ_{ij} denotes the Kronecker delta. The equations (4.81) and (4.82) can be written as

$$[A - \eta_i I, \ B]\begin{bmatrix} s_i \\ w_i \end{bmatrix} = 0 \tag{4.84}$$

and

$$[A^T - \eta_j I, \ H^T]\begin{bmatrix} t_j \\ z_j \end{bmatrix} = 0 \tag{4.85}$$

where

$$w_i = KHs_i, \quad z_j = K^T B^T t_j \tag{4.86}$$

for $i, j = 1, \ldots, n$. If it is possible to determine sets of

linearly independent vectors $\{s_1, \ldots, s_n\}$, $\{t_1, \ldots, t_n\}$ satisfying (4.83), (4.84) and (4.85), then (4.86) shows that the desired
feedback matrix is given by either

$$K = [w_1, w_2, \ldots, w_r][Hs_1, Hs_2, \ldots, Hs_r]^{-1} \qquad (4.87)$$

or

$$K^T = [z_1, z_2, \ldots, z_m][B^T t_1, B^T t_2, \ldots, B^T t_m]^{-1} \qquad (4.88)$$

provided that each of the inverse matrices in (4.87) and (4.88)
exists. In order to quote a necessary and sufficient condition for
the existence of the desired sets of eigenvectors $\{s_1, \ldots, s_n\}$ and
$\{t_1^T, \ldots, t_n^T\}$, for a given set of closed-loop eigenvalues $n_1, \ldots,$
n_n, we must define subspaces $P_1(n)$, $P_2(n)$ for a complex number n by

$$P_1(n) = \{s \mid (A - nI)s \in Range\ (B)\} \qquad (4.89)$$

$$P_2(n) = \{t \mid (A^T - nI)t \in Range\ (H^T)\} \qquad (4.90)$$

The range of a matrix is simply the subspace generated by its
columns, which in our case for both B and H^T were assumed linearly
independent. Thus (4.89) means that the column n-vector s, whose
elements are in general complex, is such that

$$(A - nI)s = \sum_{i=1}^{m} \alpha_i b_i \qquad (4.91)$$

where the b's are the columns of B, and the α's are scalars, not all
zero. The expression (4.90) is interpreted in an exactly similar
fashion. The conditions for eigenvalue assignment can now be
stated:

Theorem 4.11 There exists a real matrix K such that A + BKH has
eigenvalues $n_1, \ldots, n_n$ if and only if there exist vectors
$s_i \in P_1(n_i)$, $t_j \in P_2(n_j)$, for i, j = 1, $\ldots$, n, satisfying (4.83)
and such that

$$\bar{s}_p = s_q, \qquad \bar{t}_p = t_q, \qquad if\ \bar{n}_p = n_q$$

Example 7 With

$$A = \begin{bmatrix} 0 & 1 & 0 \\ 0 & 0 & 1 \\ 0 & 0 & 0 \end{bmatrix}, \quad B = \begin{bmatrix} 1 & 0 \\ 1 & 0 \\ 1 & 1 \end{bmatrix}, \quad H = \begin{bmatrix} 1 & 0 & 0 \\ 0 & 1 & 0 \end{bmatrix}$$

we wish to determine K so that $A_{cy} = A + BKH$ has eigenvalues -1, -2, -5. From (4.89) and (4.91), $P_1(\eta)$ is the set of vectors s such that

$$\begin{bmatrix} -\eta & 1 & 0 \\ 0 & -\eta & 1 \\ 0 & 0 & -\eta \end{bmatrix} s = \alpha_1 \begin{bmatrix} 1 \\ 1 \\ 1 \end{bmatrix} + \alpha_2 \begin{bmatrix} 0 \\ 0 \\ 1 \end{bmatrix}$$

When $\eta_1 = -1$, a simple calculation shows that s takes the form

$$[(\alpha_1 + \alpha_2), -\alpha_2, (\alpha_1 + \alpha_2)]^T = (\alpha_1 + \alpha_2)[1, 0, 1]^T$$
$$- \alpha_2[0, 1, 0]^T$$

so that a convenient basis for $P_1(-1)$ is

$$[1, 0, 1]^T, \quad [0, 1, 0]^T$$

Similarly, bases for $P_1(-2)$ and $P_1(-5)$ are, respectively,

$$[1, 0, 2]^T, \quad [0, 1, -1]^T$$

$$[1, 0, 5]^T, \quad [0, 1, -4]^T$$

Using (4.90), bases for $P_2(-1)$, $P_2(-2)$, $P_2(-5)$ are found in the same way to be

$$[1, 0, 0]^T, \quad [0, 1, -1]^T$$

$$[1, 0, 0]^T, \quad [0, -2, 1]^T$$

$$[1, 0, 0]^T, \quad [0, -5, 1]^T$$

We shall not give any attention to the problem of systematic determination of the s's and t's in Theorem 4.11, although algorithms have been developed for this purpose. The reader can check that suitable vectors in this example are

$$
s_1 = \begin{bmatrix} 1 \\ -1 \\ 1 \end{bmatrix}, \quad s_2 = \begin{bmatrix} -3 \\ 2 \\ -8 \end{bmatrix}, \quad s_3 = \begin{bmatrix} 3 \\ 1 \\ 11 \end{bmatrix}
$$

$$
t_1 = \frac{1}{4}\begin{bmatrix} 10 \\ 3 \\ -3 \end{bmatrix}, \quad t_2 = \frac{1}{3}\begin{bmatrix} 3 \\ 2 \\ -1 \end{bmatrix}, \quad t_3 = \frac{1}{12}\begin{bmatrix} 6 \\ 5 \\ -1 \end{bmatrix}
$$

From (4.84) we then obtain

$$
w_1 = \begin{bmatrix} 0 \\ -1 \end{bmatrix}, \quad w_2 = \begin{bmatrix} 4 \\ 12 \end{bmatrix}
$$

and finally the feedback matrix is given by (4.87) as

$$
K = [w_1, \ w_2][Hs_1, \ Hs_2]^{-1}
$$

$$
= \begin{bmatrix} -4 & -4 \\ -10 & -9 \end{bmatrix}
$$

An identical expression is obtained from (4.88). The reader can verify that $A + BKH$ does have the desired eigenvalues. This is easily done by checking that s_1, s_2, s_3 (respectively, t_1^T, t_2^T, t_3^T) are right (left) eigenvectors of $A + BKH$ corresponding to eigenvalues -1, -2, -5.

Problems

4.1 Repeat Example 1, but with $v_{31} = 1$, $v_{32} = 0$, to obtain a different feedback matrix K.

4.2 Show that each of the following systems is not c.c., and reduce it to the form (4.31). Hence determine whether or not each is stabilizable.

$$
(i) \quad \dot{x} = \begin{bmatrix} -1 & 1 & -1 \\ 0 & -1 & 10 \\ 0 & 1 & 3 \end{bmatrix} x + \begin{bmatrix} 0 \\ 2 \\ 1 \end{bmatrix} u
$$

(ii) $\dot{x} = \begin{bmatrix} 0 & 1 & 0 & 0 \\ -1 & 0 & 0 & 0 \\ 0 & 0 & 0 & 1 \\ 0 & 0 & -1 & 0 \end{bmatrix} x + \begin{bmatrix} 0 \\ 2 \\ 0 \\ 1 \end{bmatrix} u$

4.3 Use the method of Example 1 to find a matrix K such that A + BK has eigenvalues -1, -2, -3, when

$A = \begin{bmatrix} 0 & 1 & 0 \\ 0 & 0 & 1 \\ -3 & 1 & 3 \end{bmatrix}$, $B = \begin{bmatrix} 1 & 1 \\ 0 & 1 \\ 1 & 1 \end{bmatrix}$

4.4 Repeat the preceding problem, using Theorem 4.2.

4.5 Consider the simple predator-prey model introduced in Problem 2.1, and suppose that

$A = \begin{bmatrix} -1 & 2 \\ -3 & 2 \end{bmatrix}$

Show that without the application of control both predator and prey populations will grow exponentially. Suppose that the control term is represented by linear feedback of the form $-kx_2$. Determine the smallest value of the constant k which prevents the "population explosion" from taking place.

4.6 Return to the cattle ranching problem described in Problem 2.3. Suppose that the rancher slaughters only mature and old cattle, his aim being to maintain a constant cattle population after a sufficiently long time. The method by which this is to be done corresponds to adding a term

$-\theta x_2(k) - (1.7 - 2\theta)x_3(k)$

to the right-hand side of the appropriate state equation. Show that the desired steady state can be achieved in this way provided that the parameter θ satisfies $0.1 < \theta < 1.05$. (Use Problem 3.21.)

4.7   The approximate equations of motion in pitch and plunge for an
      aircraft in nearly steady, horizontal flight are

$$T\dot{\alpha} = \beta$$

$$\ddot{\theta} = -\omega^2(\beta - qu)$$

$$\dot{h} = V\alpha$$

where

α = flight path angle relative to the horizontal

θ = pitch angle perturbation

h = perturbation from reference altitude

$\beta = \theta - \alpha$ (angle-of-attack perturbation)

u = elevator deflection (control)

T = lift time constant (positive)

ω = undamped natural frequency in pitch-plunge (positive
      constant)

q = elevator effectiveness (positive constant)

v = magnitude of velocity with respect to ground (assumed
      constant)

Write the equations in state space form, taking α, θ, $\dot{\theta}$, and h
as the four state variables.

An autopilot is to be designed to keep $h \approx 0$ in the presence
of vertical wind distrubances. If h is measurable using an
altimeter, show (by testing the closed-loop characteristic
polynomial) that linear feedback u = -kh does not produce an
asymptotically stable closed-loop system for any value of k.
If, in addition, a gyro is used to measure θ, show that the
closed-loop system can be made asymptotically stable with
feedback $u = -k_1h - k_2\theta$, provided that $k_1 > 0$, $k_2 > Vk_1/T\omega^2$.

4.8   Show that in the scheme for constructing the matrix T in
      (4.46), if A^ib_j is linearly dependent on the previously
      selected vectors in (4.45), then so is A^sb_j, for all $s > i$.

4.9 Consider $C(A, B)$ in (4.48), and suppose that

$$A^4 b_1 = 3b_1 - Ab_1 + 2A^3 b_1$$

Use the first scheme, involving the matrix T in (4.41), to obtain the first type of canonical form (4.42) to (4.44) for the system.

4.10 Consider another type of canonical form, obtained by taking P in (4.29) to have rows

$$e_1, \ e_1 A, \ \ldots, \ e_1 A^{k_1 - 1}, \ e_2, \ e_2 A, \ \ldots, \ e_s A^{k_s - 1}$$

where e_i denotes the

$$\sigma_i = \left(\sum_{j=1}^{i} k_j \right) \text{th}$$

row of the *inverse* of the matrix in (4.46). Show that in this case the canonical form has $\hat{A} = [A_{ij}]$, where

$$A_{ii} = \begin{bmatrix} 0 & 1 & 0 & & \\ 0 & 0 & 1 & & \\ & & & \ddots & \\ & & & & 1 \\ x & x & x & \cdots & x \end{bmatrix}, \quad A_{ij} = \begin{bmatrix} 0 & \cdots\cdots & 0 \\ \vdots & & \\ 0 & \cdots\cdots & 0 \\ x & \cdots\cdots & x \end{bmatrix}, \quad i \neq j$$

and $\hat{B}$ has all rows zero except the σ_ith, which is

$$[0, \ 0, \ \ldots, \ 0, \ 0, \ 1, \ x, \ \ldots, \ x], \qquad i = 1, 2, \ldots, s$$
$$\longleftarrow (i - 1) \longrightarrow$$

Notice that when $m = 1$, this reduces precisely to the single-input case of Theorem 2.10, with P given by (2.99).

4.11 Determine the two canonical forms of Section 4.1.2 for the
pair

$$
A = \begin{bmatrix} -4 & -4 & 0 & -1 & -2 \\ 1 & 0 & 0 & 0 & 0 \\ 0 & 0 & -4 & -5 & -2 \\ 0 & 0 & 1 & 0 & 0 \\ 0 & 0 & 0 & 1 & 0 \end{bmatrix}, \quad B = \begin{bmatrix} 0 & 1 \\ 0 & 0 \\ -1 & 0 \\ 0 & 0 \\ 0 & 0 \end{bmatrix}
$$

4.12 Any $2n \times 2n$ real matrix H is called *hamiltonian* if $H = \hat{I} H^T \hat{I}$,
where

$$
\hat{I} = \begin{bmatrix} 0 & -I_n \\ I_n & 0 \end{bmatrix}
$$

(i)   Verify that M in (4.63) is hamiltonian.

(ii) Show that if θ_i is an eigenvalue of H, then so is $-\theta_i$.

4.13 Consider the linear system (4.1), with control given by
(4.57), which minimizes the performance index J in (4.56).
Use equation (4.60) and the result of Problem 3.33 to show
that the minimum value of J is $x_0^T P x_0$, where P is the positive
definite solution of the Riccati equation (4.58), and x_0 is
the initial condition (4.55).

4.14 Consider the system

$$
\dot{x}_1 = -x_1 + u_1, \quad \dot{x}_2 = x_1
$$

where $u_1(t)$ is to be chosen so as to minimize

$$
\int_0^\infty (x_2^2 + 0.1 u_1^2) \, dt
$$

Determine the optimal feedback control (4.57) by solving the
Riccati equation (4.58) directly for the elements of P.

4.15 One way to stabilize an ocean liner is to use a pair of fins, which are controlled by an actuator. The equation governing the roll motion of the liner is

$$J\ddot{\theta} + \eta\dot{\theta} + \alpha\theta = ku$$

where θ is the roll angle, and $ku(t)$ is the roll moment generated by the fins. If

$$\frac{\alpha}{J} = 0.3, \qquad \frac{\eta}{2\sqrt{\alpha J}} = 0.1, \qquad \frac{k}{\alpha} = 0.05$$

write the equations in state space form, using $x_1 = \theta$, $x_2 = \dot{\theta}$ as state variables. Determine the feedback control which minimizes the performance index

$$\int_0^\infty (100\theta^2 + u^2)\, dt$$

by solving the Riccati equation (4.58) directly for the elements of P.

4.16 Consider the Riccati equation (4.58) with

$$A = \begin{bmatrix} 1 & 0 \\ 0 & -\sqrt{2} \end{bmatrix}, \qquad B = \begin{bmatrix} 1 \\ 0 \end{bmatrix}, \qquad Q = \begin{bmatrix} 0 & 0 \\ 0 & 1 \end{bmatrix}, \qquad R = I_2$$

Show that a positive definite solution P exists. Obtain this solution using Theorems 4.7 and 4.8 (you can assume that the latter still applies in this case, despite the fact that the Jordan form of M is not diagonal).

4.17 Consider the problem of minimizing

$$J_\alpha = \int_0^\infty e^{2\alpha t}(x^T Q x + u^T R u)\, dt$$

subject to (4.1), where Q and R are as in (4.56), and α is a real nonnegative constant. By writing

$$x_\alpha(t) = e^{\alpha t} x(t), \qquad u_\alpha(t) = e^{\alpha t} u(t)$$

show that the problem is equivalent to the standard one of
Theorem 4.5, with A replaced by A + αI (use Problem 2.16).
Hence deduce that the closed-loop matrix (4.59), obtained by
minimizing J_α, will have eigenvalues with real parts all less
than $-\alpha$.

4.18 Consider equation (4.62) with Q = I, and the pair (A, H) c.o.
Let η be a column n-vector such that Pη = 0, and prove that
Hη = 0. Similarly, show that HAη = 0, HA$^2\eta$ = 0, ...,
HA$^{n-1}\eta$ = 0. Hence deduce that any solution P of (4.62) is
nonsingular.

4.19 Consider the dual system defined in (2.51), for which it is
required to minimize

$$\int_0^\infty (y^T R^{-1} y + u^T Q^{-1} u)\ dt$$

where Q and R are each positive definite. Show that the
solution of the associated Riccati equation is equal to the
inverse of the solution (assumed positive definite) of (4.62).

4.20 Consider a system for which

$$A = \begin{bmatrix} 0 & 1 & 0 & 0 \\ 1 & 1 & 0 & 0 \\ -1 & 0 & 0 & 0 \\ 0 & 0 & 0 & 0 \end{bmatrix}, \quad B = \begin{bmatrix} 0 & 0 \\ 1 & 0 \\ 0 & 0 \\ 0 & 1 \end{bmatrix}, \quad H = \begin{bmatrix} 1 & 0 & 0 & 0 \\ 0 & 0 & 1 & 0 \\ 0 & 0 & 0 & 1 \end{bmatrix}$$

It is required to determine a real matrix K such that the
eigenvalues of A + BKH are -1, -2, -3, -4. Verify that the
following vectors satisfy the conditions of Theorem 4.11

$$s_1 = \begin{bmatrix} -5 \\ 5 \\ -5 \\ -7 \end{bmatrix}, \quad s_2 = \begin{bmatrix} -2 \\ 4 \\ -1 \\ -7 \end{bmatrix}, \quad s_3 = \begin{bmatrix} -3 \\ 9 \\ -1 \\ -14 \end{bmatrix}, \quad s_4 = \begin{bmatrix} -4 \\ 16 \\ -1 \\ -23 \end{bmatrix}$$

$$t_1 = \frac{1}{3} \begin{bmatrix} -2 \\ 1 \\ 1 \\ 1 \end{bmatrix}, \quad t_2 = \frac{1}{2} \begin{bmatrix} 27 \\ -9 \\ -22 \\ -10 \end{bmatrix}, \quad t_3 = \begin{bmatrix} -16 \\ 4 \\ 13 \\ 5 \end{bmatrix}, \quad t_4 = \frac{1}{6} \begin{bmatrix} 35 \\ -7 \\ -28 \\ -10 \end{bmatrix}$$

and hence determine a suitable matrix K.

4.2 TRANSFER FUNCTION MATRICES

4.2.1 Definitions

We recall from Section 2.1 that applying the Laplace transform to the state equations (4.1) and (4.3) produces the relationship

$$\bar{y}(s) = G(s)\bar{u}(s) \tag{4.92}$$

between the transformed input and output vectors, where

$$G(s) = H(sI - A)^{-1}B \tag{4.93}$$

is the r × m transfer function matrix of the system. The corresponding expression for discrete-time systems is identical, with the trivial substitution of z for s. Each element $g_{ij}(s)$ of G(s) is a rational function of s, which by (4.92) represents the transfer function between $\bar{y}_i$ and $\bar{u}_j$, all other control variables being set equal to zero. If

$$g(s) = s^p + g_1 s^{p-1} + \cdots + g_p \tag{4.94}$$

denotes the monic least common denominator of all the elements of G(s), then we can write this *rational matrix* G(s) in the form

$$G(s) = \frac{N(s)}{g(s)} \tag{4.95}$$

where N(s) is an r × m *polynomial matrix*, i.e., each of its elements is a polynomial. We remarked in Chapter 2 [equations (2.18), (2.19)] that an alternative way of writing a polynomial matrix is to collect together terms in powers of s: If q is the highest power of s which occurs anywhere in the elements of N(s), then in general we can write

$$N(s) = N_0 s^q + N_1 s^{q-1} + \cdots + N_{q-1} s + N_q \tag{4.96}$$

where the N_i are constant $r \times m$ matrices. The integer q is defined to be the *degree* δN of $N(s)$. The matrix $G(s)$ in (4.95) is called *strictly proper* if each element $g_{ij}(s)$ is a proper rational function, so that $q < p$, and in this case $G(s) \to 0$ as $s \to \infty$.

Example 8   If

$$A = \begin{bmatrix} 1 & 0 & 0 \\ 0 & 1 & 0 \\ 0 & 0 & 2 \end{bmatrix}, \quad B = \begin{bmatrix} 1 & 0 \\ 2 & 1 \\ 0 & -8 \end{bmatrix}, \quad H = \begin{bmatrix} 1 & 1 & 0 \\ 1 & 0 & 1 \end{bmatrix} \tag{4.97}$$

it is easy to compute from (4.93) that

$$G(s) = \begin{bmatrix} \dfrac{3}{s-1} & \dfrac{1}{s-1} \\[2mm] \dfrac{1}{s-1} & \dfrac{-8}{s-2} \end{bmatrix}$$

$$= \frac{1}{(s-1)(s-2)} \begin{bmatrix} 3(s-2) & s-2 \\ s-2 & -8(s-1) \end{bmatrix} \tag{4.98}$$

$$= \frac{1}{s^2 - 3s + 2} \left\{ \begin{bmatrix} 3 & 1 \\ 1 & -8 \end{bmatrix} s + \begin{bmatrix} -6 & -2 \\ -2 & 8 \end{bmatrix} \right\} \tag{4.99}$$

$$\qquad\qquad\qquad g(s) \qquad\qquad\quad N_0 \qquad\qquad\qquad N_1$$

Notice that in (4.99), $g(s)$ is not equal to the characteristic polynomial of A.

Consider in particular a square $m \times m$ polynomial matrix $M(s)$. Then $M(s)$ is called *nonsingular* if det $M(s)$ is not identically zero, and *unimodular* if in addition det $M(s)$ is independent of s. In the latter case it follows immediately that $M^{-1}(s)$ will also be a polynomial matrix, so the term *invertible* is also used. Suppose that $M(s)$ has degree ρ and leading coefficient M_0. Then $M(s)$ is called *regular* if det $M_0 \neq 0$, and it is easy to see that det $M(s)$ has its maximum possible degree, namely $m\rho$, if and only if $M(s)$ is regular. This is a special case of a further definition. Let k_i be the highest degree occurring among the elements in the ith column of $M(s)$, so we can write

$$M(s) = M_h \, diag[\, s^{k_1}, \, \ldots, \, s^{k_m}] + M_r(s)$$

where M_h is a constant $m \times m$ matrix, and $M_r(s)$ is a polynomial matrix whose ith column has degree less than k_i. We call $M(s)$ *column proper* if M_h is nonsingular, and it is also easy to see that in this case

$$det \, M(s) = (det \, M_h)s^k + (terms \ of \ degree \ less \ than \ k)$$

where $k = k_1 + k_2 + \cdots + k_m$. Thus $det \, M(s)$ has its maximum possible degree k if and only if $M(s)$ is column proper (a similar definition can be given for a row proper matrix). When $k_i = \rho$, for all i, then the descriptions of regular and column proper coincide.

Example 9 Consider

$$M(s) = \begin{bmatrix} s^3 & s^2 + 4s \\ s^2 - s + 2 & s \end{bmatrix}$$

$$= \underbrace{\begin{bmatrix} 1 & 0 \\ 0 & 0 \end{bmatrix}}_{M_0} s^3 + \underbrace{\begin{bmatrix} 0 & 1 \\ 1 & 0 \end{bmatrix}}_{M_1} s^2 + \underbrace{\begin{bmatrix} 0 & 4 \\ -1 & 1 \end{bmatrix}}_{M_2} s + \underbrace{\begin{bmatrix} 0 & 0 \\ 2 & 0 \end{bmatrix}}_{M_3}$$

$$= \underbrace{\begin{bmatrix} 1 & 1 \\ 0 & 0 \end{bmatrix}}_{M_h} \begin{bmatrix} s^3 & 0 \\ 0 & s^2 \end{bmatrix} + \begin{bmatrix} 0 & 4s \\ s^2 - s + 2 & s \end{bmatrix} \qquad (4.100)$$

Since M_0 and M_h are both singular, $M(s)$ is neither regular nor column proper. Clearly, $det \, M(s)$ has degree 3, whereas $k_1 = 3$, $k_2 = 2$, $k = 5$, $m\rho = 6$.

For a given system, it is important to appreciate that the transfer function matrix depends *only* upon the c.c. and c.o. part.

This is made clear by combining together the decompositions of
Theorems 4.3 and 4.10, to produce:

Theorem 4.12   The system described by (4.1) and (4.3) is
algebraically equivalent to

$$
\frac{d}{dt}
\begin{bmatrix} x^{(1)} \\ x^{(2)} \\ x^{(3)} \\ x^{(4)} \end{bmatrix}
=
\begin{bmatrix}
A^{(11)} & A^{(12)} & A^{(13)} & A^{(14)} \\
0 & A^{(22)} & 0 & A^{(24)} \\
0 & 0 & A^{(33)} & A^{(34)} \\
0 & 0 & 0 & A^{(44)}
\end{bmatrix}
\begin{bmatrix} x^{(1)} \\ x^{(2)} \\ x^{(3)} \\ x^{(4)} \end{bmatrix}
+
\begin{bmatrix} B^{(1)} \\ B^{(2)} \\ 0 \\ 0 \end{bmatrix}
u
$$

$$
y = H^{(2)} x^{(2)} + H^{(4)} x^{(4)} \tag{4.101}
$$

whereby the state space is split up into four mutually exclusive
parts, such that $x^{(1)}$ is c.c. but unobservable; $x^{(2)}$ is c.c. and
c.o.; $x^{(3)}$ is uncontrollable and unobservable; $x^{(4)}$ is c.o. but
uncontrollable. Moreover, the transfer function matrix is

$$
G(s) = H^{(2)} (sI - A^{(22)})^{-1} B^{(2)} \tag{4.102}
$$

The expression (4.102) is obtained by an easy calculation from
(4.93), using the matrices in (4.101).

4.2.2 State Space Realization
As for the single-input, single-output case in Section 2.5, we now
consider a converse problem: Given $G(s)$, we wish to find a triple
$\{A, B, H\}$ such that (4.93) holds; this triple is called a *state
space realization* of $G(s)$ of *order* equal to the dimension of A.
We can assume that $G(s)$ is strictly proper, for if not, we can
divide out each offending element $g_{ij}(s)$ to produce a proper
fraction. Then instead of (4.93) we obtain

$$
G(s) = H(sI - A)^{-1} B + E(s) \tag{4.103}
$$

where $E(s)$ is a polynomial matrix. The extra term in (4.103)
represents a term $E(s)\bar{u}(s)$ in the expression for $\bar{y}(s)$ in (4.92),
which corresponds to u and its derivatives being present in the

expression for the output vector y(t). As before, we define a
minimal realization as one having least possible order. We are now
ready to state and prove the general version of Theorem 2.14.

Theorem 4.13 A realization $R = \{A, B, H\}$ of $G(s)$ is minimal if and
only if it is c.c. and c.o.

Proof.

(i) <u>Sufficiency</u>. We wish to show that if $C(A, B)$ and $O(A, H)$
each have rank n, then R is minimal. That is, if $\{A_0, B_0, H_0\}$ is
any other realization of $G(s)$, with A_0 having dimension n_0, then we
prove that $n_0 \geqslant n$. Since

$$H(sI - A)^{-1}B = H_0(sI - A_0)^{-1}B_0 \tag{4.104}$$

we can deduce [see (2.131)] that

$$HA^iB = H_0A_0^iB_0, \quad i = 0, 1, 2, \ldots \tag{4.105}$$

The reader can then easily check that (4.105) implies that

$$O(A, H) C(A, B) = O'(A_0, H_0) C'(A_0, B_0) \tag{4.106}$$

since the general terms in the ith block row and jth block column
on the left and right in (4.106) are, respectively,

$$HA^{i+j-2}B, \quad H_0A_0^{i+j-2}B_0, \quad i, j = 1, \ldots, n$$

The primes on the right in (4.106) indicate that

$$C'(A_0, B_0) = [B_0, \ldots, A_0^{n-1}B_0]$$

so this matrix has dimensions $n_0 \times m_0 n$, and similarly $O'(A_0, H_0)$
has dimensions $r_0 n \times n_0$, where r_0 and m_0 are unknown positive
integers. Therefore, the rank of their product $O'C'$ cannot be
greater than n_0. However, the matrix on the left in (4.106) has,
by assumption, rank equal to n, so we have shown that $n \leqslant n_0$, as
required.

(ii) <u>Necessity</u>. We now prove that if rank $C(A, B) = \nu < n$, then there exists a realization of $G(s)$ having order less than n.

From Theorem 4.3, R is algebraically equivalent to

$$\{\hat{A}, \hat{B}, \hat{H}\} \overset{\Delta}{=} \left\{ \begin{bmatrix} A_1 & A_2 \\ 0 & A_3 \end{bmatrix}, \begin{bmatrix} B_1 \\ 0 \end{bmatrix}, [H_1, H_2] \right\}$$

which is also a realization of $G(s)$ (see Problem 4.22). It is left as an easy exercise for the reader to prove that

$$\hat{H}(sI - \hat{A})^{-1}\hat{B} = H_1(sI - A_1)^{-1}B_1$$

showing that $\{A_1, B_1, H_1\}$ is a realization of $G(s)$ having order $\nu < n$. This contradicts the assumption that R is minimal, so that we cannot have rank $C(A, B) < n$.

The part of the proof involving $0(A, H)$ follows by duality. ■

<u>Example 10</u> It is easy to check that the realization (4.97) is both c.c. and c.o., so it is minimal for $G(s)$ in (4.98). This illustrates that the order of a minimal realization (here, 3) is not necessarily equal to the degree of $g(s)$ (here, a quadratic).

The equivalent of Theorem 2.15 also applies for the multivariable case, with the proof following very similar lines, so we merely state:

<u>Theorem 4.14</u> If $\{A, B, H\}$ is a minimal realization of $G(s)$, then $\{\hat{A}, \hat{B}, \hat{H}\}$ is also minimal if and only if the algebraic equivalence relation (4.30) holds.

The task of determining a minimal realization of a given transfer function matrix is naturally more difficult than for the scalar case of Section 2.5. One obvious starting point is to generalize the controllable canonical form realization (2.108) for a ratio of two polynomials.

<u>Theorem 4.15</u> Let $g(s)$ be the monic least common denominator, defined in (4.94), of all the elements $g_{ij}(s)$ of the $r \times m$ matrix $G(s)$, and suppose that

$$g(s)G(s) = N_0 s^{p-1} + N_1 s^{p-2} + \cdots + N_{p-1} \qquad (4.107)$$

Then a realization of $G(s)$ is

$$A = \begin{bmatrix} 0 & I_m & 0 & 0 \\ 0 & 0 & I_m & 0 \\ & & & \vdots \\ & & & \ddots I_m \\ \cdots \cdots \cdots \cdots \cdots \cdots & & \\ -g_p I_m & -g_{p-1} I_m & \cdots & -g_1 I_m \end{bmatrix}, \quad B = \begin{bmatrix} 0 \\ 0 \\ \vdots \\ 0 \\ I_m \end{bmatrix} \qquad (4.108)$$

$$H = [N_{p-1}, N_{p-2}, \ldots, N_0]$$

where A is $mp \times mp$, B is $mp \times m$, H is $r \times mp$, and (A, B) is c.c.

Proof. The easiest approach is to observe that

$$A = C \otimes I_m, \quad B = e \otimes I_m \qquad (4.109)$$

where $\otimes$ denotes Kronecker product, C is the standard $p \times p$ companion matrix for $g(s)$, and e is the last column of I_p. Then, using properties of Kronecker product [see (A2) in Appendix A], we get

$$(sI_{mp} - A)^{-1} B = [(sI_p - C) \otimes I_m]^{-1} (e \otimes I_m)$$

$$= [(sI_p - C)^{-1} \otimes I_m](e \otimes I_m)$$

$$= [(sI_p - C)^{-1} e] \otimes I_m \qquad (4.110)$$

The expression within the square brackets in (4.110) is [see (2.67)]

$$[1, s, s^2, \ldots, s^{p-1}]^T / g(s) \qquad (4.111)$$

and substituting (4.111) into (4.110) gives

$$H(sI_{mp} - A)^{-1}B = [N_{p-1}, \ldots, N_0] \left. \begin{bmatrix} I_m \\ I_m s \\ \vdots \\ I_m s^{p-1} \end{bmatrix} \right/ g(s)$$

$$= (N_{p-1} + \cdots + N_0 s^{p-1})/g(s)$$

$$= G(s)$$

as required. Finally, a direct computation produces

$$\mathcal{C}(A, B) = \begin{bmatrix} 0 & . & . & . & . & . & I_m & X & . & X \\ 0 & & & & . & . & . & . & . & . \\ . & & & I_m & . & & & & \\ . & & I_m & X & . & & & . \\ . & I_m & X & & . & & & . \\ I_m & X & X & X & . & . & . & X \end{bmatrix}$$

where X denotes a nonzero term, so $\mathcal{C}(A, B)$ has full rank mp. ∎

Remarks on Theorem 4.15   (1) When m = r = 1, the realization
(4.108) reduces to the single-input, single-output form (2.108)
(with p = n). The matrix A in (4.108) is a *block companion form*
and in view of (4.109) we have

$$\det(sI_{mp} - A) = \det[(sI_p - C) \otimes I_m]$$

$$= [\det(sI_p - C)]^m$$

$$= [g(s)]^m$$

Notice also that for any polynomial k(s) it is easy to show that

$$k(A) = k(C \otimes I_m)$$

$$= [k(C)] \otimes I_m$$

In particular, since C is nonderogatory with minimum polynomial
g(s), it follows that g(A) = 0, showing that A also has g(s) as its

minimum polynomial.

(2) The dual of (4.108) is the following realization of G(s),
which is c.o. but not in general c.c.:

$$A = \begin{bmatrix} 0 & 0 & & & -g_p I_r \\ I_r & 0 & & & \cdot \\ 0 & I_r & & & \cdot \\ \cdot & & \cdot & & \cdot \\ \cdot & & & \cdot & \\ 0 & 0 & \cdot & \cdot & I_r & -g_1 I_r \end{bmatrix}, \quad B = \begin{bmatrix} N_{p-1} \\ N_{p-2} \\ \cdot \\ \cdot \\ \cdot \\ N_0 \end{bmatrix} \quad (4.112)$$

$$H = [0, 0, \ldots, 0, I_r]$$

Notice that when m = r = 1, (4.112) reduces to the observable
canonical form (2.109) (with p = n).

(3) We can construct minimal realizations from (4.108) or
(4.112) in the following manner. We prefer to take the latter as
the starting point. By construction (4.112) is c.o., but not c.c.
Hence, by Theorem 4.3, we can transform it via algebraic
equivalence into the realization of (4.31) and (4.32), viz.,

$$\left\{ \begin{bmatrix} A_1 & A_2 \\ 0 & A_3 \end{bmatrix}, \begin{bmatrix} B_1 \\ 0 \end{bmatrix}, [H_1, H_2] \right\} \quad (4.113)$$

where the pair (A_1, B_1) is c.c. It is easy to verify that
$\{A_1, B_1, H_1\}$ is also a realization of G(s). Furthermore, since
algebraic equivalence preserves observability (Problem 2.19), the
triple (4.113) is also c.o., which implies that the pair (A_1, H_1)
is c.o. [to see this, write down the observability matrix for
(4.113)]. Thus $\{A_1, B_1, H_1\}$ is in fact minimal, by Theorem 4.13.

Notice that the order of this realization is equal to the
dimension of A_1, which by Theorem 4.3 is the rank of the
controllability matrix for the pair (A, B) in (4.112). A similar
argument applied to the c.c. realization (4.108) shows that the
least order is equal to rank $\mathcal{O}(A, H)$ where A and H are defined in
(4.108). Thus the least order of realizations of G(s) can be
calculated without actually determining a specific minimal

realization.

Example 11   Consider again the transfer function matrix in (4.98), for which g(s), N_0 and N_1 are given in (4.99). We can write down a realization using either (4.108) or (4.112). Using the latter, since m = 2, r = 2, p = 2, we obtain

$$A = \begin{bmatrix} 0 & -2I_2 \\ I_2 & 3I_2 \end{bmatrix}, \quad B = \begin{bmatrix} -6 & -2 \\ -2 & 8 \\ 3 & 1 \\ 1 & -8 \end{bmatrix}, \quad H = [0, \ I_2] \qquad (4.114)$$

By construction, the triple in (4.114) is c.o., and

$$C(A, B) = \begin{bmatrix} -6 & -2 & -6 & -2 & . \\ -2 & 8 & -2 & 16 & . \\ 3 & 1 & 3 & 1 & . \\ 1 & -8 & 1 & -16 & . \end{bmatrix} \qquad (4.115)$$

The reader can check that $C(A, B)$ has rank 3, which is therefore the order of minimal realization of G(s). We can now obtain the c.c. part of (4.114) by applying the procedure of Remark 3 above. Thus we follow the steps set out in Example 2. Columns 1, 2, and 4 in (4.115) are linearly independent, so a convenient transformation matrix is, for example,

$$P^{-1} = \begin{bmatrix} -6 & -2 & -2 & 1 \\ -2 & 8 & 16 & 0 \\ 3 & 1 & 1 & 0 \\ 1 & -8 & -16 & 0 \end{bmatrix} \qquad (4.116)$$

Next, compute P, and hence obtain $\{\hat{A}, \ \hat{B}, \ \hat{H}\}$ from (4.30). The minimal realization obtained from the resulting triple (4.113) is found to be (Problem 4.24)

$$A_1 = \begin{bmatrix} 1 & 0 & 0 \\ 0 & 0 & -2 \\ 0 & 1 & 3 \end{bmatrix}, \quad B_1 = \begin{bmatrix} 1 & 0 \\ 0 & 1 \\ 0 & 0 \end{bmatrix}, \quad H_1 = \begin{bmatrix} 3 & 1 & 1 \\ 1 & -8 & -16 \end{bmatrix}$$
$$(4.117)$$

Observe that (4.117) is different from the minimal realization (4.97), but in view of Theorem 4.14 these two realizations are algebraically equivalent (Problem 4.25).

A quite different method for computing realizations relies on a knowledge of the roots of the denominators of the elements $g_{ij}(s)$ of $G(s)$. We consider only the case when these are all distinct and simple, $s_1, \ldots, s_N$, say.

Theorem 4.16   Define

$$K_i = \lim_{s \to s_i} (s - s_i)G(s), \quad i = 1, 2, \ldots, N \qquad (4.118)$$

and matrices $L_i (r \times r_i)$, $M_i (r_i \times m)$, each having rank r_i ($\stackrel{\triangle}{=}$ rank K_i), such that

$$K_i = L_i M_i \qquad (4.119)$$

Then a minimal realization of $G(s)$ is

$$A = \mathrm{diag}[s_1 I^1, s_2 I^2, \ldots, s_N I^N], \quad I^j \stackrel{\triangle}{=} I_{r_j} \qquad (4.120)$$

$$B = \begin{bmatrix} M_1 \\ \vdots \\ M_M \end{bmatrix}, \quad H = [L_1, \ldots, L_N] \qquad (4.121)$$

Proof.   Equation (4.118) is equivalent to expressing $G(s)$ in the form

$$G(s) = \sum_{i=1}^{N} \frac{K_i}{s - s_i}$$

$$= \sum_{i=1}^{N} \frac{L_i M_i}{s - s_i}, \quad \text{by (4.119)}$$

$$= H \, \mathrm{diag}\left[\frac{I^1}{s - s_1}, \ldots, \frac{I^N}{s - s_N} \right] B$$

$$= H(sI - A)^{-1}B \qquad (4.122)$$

where A, B, H are as given in (4.120) and (4.121). It is left as
an exercise for the reader to verify that this realization is both
c.c. and c.o., and hence is minimal (by Theorem 4.13). This
entails using the facts that the s_i are distinct, and each L_i and
M_i has full rank. ∎

Example 12   We again return to G(s) in (4.98), from which we see
that $s_1 = 2$, $s_2 = 1$. From (4.118) we obtain

$$K_1 = \begin{bmatrix} 0 & 0 \\ 0 & -8 \end{bmatrix}, \quad K_2 = \begin{bmatrix} 3 & 1 \\ 1 & 0 \end{bmatrix}$$

and hence $r_1 = 1$, $r_2 = 1$. The matrices in the factorization
(4.119) are not unique, and suitable pairs here are

$$L_1 = \begin{bmatrix} 0 \\ 8 \end{bmatrix}, \quad M_1 = [0, -1]$$

$$L_2 = \begin{bmatrix} 3 & 1 \\ 1 & 0 \end{bmatrix}, \quad M_2 = \begin{bmatrix} 1 & 0 \\ 0 & 1 \end{bmatrix}$$

Equations (4.120) and (4.121) give a minimal realization of G(s) to
be

$$A = \begin{bmatrix} 2 & 0 & 0 \\ 0 & 1 & 0 \\ 0 & 0 & 1 \end{bmatrix}, \quad B = \begin{bmatrix} 0 & -1 \\ 1 & 0 \\ 0 & 1 \end{bmatrix}, \quad H = \begin{bmatrix} 0 & 3 & 1 \\ 8 & 1 & 0 \end{bmatrix}$$

In general, we see from (4.120) that the order of minimal
realizations is $\sum_{i=1}^{N} r_i$.

We close this section by pointing out two interesting links
with results in Chapters 1 and 2. First, consider a system having
a single input, so that its transfer function matrix has dimensions
r × 1. Let us write this in the form

$$G(s) = \frac{1}{g(s)} \begin{bmatrix} b_{11}s^{p-1} + b_{12}s^{p-2} + \cdots + b_{1p} \\ \cdots \cdots \cdots \cdots \cdots \cdots \\ b_{r1}s^{p-1} + b_{r2}s^{p-2} + \cdots + b_{rp} \end{bmatrix} \qquad (4.123)$$

where $g(s)$ is the pth degree polynomial in (4.94). In the notation of (4.95) and (4.96) we thus have

$$N_{i-1} = [b_{1i}, \ldots, b_{ri}]^T, \quad i = 1, 2, \ldots, p$$

Since $m = 1$, the c.c. realization (4.108) becomes, for (4.123),

$$A = C, \quad B = [0, 0, \ldots, 0, 1]^T, \quad H = \begin{bmatrix} b_{1p} & \cdots & b_{11} \\ \cdots \cdots \cdots \cdots \cdots \\ b_{rp} & \cdots & b_{r1} \end{bmatrix}$$

$$(4.124)$$

[recall that C is the companion matrix of $g(s)$]. It is interesting to interpret (4.123) and (4.124) using Theorem 2.8. In terms of our current notation, it follows that the degree of the g.c.d. of the $r + 1$ polynomials

$$g(s), \, b_j(s) = b_{j1}s^{p-1} + \cdots + b_{jp}, \quad j = 1, 2, \ldots, r \quad (4.125)$$

is equal to n-rank $\mathcal{O}(C, H)$ [where H in (4.124) is obtained from B^T in Theorem 2.8 with n, h replaced by p, r, respectively]. Now (4.124) is a minimal realization for $G(s)$ in (4.123) only if the pair (C, H) is c.o. We can conclude from Theorem 2.8 that this is equivalent to the polynomials in (4.125) being relatively prime. In other words, the realization (4.124) is minimal if and only if there is no common factor between numerators and denominator in (4.123). This represents an extension of Theorem 2.12, which applied to the case $r = 1$, when $G(s)$ is a scalar transfer function. Furthermore, it was noted in Remark 3 on Theorem 4.15 that the order of minimal realizations of (4.123) is equal to rank $\mathcal{O}(C, H)$. Therefore, this least order is equal to $n - \delta$ [g.c.d. of the polynomials in (4.125)].

The second generalization which we wish to discuss concerns Theorem 1.12. Again translating into our current notation, this states that when C is the companion matrix of $g(s)$ in (4.94), then the bezoutian matrix Z_j associated with $b_j(s)$ in (4.125) and $g(s)$ satisfies

$$Z_j = Tb_j(C) \tag{4.126}$$

where

$$T = \begin{bmatrix} g_{p-1} & g_{p-2} & \cdots & g_1 & 1 \\ g_{p-2} & g_{p-3} & \cdots & 1 & \\ \vdots & \vdots & \cdots & & \\ g_1 & 1 & \cdots & & 0 \\ 1 & & & & \end{bmatrix} \tag{4.127}$$

Furthermore, we know from Theorem 2.6 that

$$b_j(C) = O(C, h_j), \quad h_j \triangleq [b_{jp}, \ldots, b_{j1}] \tag{4.128}$$

Since T in (4.127) is nonsingular, (4.126) and (4.128) show that Z_j and $O(C, h_j)$ have the same rank, which is equal to the order of minimal realizations of $b_j(s)/g(s)$. To extend this result, define a bezoutian-type matrix Z_e of dimensions rp × mp associated with $g(s)$ and the r × m polynomial matrix $N(s)$ in (4.96), assuming that N has degree $q = p - 1$. Let Z_e be partitioned into r × m submatrices Z_{ij} (i, j = 1, ..., p) defined by the same formula (1.114) as for the elements of the ordinary bezoutian matrix, but with the a's and b's replaced by g's and N's, respectively (and n replaced by p). In this case Z_e is *block* symmetric (i.e., $Z_{ij} = Z_{ji}$, for all i, j). Otherwise, the development used to prove Theorem 1.12 carries over unaltered, leading to the following extension of (4.126):

$$Z_e = (T \otimes I_r)O_e \tag{4.129}$$

where T is defined in (4.127), and O_e has (block) rows

$$H, \; HA, \; HA^2, \; \ldots, \; HA^{p-1} \tag{4.130}$$

with A and H defined in (4.108), namely,

$$A = C \otimes I_m, \quad H = [N_{p-1}, \; N_{p-2}, \; \ldots, \; N_0] \tag{4.131}$$

We noted in Remark 1 on Theorem 4.15 that the degree of the minimum polynomial of A is p. Thus in view of Remark 3 on Theorems 2.1 and 2.2, we observe that 0_e defined in (4.130) has the same rank as $0(A, H)$, whose last block row is HA^{mp-1}. Therefore, (4.129) implies that Z_e and $0(A, H)$ have the same rank, which is equal to the order of minimal realizations of $G(s) = N(s)/g(s)$ (refer again to Remark 3 on Theorem 4.15).

It is also worth noting that

$$0_e = C^{p-1} \otimes N_0 + C^{p-2} \otimes N_1 + \cdots + C \otimes N_{p-2} + I_p \otimes N_{p-1} \tag{4.132}$$

To prove this result is a straightforward exercise which is left for the reader (Problem 4.31). The expression (4.132) can be regarded as a generalization of (4.128). This is because when $r = m = 1$, so that N(s) becomes a scalar polynomial, then (4.132) reduces to the result in Theorem 2.6, here expressible as $0(C, H) = N(C)$, where the N_i in (4.131) are now coefficients of this scalar polynomial N(s).

4.2.3 Matrix-Fraction Description

Let us rewrite the expression in (4.95) for the strictly proper transfer function matrix in the forms

$$G(s) = N(s)M_1^{-1}(s) \tag{4.133}$$

$$G(s) = M_2^{-1}(s)N(s) \tag{4.134}$$

where N(s) is an r × m polynomial matrix, and

$$M_1(s) = g(s)I_m, \quad M_2(s) = g(s)I_r \tag{4.135}$$

[recall that $g(s)$ is the monic least common denominator of all the elements of $G(s)$, and has degree p]. The expressions (4.133) and (4.134) are called, respectively, *right* and *left matrix-fraction descriptions (m.f.d.'s)* of $G(s)$, and $N(s)$ is the *numerator*, $M_1(s)$ and $M_2(s)$ are *denominators*. These names are a natural extension of the case when $N(s)$ is a scalar polynomial, and $G(s)$ is a rational function. In fact, the terms are used when $M_1(s)$ and $M_2(s)$ are *arbitrary* nonsingular m × m and r × r polynomial matrices, not restricted to the special forms in (4.135). The following example illustrates that there will not in general be a unique m.f.d. of a given transfer function matrix.

Example 13   We return to $G(s)$ in Example 8. From (4.98) the right m.f.d. (4.133), with $M_1(s)$ given by (4.135), is

$$G(s) = \begin{bmatrix} 3(s-2) & s-2 \\ s-2 & -8(s-1) \end{bmatrix} \begin{bmatrix} (s-1)(s-2) & 0 \\ 0 & (s-1)(s-2) \end{bmatrix}^{-1}$$

However, if $g_i(s)$ denotes the least common denominator of the ith column of $G(s)$, and $\tilde{N}(s)$ is the polynomial matrix of numerators of $G(s)$, relative to these denominators, then it is obvious that in general we can write

$$G(s) = \tilde{N}(s)[\operatorname{diag}(g_1(s), \ldots, g_m(s))]^{-1} \qquad (4.136)$$

For the example under discussion

$$g_1(s) = s - 1, \qquad g_2(s) = (s-1)(s-2)$$

so (4.136) gives

$$G(s) = \begin{bmatrix} 3 & s-2 \\ 1 & -8(s-1) \end{bmatrix} \begin{bmatrix} s-1 & 0 \\ 0 & (s-1)(s-2) \end{bmatrix}^{-1}$$

which is an alternative m.f.d.

When $M_1(s)$ in (4.133) is given by (4.135), we have the c.c. state space realization (4.108), which has order mp; similarly, for (4.134) and (4.135) the corresponding c.o. realization in (4.112)

has order rp. These orders coincide with the degrees of the
determinants of $M_1(s)$ and $M_2(s)$, respectively, in (4.135). It is
therefore reasonable to define the *degree* of any m.f.d., for any
nonsingular denominator $M(s)$, by the *determinantal degree*
$\delta[\det M(s)]$. The minimal realization problem is then to find
m.f.d.'s having denominators with the smallest possible determinantal degree.

We saw in Section 2.5 that to obtain a minimal realization of
a *scalar* transfer function we must remove the g.c.d. between
numerator and denominator; the order of the minimal realization is
then equal to the degree of the new denominator. We are now ready
to extend this concept to the matrix case. For any two polynomial
matrices $N(s)$, $M(s)$ having dimensions $r \times m$ and $t \times m$, an $m \times m$
polynomial matrix $D(s)$ called a *common right divisor* of N and M if

$$N(s) = \hat{N}(s)D(s), \quad M(s) = \hat{M}(s)D(s) \tag{4.137}$$

where $\hat{N}$ and $\hat{M}$ are also polynomial matrices. The matrix $D(s)$ is a
greatest common right divisor (g.c.r.d.) of N and M if, for any
other right common divisor $D_1(s)$, there exists a polynomial matrix
$X(s)$ such that $D(s) = X(s)D_1(s)$. We can demonstrate that a g.c.r.d.
is not unique. For suppose that $D(s)$ and $D_1(s)$ are two g.c.r.d.'s
of N and M. Then by definition

$$D = XD_1, \quad D_1 = YD$$

where X and Y are polynomial matrices. Hence D = XYD, which
reveals that $X^{-1}(s) = Y(s)$. In other words, from the definition in
Section 4.2.1, $X(s)$ and $Y(s)$ are unimodular. Thus a g.c.r.d. is
unique only up to premultiplication by an arbitrary unimodular
matrix.

For the realization problem we are only interested in the case
when $M(s)$ is $m \times m$. It then follows that all g.c.r.d.'s of $N(s)$
and $M(s)$ have the same determinantal degree, since with $D(s)$ and
$D_1(s)$ as above,

$$\delta(\det D) = \delta(\det XD_1) = \delta(\det X \det D_1)$$

$$= \delta(\det X) + \delta(\det D_1)$$

$$= \delta(\det D_1)$$

The matrices $N(s)$ and $M(s)$ are *relatively right prime (r.r.p.)* if their g.c.r.d. is unimodular. In this case the corresponding m.f.d. is called *irreducible*, because of the following argument. With $M(s)$ $m \times m$, let $D(s)$ in (4.137) be a g.c.r.d. of N and M. Substitution of (4.137) into

$$G(s) = N(s)M^{-1}(s) \tag{4.138}$$

produces the m.f.d. $G(s) = \hat{N}(s)\hat{M}^{-1}(s)$. The degrees of these two m.f.d.'s of $G(s)$ are related by

$$\delta(\det M) = \delta(\det \hat{M}D), \quad \text{from (4.137)}$$

$$= \delta(\det \hat{M}) + \delta(\det D)$$

This shows that the m.f.d. (4.138) of $G(s)$ has the least degree when it has been put into irreducible form, since $\delta(\det D) > 0$ unless $D(s)$ is unimodular.

When $N(s)$ is $m \times r$ and $M(s)$ is $m \times t$ there is a corresponding definition of left divisors of $N(s)$ and $M(s)$. In particular, in this case N and M are relatively left prime if and only if N^T and M^T are relatively right prime.

We want to show that the minimal degree is the same for all possible m.f.d.'s of $G(s)$, and to achieve this we need some fundamental results on polynomial and rational matrices. Let $P(s)$ be a polynomial matrix, and define its *rank* R to be the order of the largest nonvanishing minor of $P(s)$.

Theorem 4.17 (Smith Canonical Form) There exist unimodular polynomial matrices $X_1(s)$, $X_2(s)$ such that

$$X_1(s)P(s)X_2(s) = S(s) \tag{4.139}$$

where the *Smith form* of P(s) is

$$
S(s) = \begin{bmatrix} \text{diag}\{i_1(s), \ldots, i_R(s)\} & \vdots & 0 \\ - - - - - - - - - - - - - & - & - \\ 0 & \vdots & 0 \end{bmatrix}
\tag{4.140}
$$

and the unique polynomials

$$
i_j(s) \triangleq \frac{\Delta_j(s)}{\Delta_{j-1}(s)}, \quad j = 1, \ldots, R
$$

are the *invariant factors* of P, where $\Delta_j(s)$ is the jth
determinantal divisor of P, being the monic g.c.d. of all minors of
order j of P (with $\Delta_0 = 1$).

Remarks on Theorem 4.17   (1) Equation (4.139) defines an
equivalence relation between P(s) and S(s). Pre- and postmultiplication by $X_1(s)$ and $X_2(s)$ corresponds to performing elementary row
and column operations, respectively, on P(s). If two polynomial
matrices are equivalent in this sense, then they have the same
Smith form.

(2) The definition (4.140) implies that $i_j(s)$ is a factor of
$i_{j+1}(s)$, for j = 1, 2, ..., R - 1.

(3) The matrix S(s) has the same dimensions as P(s), so the
specific pattern of zeros depends upon the value of R, and whether
P has the same, fewer, or more rows than columns.

(4) When P(s) is m × m, then taking determinants of both sides
of (4.139) shows that det P = k det S, where k is a constant,
independent of s. Moreover, if P(s) is unimodular, then since
$\Delta_m = 1$, Remark 2 implies that the Smith form of P(s) is I_m.

(5) It is instructive to apply the theorem to

$$
P(s) = \begin{bmatrix} N(s) \\ M(s) \end{bmatrix} \begin{matrix} r \\ m \end{matrix}
\tag{4.141}
$$

where N and M are numerator and denominator in a right m.f.d.
(4.138). In this case we can write (4.139) as

$$\begin{bmatrix} N \\ M \end{bmatrix} = X_1^{-1}(SX_2^{-1})$$

$$= \begin{matrix} & \overset{m}{} & \overset{r}{} \\ & \begin{bmatrix} Y_{11} & Y_{12} \\ Y_{21} & Y_{22} \end{bmatrix} & \begin{bmatrix} S_1 \\ 0 \end{bmatrix} \begin{matrix} m \\ r \end{matrix} \\ & X_1^{-1}(s) & \end{matrix} \qquad (4.142)$$

where X_1^{-1}, X_2^{-1}, and S_1 are polynomial matrices. Expanding (4.142) gives

$$N = Y_{11}S_1, \qquad M = Y_{21}S_1$$

showing that S_1 is a common right divisor of M and N. Furthermore, premultiply both sides of (4.142) by X_1, and write the latter matrix in partitioned form, to obtain

$$\begin{matrix} & \overset{r}{} & \overset{m}{} & & \\ \begin{matrix} m \\ r \end{matrix} & \begin{bmatrix} X_{11} & X_{12} \\ X_{21} & X_{22} \end{bmatrix} & \begin{bmatrix} N \\ M \end{bmatrix} & = & \begin{bmatrix} S_1 \\ 0 \end{bmatrix} \\ & X_1(s) & & & \end{matrix} \qquad (4.143)$$

from which

$$X_{11}N + X_{12}M = S_1 \qquad (4.144)$$

If S_2 is another common right divisor of M and N, i.e., $M = M_1S_2$, $N = N_1S_2$, then (4.144) becomes

$$(X_{11}N_1 + X_{12}M_1)S_2 = S_1 \qquad (4.145)$$

Hence, by definition, the matrix S_1 is a g.c.r.d. of M(s) and N(s). This shows that one way of obtaining a g.c.r.d. is to reduce P(s) by elementary row operations according to (4.143).

It is also interesting to notice that in (4.143), $-X_{21}^{-1}X_{22}$ is a *left* m.f.d. for NM^{-1} (see Problem 4.34).

(6) An important special case of the preceding is when M(s) and N(s) are relatively prime. Their g.c.r.d. is then unimodular, so that in this case the Smith form of (4.141) is $[I_m, \ 0]^T$.

However, a knowledge of the invariant factors of each of M(s), N(s) is not sufficient to determine whether or not M and N are relatively prime. This is easily seen from the examples:

$$M(s) = \begin{bmatrix} 1 & 0 \\ 0 & s \end{bmatrix}, \qquad N(s) = \begin{bmatrix} 1 & 0 \\ 0 & s \end{bmatrix} \tag{4.146}$$

$$M_1(s) = \begin{bmatrix} 1 & 0 \\ 0 & s \end{bmatrix}, \qquad N_1(s) = \begin{bmatrix} s & 0 \\ 0 & 1 \end{bmatrix} \tag{4.147}$$

for which the Smith forms of P(s) in (4.141) are easily found (using the definition (4.140)) to be, respectively,

$$\begin{bmatrix} 1 & 0 & 0 & 0 \\ 0 & s & 0 & 0 \end{bmatrix}^T, \quad \begin{bmatrix} 1 & 0 & 0 & 0 \\ 0 & 1 & 0 & 0 \end{bmatrix}^T$$

Thus the matrices in (4.147) are relatively prime, but those in (4.146) are not. However, all four matrices in (4.146) and (4.147) have the same invariant factors 1, s.

(7) When M, N are relatively prime, we can premultiply both sides of (4.144) by the polynomial matrix $S_1^{-1}(s)$ to obtain

$$X(s)N(s) + Y(s)M(s) = I_m \tag{4.148}$$

<u>Theorem 4.18</u> The matrices N(s), M(s) are relatively right prime if and only if there exist polynomial matrices X(s), Y(s) satisfying (4.148).

Proof. It remains only to establish the converse, namely that if there exist X, Y satisfying (4.148) then N, M are relatively prime. Suppose that D is any g.c.r.d. in (4.137). Then (4.148) becomes

$$(X\hat{N} + Y\hat{M})D = I$$

showing that the polynomial matrix $(X\hat{N} + Y\hat{M})$ is the inverse of D, i.e., D is unimodular. ∎

Remarks on Theorem 4.18   (1) When $r = m = 1$, equation (4.148)
reduces to the diophantine equation (1.104), so the first part of
Theorem 1.11, for two scalar polynomials, now follows as a special
case of the theorem, except that we have no conditions on the
degrees of $X(s)$, $Y(s)$ in (4.148).

(2) Applying the theorem to (4.144) shows that if the Smith
form of $P(s)$ in (4.141) is $[I_m, 0]^T$, then $N(s)$ and $M(s)$ are
relatively prime (this is the converse of Remark 6 on Theorem
4.17).

(3) It can be deduced from the theorem (see Problem 4.32) that
a necessary and sufficient condition for the system $\dot{x} = Ax + Bu$ to
be c.c. is that $sI - A^T$, B^T are relatively right prime. By a
different argument, it can be shown that the discrete system
$x(k + 1) = Ax(k) + Bu(k)$ is c.c. if and only if $I - sA^T$, B^T are
r.r.p.

(4) The theorem enables us to give the following relationship
between any two (right) irreducible m.f.d.'s of the same transfer
function matrix. This corresponds to Theorem 4.14 for state space
minimal realizations.

Theorem 4.19   Consider any two irreducible right m.f.d.'s

$$G(s) = N(s)M^{-1}(s)$$

$$= N_1(s)M_1^{-1}(s) \qquad\qquad (4.149)$$

where each pair of polynomial matrices N, M and N_1, M_1 is
relatively right prime. Then there exists a unimodular matrix $E(s)$
such that

$$N = N_1 E, \quad M = M_1 E \qquad\qquad (4.150)$$

Proof. From (4.149) we have

$$N = N_1 M_1^{-1} M$$

and substituting this into (4.148) (which holds since N, M are relatively prime) produces

$$(XN_1 + YM_1)M_1^{-1}M = I$$

This shows that the matrix $M_1^{-1}M$ has a polynomial inverse, i.e., $M_1^{-1}M$ is unimodular. We can therefore set $E = M_1^{-1}M$, since this trivially satisfies the second part of (4.150). ∎

Remarks on Theorem 4.19   (1) We saw earlier that the degree of a particular m.f.d. will be least when it has been put into irreducible form, by removing a nonunimodular g.c.r.d. The theorem shows that *all* irreducible right m.f.d.'s of a given transfer function matrix have the *same* degree, since from (4.150)

$$\delta(\det M) = \delta(\det M_1) + \delta(\det E) = \delta(\det M_1)$$

   (2) It can be shown that this minimal degree of irreducible right m.f.d.'s is equal to the order of minimal state space realizations of the same transfer function matrix. It follows that this minimal degree must also equal the degree of irreducible *left* m.f.d.'s, since both ways of factorizing G(s) must correspond to minimal state space realizations having the *same* order. An analogue of Theorem 2.12 can therefore be stated: A state space realization {A, B, H} of an m.f.d. $N(s)M^{-1}(s)$, for which the dimension of A is equal to the degree of det M(s), is minimal if and only if N(s) and M(s) are relatively prime.

   (3) It may be the case that the differential (or difference) equations describing a linear system do not originate in the first-order forms (4.1) [or (4.2)]. The most general represen-tation of such a system, after taking Laplace (or z) transforms, and assuming as before that all initial conditions are zero, is

$$T(s)\bar{\xi}(s) = U(s)\bar{u}(s) \tag{4.151}$$

$$\bar{y}(s) = V(s)\bar{\xi}(s) + W(s)\bar{u}(s) \tag{4.152}$$

The transformed state vector $\bar{\xi}(s)$ has ℓ components, and the polynomial matrices $T(s)$, $U(s)$, $V(s)$, and $W(s)$ have dimensions $\ell \times \ell$, $\ell \times m$, $r \times \ell$, and $r \times m$, respectively. The extra term in (4.152) allows the output vector $y(t)$ to depend upon the control variables and their derivatives. It is trivial to obtain $\bar{y}(s) = G(s)\bar{u}(s)$ from (4.151) and (4.152), where

$$G(s) = V(s)T^{-1}(s)U(s) + W(s) \qquad (4.153)$$

assuming that $\det T(s) \neq 0$. When the equations are in state space form, then

$$T(s) = sI_n - A, \quad V(s) = H, \quad U(s) = B \qquad (4.154)$$

and (4.153) reduces to the familiar form (4.103). It can be seen that the m.f.d.'s in (4.133) and (4.134) are each special cases of (4.153), which is termed the *polynomial system matrix representation* of $G(s)$. Using a similar argument to the m.f.d. case, it can be shown that the representation $\{T, U, V, W\}$ of $G(s)$ has least order if and only if each pair T, V and T^T, U^T is r.r.p.

(4) An alternative characterization of the minimal degree of an irreducible m.f.d. is obtained by returning to (4.95), from which we have $gG = N$. Applying Theorem 4.17 to the polynomial matrix $N(s)$ produces:

Theorem 4.20 (Smith-McMillan Canonical Form) Any rational $r \times m$ matrix $G(s)$ can be expressed as

$$G(s) = X_3(s)S_M(s)X_4(s) \qquad (4.155)$$

where X_3, X_4 are unimodular polynomial matrices, and the Smith-McMillan form of $G(s)$ is

$$S_M(s) = \left[\begin{array}{c|c} \mathrm{diag}[\varepsilon_1(s)/\psi_1(s), \ \ldots, \ \varepsilon_R(s)/\psi_R(s)] & 0 \\ \hline 0 & 0 \end{array} \right]$$

R is the rank of $G(s)$, and

(i) $\varepsilon_i(s)$ and $\psi_i(s)$ are relatively prime monic polynomials.

(ii) $\varepsilon_i(s)|\varepsilon_{i+1}(s)$, i = 1, 2, ..., R - 1.

(iii) $\psi_{i+1}(s)|\psi_i(s)$, i = 1, 2, ..., R - 1, and $\psi_1(s)$ = g(s), where g(s) is defined in (4.94).

Remarks on Theorem 4.20   (1) The roots of $\psi_1(s)$, ..., $\psi_R(s)$ are called the *poles* of G(s), and those of $\varepsilon_1(s)$, ..., $\varepsilon_R(s)$ the *zeros* of G(s), but the term *transmission zeros* is often used for the latter in the control literature. Various other types of zeros of a system have been defined. For example, suppose that G(s) arises from (4.153), and let $D_\ell(s)$ be a greatest common left divisor of T(s) and U(s), and $D_r(s)$ a greatest common right divisor of T(s) and V(s). Then the roots of det $D_\ell(s)$ and det $D_r(s)$ are, respectively, the *input decoupling (i.d.)* and *output decoupling (o.d.)* *zeros* of the system. In particular, when G(s) has a state space realization, with T, U, V given by (4.154), then the numbers of i.d. and o.d. zeros are, respectively, equal to n-rank C(A, B) and n-rank O(A, H).

(2) We can write (4.155) in the form of a right m.f.d.:

$$G(s) = [X_3(s)\varepsilon(s)][X_4^{-1}(s)\psi(s)]^{-1} \qquad (4.156)$$

where

$$\varepsilon(s) \quad \begin{bmatrix} \text{diag}(\varepsilon_1, \ldots, \varepsilon_R) & | & 0 \\ \text{-----------} & | & \text{--} \\ 0 & | & 0 \end{bmatrix} \quad (r \times m) \qquad (4.157)$$

$$\psi(s) = \begin{bmatrix} \text{diag}(\psi_1, \ldots, \psi_R) & | & 0 \\ \text{------------} & | & \text{---} \\ 0 & | & I_{m-R} \end{bmatrix} \quad (m \times m) \qquad (4.158)$$

By condition (i) of Theorem 4.20 it follows that the matrices $\varepsilon(s)$ and $\psi(s)$ are relatively right prime, and hence so are $X_3\varepsilon$ and $X_4^{-1}\psi$. Thus (4.156) is an irreducible m.f.d., whose degree is

$$\delta(\det\ X_4^{-1}\psi)\ =\ \delta(\det\ X_4^{-1})\ +\ \delta(\det\ \psi)$$

$$=\ \sum_{i=1}^{R}\ \delta\psi_i \tag{4.159}$$

using (4.158) and the fact that $X_4(s)$ is unimodular. The
expression (4.159) is called the *McMillan degree* (written $\delta[G(s)]$)
of a rational matrix. We noted in Remark 1 on Theorem 4.19 that
all irreducible m.f.d.'s of a given $G(s)$ have the same degree, so
the McMillan degree coincides with the degree as defined at the
beginning of this section.

Example 14   Consider the strictly proper matrix

$$G(s)\ =\ \begin{bmatrix} \dfrac{1}{(s+1)^2} & \dfrac{1}{(s+1)(s+2)} \\[3mm] \dfrac{-6}{(s+1)(s+2)^2} & \dfrac{s-3}{(s+2)^2} \end{bmatrix} \tag{4.160}$$

Since det $G(s) \neq 0$, we have R = 2, and

$$g(s)\ =\ (s+1)^2(s+2)^2$$

so that

$$g(s)G(s)\ =\ \begin{bmatrix} (s+2)^2 & (s+1)(s+2) \\ -6(s+1) & (s-3)(s+1)^2 \end{bmatrix} \tag{4.161}$$

It is easy to compute that the Smith form of the polynomial matrix
on the right in (4.161) is

$$S(s)\ =\ \begin{bmatrix} 1 & 0 \\ 0 & s(s-1)(s+2)(s+1)^2 \end{bmatrix}$$

Hence the Smith-McMillan form of (4.160) is

$$\frac{S(s)}{g(s)}\ =\ \begin{bmatrix} \dfrac{1}{(s+1)^2(s+2)^2} & 0 \\[3mm] 0 & \dfrac{s(s-1)}{s+2} \end{bmatrix}$$

so that

$$\varepsilon_1 = 1, \quad \varepsilon_2 = s(s - 1), \quad \psi_1 = (s + 1)^2(s + 2)^2, \quad \psi_2 = s + 2$$

The zeros of $G(s)$ are thus at $s = 0$, $s = 1$, and the poles are at
$s = -1$ (twice) and $s = -2$ (three times). The McMillan degree of
$G(s)$ is $\delta\psi_1 + \delta\psi_2 = 5$. Notice that this example illustrates that
even though $G(s)$ is strictly proper, its Smith-McMillan form need
not be.

(3) Return to the irreducible m.f.d. in (4.156). From Theorem
4.19 it follows that for any other m.f.d. in (4.149) we have

$$N(s) = X_3(s)\varepsilon(s)E(s) \tag{4.162}$$

with $E(s)$ unimodular. As is apparent from (4.157), $\varepsilon(s)$ is itself
in Smith form, so (4.162) shows that all numerators of irreducible
right m.f.d.'s of $G(s)$ have the same Smith form. A similar
argument applies for irreducible left m.f.d.'s, showing that the
zeros of $G(s)$ are the roots of the invariant polynomials of the
numerator of *any* irreducible m.f.d. $N(s)M^{-1}(s)$ of $G(s)$. Similarly,
the poles are the roots of det $M(s) = 0$.

We have assumed throughout that $G(s)$ is strictly proper; in
the scalar case this simply means that the degree of the numerator
is less than that of the denominator. For the matrix case, we need
the concept of a column proper matrix $M(s)$, introduced in Section
4.2.1; recall also that k_i denotes the degree of the ith column of
$M(s)$, and let ℓ_i denote the degree of the ith column of $N(s)$.

Theorem 4.21 Consider

$$G(s) = N(s)M^{-1}(s) \tag{4.163}$$

(i) If $G(s)$ is strictly proper, then $\ell_i < k_i$, $i = 1, 2, \ldots, m$.
(ii) Conversely, if $\ell_i < k_i$, $i = 1, 2, \ldots, m$ and $M(s)$ is column
     proper then $G(s)$ is strictly proper.

Proof.

(i) Since $N(s) = G(s)M(s)$, the (j, i) element of $N(s)$ can be written as

$$n_{ji}(s) = \sum_{p=1}^{m} g_{jp} m_{pi}, \quad j = 1, 2, \ldots, r$$

If $G(s)$ is strictly proper, then every element $g_{jp}(s)$ must be a strictly proper rational function. Hence the degree of the polynomial n_{ji} must be less than the highest degree among the polynomial elements m_{pi} — in other words, $\ell_i < k_i$.

(ii) For the converse, write (4.163) in the form

$$G(s) = \frac{N(s) \text{ adj } M(s)}{\det M(s)}$$

If M_{ij} denotes the i, j element of adj M, then

$$g_{ij}(s) = \sum_{p=1}^{m} n_{ip} M_{pj} / \det M$$

$$= \det \tilde{M} / \det M \tag{4.164}$$

where $\tilde{M}$ is the matrix obtained by replacing the jth row of M by $n_{i1}, \ldots, n_{im}$. Recall that the coefficient matrix M_h of highest-degree column terms of M is nonsingular, so $\delta(\det M)$ takes its maximum possible value $k = k_1 + k_2 + \cdots + k_m$. However, by the construction above, since $\delta n_{ij} \leq \ell_j < k_j$, it follows that the highest-degree column coefficient matrix for $\tilde{M}$ is M_h with its jth row replaced by zeros. Hence $\tilde{M}$ is not column proper, so $\delta(\det \tilde{M}) < k$. Equation (4.164) therefore shows that $g_{ij}(s)$ is strictly proper, for all i and j. ∎

Example 15 Consider the matrix $M(s)$ in (4.100), which we saw is *not* column proper, and has column degrees $k_1 = 3$, $k_2 = 2$. Take

$$N(s) = [2s^2 - 1, \ s]$$

so that $\ell_1 = 2 < k_1$, $\ell_2 = 1 < k_2$. However, a direct calculation gives

$$G(s) = N(s)M^{-1}(s) = \left[\frac{s^2 + s - 3}{-3s^2 + 2s - 8}, \quad \frac{s^3 + 8s^2 - s - 4}{3s^2 - 2s + 8} \right]$$

which is certainly *not* strictly proper.

The notion of a row proper matrix can be developed in a corresponding fashion. An interesting application of Theorem 4.21 is to obtain the following generalization of the elementary division formula (1.3) for scalar polynomials.

Theorem 4.22 (Division of Polynomial Matrices)

    (i) Let $M(s)$ be a column proper nonsingular $m \times m$ polynomial matrix. Then for any $r \times m$ polynomial matrix $N(s)$, there exist polynomial matrices $Q(s)$, $R(s)$ such that

$$N(s) = Q(s)M(s) + R(s) \qquad (4.165)$$

      and $R(s)M^{-1}(s)$ is strictly proper.

    (ii) Furthermore, if

$$\delta(\text{ith column of } R) < \delta(\text{ith column of } M),$$
$$i = 1, 2, \ldots, m \qquad (4.166)$$

      then $Q(s)$ and $R(s)$ are unique.

Proof.

    (i) Form the rational matrix $G = NM^{-1}$. By dividing out elements of G, if necessary, write

$$G(s) = Q(s) + G_1(s) \qquad (4.167)$$

where $G_1(s)$ is strictly proper, and $Q(s)$ is a polynomial matrix. Postmultiplying both sides of (4.167) by M gives

$$N(s) = Q(s)M(s) + R(s) \qquad (4.168)$$

where

$$R(s) \overset{\Delta}{=} G_1(S)M(s)$$

Since (4.168) shows that $R = N - QM$, it must be a polynomial matrix, and by construction RM^{-1} is strictly proper.

It follows from part (i) of Theorem 4.21 that the columns of $R(s)$ and $M(s)$ satisfy (4.166).

(ii) Suppose that there exists another pair of polynomial matrices Q_1, R_1 satisfying (4.165) and such that R_1 also satisfies (4.166) (we cannot assume that R_1M^{-1} is strictly proper). Then we have

$$QM + R = Q_1M + R_1$$

or

$$Q - Q_1 = (R_1 - R)M^{-1} \qquad\qquad (4.169)$$

The left side of (4.169) is a polynomial matrix. However, on the right side of (4.169), M is a column proper, and since R_1 and R both satisfy (4.166), so does their difference, i.e.,

$$\delta[\text{ith column of } (R_1 - R)] < \delta(\text{ith column of M})$$

Thus the conditions of part (ii) of Theorem 4.21 hold for $(R_1 - R)$ and M, showing that $(R_1 - R)M^{-1}$ is a strictly proper rational matrix. Hence, for (4.169) to hold we must have

$$Q - Q_1 = 0 = R_1 - R \blacksquare$$

Remarks on Theorem 4.22   (1) In the special case when $M(s)$ is regular, the theorem reduces to a classical result that there exists a unique pair of matrices Q, R satisfying (4.165) such that $\delta R < \delta M$.

(2) The matrices Q, R in (4.165) are called the *right* quotient and remainder, respectively. When $N(s)$ is $m \times r$, corresponding results can be written down for left division of $N(s)$ by $M(s)$.

(3) See Problem 4.37 for a result when $M(s)$ is nonsingular, but not necessarily regular or column proper.

332   Feedback, Realization, and Polynomial Matrices

Problems

4.21 Consider a system for which $r = m$, and take $K = I_m$ in the
output feedback (4.72). Show that if A_{cy} is the resulting
closed-loop matrix in (4.73), then

$$\frac{\det(sI_n - A_{cy})}{\det(sI_n - A)} = \det[I_m - G(s)]$$

where $G(s)$ is defined in (4.93) [the result

$$\det(I_n - XY) = \det(I_m - YX)$$

for any $n \times m$ matrix X and $m \times n$ matrix Y, can be assumed].
This is a generalization of Problem 2.47.

4.22 If $\{A, B, H\}$ is a state space realization of $G(s)$, show that
the algebraically equivalent system $\hat{A}$, $\hat{B}$, $\hat{H}$ defined by (4.30)
is also a realization.

4.23 Verify that (4.112) is indeed a c.o. realization of $G(s)$ in
(4.107).

4.24 Compute the inverse of the matrix in (4.116), and hence obtain
the realization in (4.117). By computing $H_1(sI - A_1)^{-1}B_1$,
verify that it gives the correct expression for $G(s)$ in (4.98).

4.25 The triples (4.97) and (4.117) are both minimal realizations
of the same transfer function matrix, and are therefore
algebraically equivalent (Theorem 4.14). Determine a
transformation matrix satisfying (4.30).

4.26 Use Theorem 4.16 to obtain a minimal state space realization
of

$$\frac{1}{s^3 + 2s^2 - s - 2} \begin{bmatrix} s^2 + s + 4 & s^2 + 6 \\ s^2 - 5s - 2 & 2s^2 - 7s - 2 \end{bmatrix}$$

4.27 Determine a minimal realization of the transfer function
matrix in the preceding problem by applying the method of
Example 11.

4.28 If $\{A_1, B_1, H_1\}$, $\{A_2, B_2, H_2\}$ are state space realizations of $G_1(s)$ and $G_2(s)$, respectively, show that

$$\left\{ \begin{bmatrix} A_1 & B_1 H_2 \\ 0 & A_2 \end{bmatrix}, \begin{bmatrix} 0 \\ B_2 \end{bmatrix}, [H_1, 0] \right\}$$

is a realization of $G_1(s)G_2(s)$, assuming that this product exists.

4.29 If C is a companion matrix in any of the forms given in Section 1.2, show that the Smith form of $sI - C$ is

diag[1, 1, $\ldots$, 1, det($sI - C$)]

4.30 Consider a pair of unimodular matrices

$$A(s) = A_0 s^p + A_1 s^{p-1} + \cdots + A_p, \quad A^{-1}(s) = B_0 s^q + \cdots + B_q$$

Show that if $p \geqslant 2$, $q \geqslant 2$, and A_1, B_1 are both nonsingular, then $A_0 s + A_1$ is unimodular with inverse $(B_0 s + B_1) B_1^{-1} A_1^{-1}$.

4.31 In (4.132) let the block rows of the matrix on the right be denoted by Y_1, Y_2, $\ldots$, Y_p (each is an $r \times mp$ matrix). By writing e_i for the ith row of I_p, use an argument like that for the scalar case in Section 1.2 to show that $Y_{i+1} = Y_i A$, where A is defined in (4.131), and hence verify that the identity (4.132) holds.

4.32 Use Theorem 4.18 to prove that N(s), M(s) are relatively right prime if and only if the matrix P(s) defined in (4.141) has rank m for each value $s = s_i$, where the s_i are the roots of the equation det M(s) = 0. Hence deduce from Theorem 2.2 that the matrices $sI - A^T$, B^T are relatively right prime if and only if the pair (A, B) is c.c.

4.33 Use the result of the first part of the preceding problem to test whether the following pairs of matrices are relatively right prime.

(i) $M(s) = \begin{bmatrix} s & 0 \\ s^2 & s \end{bmatrix}$, $N(s) = \begin{bmatrix} 0 & (s+1)^2 \\ (s+2)^2 & s+3 \end{bmatrix}$

(ii) $M(s) = \begin{bmatrix} (s+1)(s+2) & 0 \\ 0 & s+2 \end{bmatrix}$, $N(s) = \begin{bmatrix} s & s+1 \\ -s^2 & s \\ s^2+2s & s+2 \end{bmatrix}$

4.34 Consider equations (4.142) and (4.143), and suppose that $M(s)$ is nonsingular.

(i)   Show that $Y_{21}(s)$ is nonsingular. By multiplying together the partitioned form of $X_1(s)$ and $X_1^{-1}(s)$, obtain an expression for $X_{21}(s)$, and hence deduce that X_{21} is nonsingular.

(ii)   Show that $-X_{21}^{-1}(s)X_{22}(s)$ is an irreducible left m.f.d. for $N(s)M^{-1}(s)$.

4.35 Suppose that in Theorem 4.19 N_1, M_1 are relatively right prime, but M and N are not. Show that there still exists a polynomial matrix E satisfying (4.150), but E is no longer unimodular.

4.36 If $N(s)$ is defined by (4.96), and $M(s) = sI_m - A$, show that the right remainder on division of $N(s)$ by $M(s)$ is

$$N_0 A^q + N_1 A^{q-1} + \cdots + N_q$$

4.37 Let $P(s)$, $T(s)$ be $r \times m$ and $m \times m$ polynomial matrices, with $t(s) = \det T(s) \ne 0$. Take $M(s) = t(s)I_m$, $N(s) = P(s) \text{ adj } T(s)$ in Theorem 4.22, and hence deduce that there exists a unique pair of polynomial matrices Q, R such that $P = QT + R$, with $\delta(R \text{ adj } T) < \delta t$, $\delta R < \delta T$.

4.3   RELATIVE PRIMENESS AND G.C.D. FOR POLYNOMIAL MATRICES

4.3.1 Definitions

In Section 4.2.3 we have already given the definition of a greatest common right divisor (g.c.r.d.) $D(s)$ of two polynomial matrices

N(s) and M(s), having the same number of columns. Recall from
(4.137) that D(s) satisfies

$$N(s) = \hat{N}(s)D(s), \quad M(s) = \hat{M}(s)D(s) \qquad (4.170)$$

and that D(s) is unique up to premultiplication by a unimodular
matrix. In particular, N(s) and M(s) are relatively right prime
(r.r.p.) if and only if D(s) is unimodular. Corresponding
definitions and properties hold for *left* divisors and primeness,
simply by using the transposes of N(s) and M(s), so there is no
need to consider this aspect separately.

It should be noted that provided N(s) and M(s) satisfy the
conditions of the division algorithm in Theorem 4.22, then there is
no loss of generality in assuming that $\delta N < \delta M$. For if $\delta N \geqslant \delta M$, we
can obtain the remainder R(s) in (4.165) for which $\delta R < \delta M$, and it
is easy to show that D(s) is a common right divisor of N(s) and
M(s) if and only if it is a common right divisor of N(s) and R(s)
(see Problem 4.38).

The definitions can be extended in an obvious way to more than
two matrices. For example, if $N_i(s)$ is a set of polynomial
matrices, each having the same number of columns, then D(s) is a
g.c.r.d. if

$$N_i(s) = \hat{N}_i(s)D(s), \quad \text{for all } i$$

and, in addition, if $D_1(s)$ is any other common right divisor, then
$D = XD_1$ for some polynomial matrix X.

The importance of these concepts for matrix-fraction
descriptions, and polynomial system matrix representations, was
indicated in Section 4.2.3. In particular, we shall devote most
attention to the case when M(s) is m × m, because of its relevance
for the problem of obtaining irreducible m.f.d.'s of a rational
transfer function matrix G(s). For example, in Theorem 4.18 we
gave a relative primeness test which is an interesting
generalization of an equation for scalar polynomials (see Remark 1
on Theorem 4.18). Two further criteria have also been given
previously: In Remark 2 on Theorem 4.18 we noted that relative

primeness of $N(s)$ and $M(s)$ is equivalent to the matrix $P(s)$ defined in (4.141) having all its invariant factors equal to unity; and the reader was asked to show in Problem 4.32 that this is equivalent to $P(s)$ having full rank for those values of s which are roots of det $M(s) = 0$. It was also noted that in Remark 5 on Theorem 4.17 that a g.c.r.d. of $N(s)$ and $M(s)$ can be obtained by reducing $P(s)$ using elementary row operations according to (4.143). Unfortunately, none of these is very convenient for practical application. In the following sections we extend some of the methods developed for scalar polynomials in Chapter 1. It is remarkable that two different types of extensions can be obtained, one arising from the problem of relative primeness as defined above, and the other by considering relative primeness of the determinants of two square polynomial matrices; it is with the latter that we begin. This is followed in Section 4.3.3 by Sylvester- and bezoutian-type resultant matrices for the original primeness property. Finally, some procedures are described in Section 4.3.4 for actually determining a g.c.r.d. of two polynomial matrices.

4.3.2 Relatively Prime Determinants

In this section we study the problem when $N(s)$, $M(s)$ are *both* square, but have different dimensions, say $r \times r$ and $m \times m$, respectively, and their *determinants* are required to be relatively prime. As in (4.96),

$$N(s) = N_0 s^q + N_1 s^{q-1} + \cdots + N_q \tag{4.171}$$

where now each N_i is $r \times r$, and assume that $M(s)$ is *regular*, so we can write

$$M(s) = I_m s^p + M_1 s^{p-1} + \cdots + M_p \tag{4.172}$$

A necessary first step is the following generalization of the basic property of companion matrices, given in Theorem 1.1:

Theorem 4.23   The mp × mp *block companion* matrix

$$
C_M \triangleq \begin{bmatrix}
0 & I_m & 0 & \cdots & 0 \\
0 & 0 & I_m & \cdots & \cdot \\
\cdot & \cdot & \cdot & \cdots & \cdot \\
\cdot & \cdot & \cdot & \cdots & I_m \\
-M_p & -M_{p-1} & -M_{p-2} & \cdots & -M_1
\end{bmatrix}
\tag{4.173}
$$

associated with M(s) in (4.172), satisfies the condition

$$
\det(sI_{mp} - C_M) = \det M(s)
\tag{4.174}
$$

Proof. This readily follows by analogy with the case when m = 1, dealt with in Section 1.2, and is left as an exercise for the reader. ∎

Armed with this result, we can now generalize the resultant property of the ordinary companion matrix in Theorem 1.3:

Theorem 4.24   The determinants of N(s) and M(s) in (4.171) and (4.172) are relatively prime if and only if the mpr × mpr matrix

$$
R_{MN} \triangleq N_0 \otimes C_M^q + N_1 \otimes C_M^{q-1} + \cdots + N_{q-1} \otimes C_M + N_q \otimes I_{mp} \tag{4.175}
$$

is nonsingular.

Proof. Let $n_{ij}^{(t)}$ denote the i, j element of N_t, so that the i, j element of N(s) is

$$
n_{ij}^{(0)} s^q + n_{ij}^{(1)} s^{q-1} + \cdots + n_{ij}^{(q)}
$$

Hence the i, j submatrix of R_{MN} is

$$
n_{ij}^{(0)} C_M^q + n_{ij}^{(1)} C_M^{q-1} + \cdots + n_{ij}^{(q-1)} C_M + n_{ij}^{(q)} I_{mp}
\tag{4.176}
$$

If we write

$$
\det N(s) = a(s - \mu_1)^{t_1}(s - \mu_2)^{t_2} \cdots (s - \mu_k)^{t_k}
\tag{4.177}
$$

for some $a \neq 0$, then since all the submatrices in (4.176) commute
with each other, it follows that the operations involved in
evaluating det R_{MN} are the same as those for det $N(s)$. It
therefore follows from (4.177) that

$$\det R_{MN} = a \det(C_M - \mu_1 I_{mp})^{t_1} \cdots \det(C_M - \mu_k I_{mp})^{t_k}$$

Thus R_{MN} is nonsingular if and only if no μ_i is an eigenvalue of
C_M. By Theorem 4.23, this is equivalent to no μ_i being a root of
det $M(s) = 0$. In other words, in view of (4.177), R_{MN} is nonsingular if and only if det $N(s)$ and det $M(s)$ have no common
roots. ∎

Remarks on Theorem 4.24   (1) When $N(s)$ and $M(s)$ have the *same*
dimensions, it can be shown that if det $N(s)$ and det $M(s)$ are
relatively prime, then

Smith form of NM = (Smith form of N)(Smith form of M)

(2) It is known [see (A4) in Appendix A] that

$$N_i \otimes C_M^{q-i} = P_1(C_M^{q-i} \otimes N_i)P_2, \quad i = 0, 1, \ldots, q$$

where P_1 and P_2 are nonsingular. Substituting this into (4.175)
shows that R_{MN} can be replaced by

$$R_{MN}' \overset{\Delta}{=} C_M^q \otimes N_0 + C_M^{q-1} \otimes N_1 + \cdots + I_{mp} \otimes N_q \tag{4.178}$$

(3) When $r = 1$, so that $N(s)$ is a scalar polynomial, then R_{MN}
reduces to our (by-now) familiar polynomial in a companion matrix
[e.g., see (1.37)] and Theorem 4.24 reduces to Theorem 1.3.
Moreover, R_{MN}' can be expressed as an observability-type matrix,
extending the scalar result in Theorem 2.6 (see Problem 4.41).

The last remark leads us to hope that generalizations can be
obtained of other forms of the resultant of two scalar polynomials,
as developed in Chapter 1. In fact, in order to derive a
Sylvester-type form, a necessary preliminary is to obtain another

extension of the diophantine equation (1.104), different from that
in Theorem 4.18. It is necessary to constrain $N(s)$ to be regular
also, i.e., N_0 in (4.171) is nonsingular.

<u>Theorem 4.25</u> Two *regular* polynomial matrices $N(s)$ and $M(s)$ in
(4.171) and (4.172) have relatively prime determinants if and only
if the equation

$$N(s)X(s) + Y(s)M(s) = E \qquad (4.179)$$

for a given $r \times m$ matrix E with constant elements, has a unique
solution $X(s)$, $Y(s)$ with $\delta X < \delta M$, $\delta Y < \delta N$.

Proof. Write

$$X(s) = X_0 s^{p-1} + X_1 s^{p-2} + \cdots + X_{p-1}$$

$$Y(s) = Y_0 s^{q-1} + Y_1 s^{q-2} + \cdots + Y_{q-1} \qquad (4.180)$$

and let C_N be the $rq \times rq$ block companion matrix associated with
$N(s)$, but in the form corresponding to C_I in (1.30). Specifically,
on assuming for convenience that in (4.171) $N_0 = I_r$, we have

$$C_N = \begin{bmatrix} 0 & 0 & . & -N_q \\ I_r & 0 & . & -N_{q-1} \\ 0 & I_r & . & -N_{q-2} \\ . & . & . & . & . & . \\ 0 & 0 & . & -N_1 \end{bmatrix} \qquad (4.181)$$

As in Theorem 4.23, the characteristic polynomial of C_N is equal to
$\det N(s)$. The essence of the proof is to establish a relationship
between the solution $X(s)$, $Y(s)$ of (4.179), and the solution for
the matrix F of the equation

$$C_N F - F C_M = G \qquad (4.182)$$

In (4.182), both F and G are $rq \times mp$, partitioned into $r \times m$ submatrices F_{ij}, G_{ij}, respectively ($i = 1, \ldots, q$, $j = 1, \ldots, p$).
Substitute (4.180) into (4.179), and determine the coefficient of a

general power of s; similarly, substitute (4.173) and (4.181) into
(4.182) to obtain an expression for G_{ij}. On setting

$$X_{p-j} = -F_{qj}, \quad Y_{q-i} = F_{ip}, \quad i = 1, \ldots, q, \quad j = 1, \ldots, p$$

some easy manipulations then establish the equivalence of the two
equations (4.179) and (4.182), when $G_{11} = E$, all other $G_{ij} = 0$.
The desired result then follows from the fact (Problem 4.40) that
(4.182) has a unique solution if and only det N(s) and det M(s) are
relatively prime. ∎

Example 16   We illustrate Theorems 4.24 and 4.25 with

$$N(s) = I_2 s + \underbrace{\begin{bmatrix} 1 & 1 \\ 0 & 1 \end{bmatrix}}_{N_1}, \quad M(s) = I_2 s + \underbrace{\begin{bmatrix} 0 & 2 \\ 2 & 0 \end{bmatrix}}_{M_1} \qquad (4.183)$$

so that $p = q = 1$, $r = m = 2$. Clearly, $\det N(s) = (s + 1)^2$,
$\det M(s) = s^2 - 4$, and these are relatively prime. It is straightforward to determine that (4.179), with $E = I_2$, has the unique
solution

$$X(s) = -\frac{1}{9}\begin{bmatrix} 7 & 11 \\ 6 & 3 \end{bmatrix}, \quad Y(s) = \frac{1}{9}\begin{bmatrix} 7 & 11 \\ 6 & 3 \end{bmatrix}$$

Also, in (4.173) we have simply $C_M = -M_1$, and from (4.178)

$$R'_{MN} = -M_1 \otimes N_0 + I_2 \otimes N_1$$

$$= \begin{bmatrix} N_1 & -2I_2 \\ -2I_2 & N_1 \end{bmatrix} \qquad (4.184)$$

and it is easy to check that R'_{MN} is nonsingular.

We can now state a Sylvester-type form of the resultant
matrix, generalizing that in Theorem 1.7. As this is based on
Theorem 4.25, we again require N(s) to be regular.

Theorem 4.26 Two *regular* polynomial matrices $N(s)$ and $M(s)$ in
(4.171) and (4.172) have relatively prime determinants if and only
if the $[mr(p + q)]$th-order matrix

$$
\begin{bmatrix}
\tau_0 & \tau_1 & \cdots & & & \tau_q & 0 & \cdots & 0 \\
0 & \tau_0 & \tau_1 & \cdots & & \tau_{q-1} & \tau_q & \cdots & 0 \\
\multicolumn{9}{c}{\cdots \cdots \cdots \cdots \cdots \cdots \cdots \cdots} \\
0 & 0 & 0 & \tau_0 & \cdots & \cdots & & \tau_{q-1} & \tau_q \\
\mu_0 & \mu_1 & \cdots & & \mu_p & 0 & & 0 & 0 \\
0 & \mu_0 & \cdots & & \mu_{p-1} & \mu_p & & 0 & 0 \\
\multicolumn{9}{c}{\cdots \cdots \cdots \cdots \cdots \cdots \cdots} \\
0 & 0 & \mu_0 & \cdots & & \cdots & & \mu_{p-1} & \mu_p
\end{bmatrix}
\begin{matrix} \\[-1ex] \\ \left. \vphantom{\begin{matrix}a\\a\\a\\a\end{matrix}}\right\} p \text{ rows} \\[2ex] \\ \left.\vphantom{\begin{matrix}a\\a\\a\end{matrix}}\right\} q \text{ rows} \end{matrix}
\qquad (4.185)
$$

where $\tau_i = N_i \otimes I_m$, $\mu_j = I_r \otimes M_j$ ($i = 0, \ldots, q$, $j = 0, \ldots, p$;
$N_0 = I_r$, $M_0 = I_m$), is nonsingular.

Proof. This essentially parallels the derivation of equation
(1.106) involving the ordinary Sylvester matrix. First, replace
$M(s)$ in Theorem 4.25 by its transpose — this is allowable since
$\det M^T \equiv \det M$. Then equate powers of s in the corresponding
equation (4.179) to obtain

$$
N_q X_{p-1} + Y_{q-1} M_p^T = E \qquad (4.186)
$$

$$
\sum_{i+j=k} (N_i X_j + Y_i M_j^T) = 0, \quad k = 0, 1, \ldots, p + q - 2 \qquad (4.187)
$$

Applying the stacking operator $v(\cdot)$ [see (A3) in Appendix A] to the
set of equations (4.186) and (4.187) produces

$$
(N_q \otimes I_m) v(X_{p-1}) + (I_r \otimes M_p) v(Y_{q-1}) = v(E) \qquad (4.188)
$$

$$
\sum_{i+j=k} [(N_i \otimes I_m) v(X_j) + (I_r \otimes M_j) v(Y_j)] = 0
$$

Finally, the condition for the equations (4.188) to have a unique solution $v(X_0), \ldots, v(X_{p-1}), v(Y_0), \ldots, v(Y_{q-1})$ is that the matrix of block coefficients be nonsingular. This matrix, after a simple rearrangement, takes the form (4.185). ■

Remarks on Theorems 4.25 and 4.26   (1) When $r = m = 1$, the approach used in Theorem 4.25 provides a simple proof of Theorem 1.11 concerning the diophantine equation (1.104). It is intriguing that Theorems 4.18 and 4.25 are quite different, yet both reduce to this same equation (1.104) in the scalar case.

(2) Theorem 4.25 still applies if E in (4.179) is a polynomial matrix of degree not greater than $p + q - 2$.

(3) It can also be shown that Theorem 4.25 still holds if only one of $N(s)$, $M(s)$ is regular (the other being nonsingular). Furthermore, if a solution of (4.179) exists with $\delta X < \delta M$, $\delta Y < \delta N$, then it is not unique if and only if both $N(s)$ and $M(s)$ are not regular.

(4) When $r = m = 1$, the matrix (4.185) has precisely the usual Sylvester form, displayed in (1.84).

(5) When $r = m$ and $E = I_m$, then $N(s)$ and $M(s)$ in (4.179) are examples of a class of matrices which have been termed *skew prime*.

The third method in Chapter 1 for expressing resultants which we now extend is that using bezoutians. We assume, without loss of generality, that $N(s)$ in (4.171) has the same degree p as $M(s)$ in (4.172), but we no longer need to impose the condition that $N(s)$ be regular. The bezoutian-type matrix Z associated with $N(s)$ and $M(s)$ has dimensions $mpr \times mpr$, with block partitions Z_{ij} (each $mr \times mr$) defined by

$$\sum_{i=1}^{p} \sum_{j=1}^{p} Z_{ij} \lambda^{i-1} \mu^{j-1} = \frac{M(\lambda) \, \theta \, N(\mu) - M(\mu) \, \theta \, N(\lambda)}{\lambda - \mu} \qquad (4.189)$$

This parallels the ordinary bezoutian matrix defined in Chapter 1 by equations (1.110) and (1.111). The blocks Z_{ij} will therefore be given by a formula like (1.114), where a term $|M_i N_j|$ will be interpreted as $M_i \otimes N_j - M_j \otimes N_i$. Furthermore, the factorization of Theorem 1.12 also carries over:

Theorem 4.27   The bezoutian matrix associated with $N(s)$ and $M(s)$ in (4.171) and (4.172) satisfies

$$Z = TR'_{MN} \tag{4.190}$$

where R'_{MN} is defined in (4.178) (with $q = p$) and T is the block triangular $mpr \times mpr$ matrix

$$T = \begin{bmatrix} M_{p-1} & M_{p-2} & \cdots & M_1 & I_m \\ M_{p-2} & M_{p-3} & \cdots & I_m & 0 \\ \cdots & \cdots & \cdots & \ddots & \\ M_1 & I_m & & \ddots & 0 \\ I_m & 0 & & & \end{bmatrix} \otimes I_r \tag{4.191}$$

Moreover, Z is nonsingular if and only if $\det N(s)$ and $\det M(s)$ are relatively prime.

Proof.   Verification of (4.190) closely follows that of the corresponding result in Theorem 1.12, using the fact that Z is block symmetric, i.e., $Z_{ij} = Z_{ji}$. The second part follows immediately from Theorem 4.24, since T in (4.191) is nonsingular. ∎

Example 16 (continued)   Using $N(s)$ and $M(s)$ in (4.183) gives the Sylvester-type matrix (4.185):

$$\begin{bmatrix} I_2 \otimes I_2 & N_1 \otimes I_2 \\ I_2 \otimes I_2 & I_2 \otimes M_1 \end{bmatrix}$$

and this is easily verified to be nonsingular. Because our example is such a simple one, T in (4.191) reduces to $I_2 \otimes I_2$, so $Z \equiv R'_{MN}$ for this problem. The block symmetry of R'_{MN} is apparent in (4.184).

Remarks on Theorem 4.27   (1) The recurrence formula for the block rows $Z_1, \ldots, Z_p$ of Z, corresponding to (1.117), is

$$Z_i = Z_{i+1}(C_M \otimes I_r) + (M_{p-i} \otimes I_r)Z_p, \quad i = p - 1, p - 2, \ldots, 1$$

where C_M is defined in (4.173), and

$$Z_p = [|M_0 N_p|, |M_0 N_{p-1}|, \ldots, |M_0 N_1|] \tag{4.192}$$

with $|M_0 N_j| \triangleq I_m \otimes N_j - M_j \otimes N_0$.

(2) The formula (4.190) can be expressed in terms of appropriate controllability/observability-type matrices, as was done for the standard case in Section 2.4. Define a matrix D (mpr × mr) having all rows zero except the last mr, which consist of I_{mr}, and write

$$\tilde{C}_M \triangleq C_M \otimes I_r \tag{4.193}$$

Define also an mpr × mpr matrix

$$C_I \triangleq C_p(\tilde{C}_M, D) \triangleq [D, \tilde{C}_M D, \tilde{C}_M^2 D, \ldots, \tilde{C}_M^{p-1} D] \tag{4.194}$$

and note that we have introduced the notation C_p to indicate that only the first p block columns of the complete controllability matrix $C(\tilde{C}_M, D)$ are taken. Using 0_p to similarly denote the first p block rows of an observability matrix, we can also define

$$C_{II} = C_p(\tilde{C}_M^t, Z_p^t), \quad 0_I = 0_p(\tilde{C}_M, Z_p), \quad 0_{II} = 0_p(\tilde{C}_M^t, D^t)$$

where the superscript t denotes *blockwise* transposition. Then it is easy to show (Problem 4.42) that

$$Z = C_I^{-1} 0_I = C_{II} 0_{II}^{-1} = 0_{II}^{-1} 0_I = C_{II} C_I^{-1} \tag{4.195}$$

The expressions (4.195) correspond exactly to those in Theorem 2.11.

(3) The bezoutian matrix Z can be related to the Sylvester-type matrix in (4.185) in an analogous way to the scalar case in Theorem 1.13. In view of (4.190), it would also be possible to relate R'_{MN} to (4.185), thereby generalizing Theorem 1.7.

(4) Equation (4.190) shows that Z and R_{MN} have the same rank. It can be proved that when det M(s) and det N(s) are not relatively prime, then (with q = p)

$$\sum_{i=1}^{m} \sum_{j=1}^{r} \delta_{ij} = mrp - \text{rank } R_{MN} \qquad (4.196)$$

where δ_{ij} is the degree of the g.c.d. of the ith invariant factor of N(s) and the jth invariant factor of M(s). In particular, when r = m = 1, this result reduces to Theorem 1.4, since R_{MN} becomes a polynomial in the ordinary companion matrix. It is worth mentioning that (4.196) still holds when N(s) is rectangular, and this permits a convenient proof of the first part of Theorem 1.6.

(5) If in the definition of Z in (4.189), M(s) is a scalar polynomial (i.e., m = 1) then we obtain the matrix Z_e in Section 4.2.2, for the case when N(s) is square. The relationship (4.190) then reduces to (4.129), in which 0_e is given by (4.132). Indeed, the latter expression is a special case of (4.178).

4.3.3 Resultants for Prime Matrices

We now return to the definition of relative right primeness of two polynomial matrices, as given in Section 4.3.1, and present generalized Sylvester-type and bezoutian-type matrices different from those in Section 4.3.2 (Theorems 4.26 and 4.27, respectively). Note that when N(s) and M(s) are both m × m, then the condition of the preceding section, that their determinants are relatively prime, is only a *sufficient* condition for N(s) and M(s) to be r.r.p. (see Problem 4.44). The converse, however, does not hold; for example, the two matrices in (4.147) are relatively prime, but their determinants are not.

For the general problem, write

$$N(s) = \sum_{i=0}^{p} N_i s^{p-i}, \qquad M(s) = \sum_{i=0}^{p} M_i s^{p-i} \qquad (4.197)$$

where N(s) is now r × m, and M(s) is m × m with $M_0 \neq 0$, but M(s) need only be nonsingular [i.e., det M(s) $\neq$ 0], not necessarily

regular. The matrices in (4.197) will be regarded as a *right*
m.f.d. NM^{-1} of a *proper* transfer function matrix $G(s)$. Let
$M_L^{-1}(s)N_L(s)$ be any left m.f.d. for $G(s)$. In particular, it follows
(see Remark 2 on Theorem 4.19) that $N(s)$, $M(s)$ are r.r.p. if and
only if there exists a pair $N_L(s)$, $M_L(s)$ which are relatively left
prime, with $\delta(\det M) = \delta(\det M_L)$. Equating the two expressions for
$G(s)$ gives

$$M_L(s)N(s) - N_L(s)M(s) = 0 \qquad (4.198)$$

On writing

$$N_L(s) = \sum_{i=0}^{\pi} N_{Li}s^{\pi-i}, \quad M_L(s) = \sum_{i=0}^{\pi} M_{Li}s^{\pi-i} \qquad (4.199)$$

and equating coefficients of powers of s in (4.198), we obtain

$$[-N_{L0}, \; -N_{L1}, \; -N_{L2}, \; \ldots, \; -N_{L\pi}, \; M_{L0}, \; M_{L1}, \; \ldots, \; M_{L\pi}]S_\pi = 0$$
$$(4.200)$$

In (4.200), S_π is a particular case of what we may call a
generalized Sylvester matrix S_k of order k associated with $N(s)$ and
$M(s)$ in (4.197) defined by

$$S_k = \begin{bmatrix} M_0 & M_1 & \cdots & M_p & 0 & \cdots & 0 \\ 0 & M_0 & \cdots & M_{p-1} & M_p & \cdots & 0 \\ \cdot & \cdot & \cdot & \cdot & \cdot & \cdot & \cdot \\ 0 & \cdots & M_0 & \cdots & \cdots & M_{p-1} & M_p \\ N_0 & N_1 & \cdots & N_p & 0 & \cdots & 0 \\ 0 & N_0 & \cdots & N_{p-1} & N_p & \cdots & 0 \\ \cdot & \cdot & \cdot & \cdot & \cdot & \cdot & \cdot \\ 0 & \cdots & N_0 & \cdots & \cdots & N_{p-1} & N_p \end{bmatrix}, \quad k = 1, 2, \ldots, \pi$$
$$(4.201)$$

In (4.201) there are k block rows of each type, so S_k has
dimensions $k(m + r) \times 2km$. Clearly, when $r = m = 1$, then S_p
reduces to the ordinary Sylvester matrix in (1.84). If we define

$$\alpha_k = \text{rank } S_k - \text{rank } S_{k-1}, \quad k = 1, 2, \ldots \quad (\text{rank } S_0 = 0)$$
(4.202)

then the structure of S_k implies that α_k is a nonincreasing function of k. The following result can then be proved.

<u>Theorem 4.28</u> If ν denotes the smallest integer for which $\alpha_{\nu+1} = m$, then N(s), M(s) in (4.197) are r.r.p. if and only if

$$\text{rank } S_\nu = \delta(\det M(s)) + m\nu$$
(4.203)

<u>Example 17</u> Let

$$N(s) = 0s^3 + \begin{bmatrix} 0 & 1 \\ 1 & 1 \end{bmatrix} s^2 + \begin{bmatrix} 2 & 0 \\ 2 & 2 \end{bmatrix} s + \begin{bmatrix} 1 & 1 \\ 1 & 0 \end{bmatrix}$$
(4.204)
$$\qquad\quad N_0 \qquad\quad N_1 \qquad\qquad N_2 \qquad\quad N_3$$

$$M(s) = \begin{bmatrix} 0 & 1 \\ 0 & 0 \end{bmatrix} s^3 + \begin{bmatrix} 2 & 0 \\ 1 & 1 \end{bmatrix} s^2 + \begin{bmatrix} 3 & 4 \\ 1 & 1 \end{bmatrix} s + \begin{bmatrix} 5 & 2 \\ -1 & -1 \end{bmatrix}$$
(4.205)
$$\qquad\quad M_0 \qquad\qquad M_1 \qquad\qquad M_2 \qquad\quad M_3$$

Using, for example, gaussian elimination, the matrices generated by (4.201) can be found to have rank $S_1 = 4$, rank $S_2 = 7$, rank $S_3 = 9$. Thus from (4.202) we obtain $\alpha_1 = 4$, $\alpha_2 = 3$, $\alpha_3 = 2 = m$, so that $\nu = 2$. However, since det M(s) has degree 5, the condition (4.203) is *not* satisfied, so that N(s) and M(s) in (4.204) and (4.205) are not r.r.p.

We now present yet another bezoutian-type matrix, denoted by Γ, having dimensions $r\pi \times mp$ and block partitions Γ_{ij} (each $r \times m$) defined by

$$\sum_{i=1}^{\pi} \sum_{j=1}^{p} \Gamma_{ij}\lambda^{i-1}\mu^{j-1} = \frac{M_L(\lambda)N(\mu) - N_L(\lambda)M(\mu)}{\lambda - \mu}$$
(4.206)

This provides a generalization, different from (4.189), of the ordinary bezoutian in (1.110) and (1.111). We assume from now on that $G(s) = NM^{-1} = M_L^{-1}N_L$ is strictly proper, and that in (4.197)

$\delta N < p$, and in (4.199) $\delta N_L < \pi$. A formula for Γ_{ij} in terms of the
coefficients of the polynomial matrices in (4.206), corresponding
to that for the scalar case in (1.114), can be developed (see
Problem 4.45). It should be noted that not only is Γ not in
general symmetric, but [unlike Z defined by (4.189)] neither is it
block symmetric. However, it is once again possible to generalize
the relationship in Theorem 1.13 between the ordinary Sylvester and
bezoutian matrices. Notice that S_π in (4.201) now has only $\pi - 1$
block rows involving the coefficients M_i, since by assumption the
lower block rows each begin with N_1. Partition S_π according to

$$
S_\pi = \begin{bmatrix} S_{\pi 1} & S_{\pi 2} \\ S_{\pi 3} & S_{\pi 4} \end{bmatrix} \begin{matrix} \pi - 1 \\ \\ \pi \end{matrix} \quad \begin{matrix} \text{block} \\ \text{rows} \end{matrix} \tag{4.207}
$$
$$
\quad\quad \pi - 1 \quad p
$$
$$
\text{block columns}
$$

The first part of the following result is then obtained directly
from (4.198) and (4.206).

Theorem 4.29

(i) The Sylvester matrix in (4.207) and the bezoutian matrix Γ
     defined by (4.206) are related by

$$
\begin{bmatrix} I & 0 \\ R_1 & R_2 \end{bmatrix} \begin{bmatrix} S_{\pi 1} & S_{\pi 2} \\ S_{\pi 3} & S_{\pi 4} \end{bmatrix} = \begin{bmatrix} S_{\pi 1} & S_{\pi 2} \\ 0 & \Gamma J \end{bmatrix} \tag{4.208}
$$

     where

$$
R_1 = -\begin{bmatrix} N_{L1} & N_{L2} & \cdots & N_{L,\pi-1} \\ 0 & N_{L1} & \cdots & N_{L,\pi-2} \\ \cdot & & \ddots & \\ \cdot & & & \cdot \\ \cdot & & & \quad N_{L1} \\ \cdot & & & \\ 0 & 0 & \cdots & 0 \end{bmatrix}
$$

$$
R_2 = \begin{bmatrix}
M_{L0} & M_{L1} & \cdots & & M_{L,\pi-1} \\
0 & M_{L0} & & & M_{L,\pi-2} \\
\vdots & & \ddots & & \vdots \\
\vdots & 0 & & \ddots & \vdots \\
\vdots & & & \ddots & \ddots \\
0 & 0 & \cdots & \cdots & M_{L0}
\end{bmatrix}
$$

and I, J in (4.208) have appropriate dimensions.

(ii) $N(s)$, $M(s)$ are r.r.p. if and only if rank $\Gamma = \delta(\det M)$.

Example 18   Suppose that $p = \pi = 2$; then (4.208) becomes

$$
\begin{bmatrix}
I & 0 & 0 \\
\hline
-N_{L1} & M_{L0} & M_{L1} \\
0 & 0 & M_{L0}
\end{bmatrix}
\begin{bmatrix}
M_0 & M_1 & M_2 \\
N_1 & N_2 & 0 \\
0 & N_1 & N_2
\end{bmatrix}
=
\begin{bmatrix}
M_0 & M_1 & M_2 \\
\hline
0 & \Gamma_{12} & \Gamma_{11} \\
0 & \Gamma_{22} & \Gamma_{21}
\end{bmatrix}
$$

Notice that this gives a convenient way of obtaining the formulas for the block elements Γ_{ij} (see Problem 4.45).

Remarks on Theorems 4.28 and 4.29   (1) In the case when $N(s)$ and $M(s)$ are scalar polynomials, ν in Theorem 4.28 is equal to the degree p of $M(s)$, and the result reduces to the second part of Theorem 1.7.

(2) In this scalar case, Theorem 4.29 reduces to Theorem 1.13, relating the standard Sylvester and bezoutian matrices.

(3) Assuming that M_{L0} is nonsingular, it follows from (4.208) that rank $S_\pi = (\pi - 1)$ rank M_0 + rank Γ. This shows that the rank of Γ is independent of the choice of $M_L(s)$ and $N_L(s)$.

(4) In the special case when $M(s) = g(s)I_m$, $M_L(s) = g(s)I_r$, where $g(s)$ is a scalar polynomial, then Γ reduces to the matrix Z_e defined in Section 4.2.2 [see (4.129)].

(5) It is now relevant to recall that in Section 3.4 we pointed out the relationship between the ordinary bezoutian matrix and the Cauchy index of a rational function. What is interesting here is that the matrix Γ defined by (4.206) can be related in an exactly similar way to a *matrix Cauchy index* of a *real* rational

matrix G(s), now restricted to be *symmetric*. Using the same
notation as in (3.192), this is defined by

$$I_\alpha^\beta G(s) = \text{(number of eigenvalues of G(s) which jump from } -\infty \text{ to}$$
$$+\infty)$$
$$- \text{(number of eigenvalues of G(s) which jump from } +\infty$$
$$\text{to } -\infty)$$
$$(4.209)$$

as s moves along the real axis from α to β (ignoring any jumps at
these end points). Suppose that G(s) is strictly proper, and is
expressed as

$$G(s) = \frac{G_0}{s} + \frac{G_1}{s^2} + \frac{G_2}{s^3} + \cdots \qquad (4.210)$$

where each G_i is a constant m × m matrix [compare with (3.208)] and
let the McMillan degree $\delta[G(s)]$ of G(s) be K. As before, let $M_L^{-1}N_L$
be a left m.f.d. of G(s); because of the symmetry, we can take
$N_L^T(M_L^{-1})^T$ as a right m.f.d., and use this in the definition of Γ in
(4.206). We then have the following generalization of Theorems
3.28 and 3.29:

Theorem 4.30 For the real symmetric strictly proper rational
matrix G(s) in (4.210)

$$I_{-\infty}^\infty G(s) = \text{signature}(S) = \text{signature}(\Gamma) = \text{signature (U)}$$

where

(i) S is the block orthosymmetric matrix

$$S \triangleq \begin{bmatrix} G_0 & G_1 & G_2 & \cdot & \\ G_1 & G_2 \cdot & \cdot & \cdot & \cdot \\ G_2 & \cdot & \cdot & \cdot & \cdot \\ & \cdot & \cdot & \cdot & \cdot \cdot & G_{2K-2} \end{bmatrix} \qquad (4.211)$$

(ii) Γ is the bezoutian matrix for G(s) = $M_L^{-1}N_L = N_L^T M_L^{-T}$, defined by (4.206).

(iii) If {A, B, H} is a real minimal realization of G(s), then U is the unique symmetric matrix such that

$$UA = A^T U, \quad UB = H^T \tag{4.212}$$

The expression (4.212) is the direct analogue of (3.211). A Sturm sequence for polynomial matrices has also been developed, leading to an expression for the matrix Cauchy index, corresponding to Theorem 3.27 for the rational function case.

(6) It can be shown that the McMillan degree $\delta[G(s)]$ of the strictly proper rational matrix G(s) is equal to the rank of Γ. Part (ii) of Theorem 4.29 then follows from Remark 2 on Theorem 4.20, since the McMillan degree is equal to $\delta(\det M(s))$ if and only if NM^{-1} is irreducible.

(7) Suppose that G(s) is rectangular, but still expanded in the form (4.210). Then it can also be shown that $\delta[G(s)]$ is equal to the rank of a block orthosymmetric matrix having the form (4.211), but with p block rows and columns, where p is the degree of the least common denominator g(s) of the elements of G(s). Moreover, the controllability indices can be obtained from this matrix.

We close this section with a brief look at the original, most general, form of the relative primeness problem, when M(s) is not necessarily square. Recall a fact quoted earlier, namely that N(s) and M(s) are r.r.p. if and only if the matrix

$$P(s) = \begin{bmatrix} N(s) \\ M(s) \end{bmatrix} \overset{m}{\underset{l}{}}$$
$$\triangleq P_0 s^p + P_1 s^{p-1} + \cdots + P_p \tag{4.213}$$

has all its invariant factors equal to unity. Associate with (4.213) a proper rational matrix $G(s) = P(s)/s^p$, for which we can write down at once from (4.108) a realization

$$A = \begin{bmatrix} 0 & I_m & 0 & \cdots & 0 \\ 0 & 0 & I_m & \cdots & 0 \\ & & & & I_m \\ 0 & 0 & \cdots & \cdots & 0 \end{bmatrix}, \quad B = \begin{bmatrix} 0 \\ 0 \\ \vdots \\ 0 \\ I_m \end{bmatrix},$$

$$H = [P_p, \ P_{p-1}, \ \ldots, \ P_1] \tag{4.214}$$

so that $G(s) = H(sI_{mp} - A)^{-1}B + P_0$. The system equations corresponding to (4.214) are

$$\dot{x} = Ax + Bu, \quad y = Hx + P_0 u \tag{4.215}$$

and if we apply linear feedback $u = Fx$, (4.215) becomes

$$\dot{x} = (A + BF)x, \quad y = (H + P_0 F)u$$

The system (4.215) is said to be *strongly observable* if the pair $(A + BF, H + P_0 F)$ is c.o. for all matrices F (of appropriate dimensions). It can be shown that $N(s)$, $M(s)$ are r.r.p. if and only if the system (4.215) is strongly observable. Various explicit criteria for testing this property have been developed. One of the simplest to state is as follows. Because of the form of A and H in (4.214) we can write

$$\mathcal{O}(A, \ H) = \begin{bmatrix} Q_0 \\ 0 \end{bmatrix} p \text{ block rows} \tag{4.216}$$

where Q_0 is easily written down (see Problem 4.48). Let us also define

$$Q_1 = \begin{bmatrix} P_0 & 0 & & & \\ P_1 & P_0 & & 0 & \\ \vdots & & \ddots & & \\ P_{p-1} & P_{p-2} & \cdots & \cdots & P_0 \end{bmatrix}$$

Then a necessary and sufficient condition for $N(s)$ and $M(s)$ to be r.r.p. is that

$$
\text{rank}
\begin{bmatrix}
Q_0 & Q_1 & 0 & & \\
0 & Q_0 & Q_1 & & \Large 0 \\
 & & \ddots & \ddots & \\
\Large 0 & & & & Q_0 Q_1
\end{bmatrix}
= \text{rank}
\begin{bmatrix}
Q_0 \\
0 \\
\vdots \\
0
\end{bmatrix}
+ \text{rank}
\begin{bmatrix}
Q_1 & 0 & & & \\
Q_0 & Q_1 & & \Large 0 & \\
0 & Q_0 & \ddots & & \\
 & & \ddots & \ddots & \\
\Large 0 & & & & Q_0 Q_1
\end{bmatrix}
$$

$$(4.217)$$

where in (4.217) each of the matrices has m block rows.   Notice that the matrix on the left in (4.217) has the Sylvester-type form of (4.201).

4.3.4 G.C.D. of Polynomial Matrices

We noted in Section 4.3.1 that one method of obtaining a g.c.r.d. of of $N(s)$ ($r \times m$) and $M(s)$ ($m \times m$), as given in (4.197), is to reduce $P(s)$ in (4.213) using elementary row operations.   In fact, this process can be used to obtain a g.c.r.d. of more than two matrices (see Problem 4.46).   However, a major disadvantage is the need to carry out polynomial manipulations.   An attractive procedure for avoiding this work involves the Sylvester-type matrices S_k in Theorem 4.28.   Recall that it is necessary to determine the rank of the matrices S_k defined in (4.201), for increasing values of k. A natural way of computing these ranks would be to use gaussian elimination.   It turns out that if this reduction is carried out in a specified manner, which we shall now describe, then it also produces the coefficients of a g.c.r.d.   Begin with a trivial permutation of the block rows of S_k in (4.201) to give the equivalent matrix

$$
S_k' = \begin{bmatrix}
N_0 & N_1 & \cdots & N_p & 0 & \cdots & 0 \\
M_0 & M_1 & \cdots & M_p & 0 & \cdots & 0 \\
0 & N_0 & \cdots & N_{p-1} & N_p & \cdots & 0 \\
0 & M_0 & \cdots & M_{p-1} & M_p & \cdots & 0 \\
\cdot & \cdot & \cdot & \cdot & \cdot & \cdot & \cdot \\
0 & 0 & \cdots & \cdots & \cdots & \cdots & N_p \\
0 & 0 & \cdots & \cdots & \cdots & \cdots & M_p
\end{bmatrix} \quad \begin{array}{c} \\ \\ \text{2k block} \\ \text{rows} \\ \\ \end{array} \qquad (4.218)
$$

Reduce S_1' to row echelon form using only elementary row transformations, to produce a matrix E_1, say (of course, rank S_1' = rank E_1). To develop the next step, consider the form of S_2', which we see consists of the two block rows of S_1', extended by these two rows shifted by one block column to the right. Because of this property, instead of applying row transformations to S_2', we can use S_2'', which is formed in exactly the same way from E_1 as S_2' is from S_1', by appending a pair of block rows shifted to the right. Then reduce S_2'' by elementary row transformations to so-called "shifted" row echelon form E_2 (say), which has the same rank as S_2'. Note that row interchanges are not allowed in the procedure. A matrix S_3'' is then constructed by extending E_2 with its lowest pair of block rows shifted one block column to the right. Row transformations applied to S_3'' produce E_3, and the scheme is continued until $E_{\nu+1}$ is obtained, where ν is defined in Theorem 4.28.

<u>Theorem 4.31</u> If the Sylvester-type matrix S_k' in (4.218) is transformed according to the scheme above into shifted row echelon form, then the last nonzero pair of block rows of $E_{\nu+1}$ gives the coefficients of a g.c.r.d. of $N(s)$ and $M(s)$ in (4.197).

<u>Example 19</u> Return to $N(s)$ and $M(s)$ in (4.204) and (4.205), where $p = 3$. From (2.218) we can write down

$$S_1' = \begin{bmatrix} 0 & 0 & 0 & 1 & 2 & 0 & 1 & 1 \\ 0 & 0 & 1 & 1 & 2 & 2 & 1 & 0 \\ 0 & 1 & 2 & 0 & 3 & 4 & 5 & 2 \\ 0 & 0 & 1 & 1 & 1 & 1 & -1 & -1 \end{bmatrix}$$

A simple calculation gives the row echelon form of S_1':

$$E_1 = \begin{bmatrix} 0 & 1 & 2 & 0 & 3 & 4 & 5 & 2 \\ 0 & 0 & 1 & 1 & 1 & 1 & -1 & -1 \\ 0 & 0 & 0 & 1 & 2 & 0 & 1 & 1 \\ 0 & 0 & 0 & 0 & 1 & 1 & 2 & 1 \end{bmatrix}$$

and rank $E_1 = 4$. We construct

$$S_2'' = \begin{bmatrix} E_1 & 0_{42} \\ 0_{42} & E_1 \end{bmatrix} \begin{matrix} 4 \\ 4 \end{matrix}$$

where 0_{mn} denotes an m × n block of zeros. Then reduce S_2'' to shifted row echelon form by subtracting suitable multiples of the first four rows from the last four rows, and applying further appropriate transformations within the last four rows. This produces

$$E_2 = \begin{bmatrix} & & E_1 & & & & 0_{42} \\ & & 0 & 0 & 2 & 3 & 5 & 2 \\ 0_{44} & & 0 & 0 & 0 & 1 & 1 & 0 \\ & & 0 & 1 & 2 & 0 & 1 & 1 \\ & & 0 & 0 & 0 & 0 & 0 & 0 \end{bmatrix} \triangleq \begin{bmatrix} E_1 & 0_{42} \\ 0_{44} & E_2' \end{bmatrix} \qquad (4.219)$$

so that rank $E_2 = 7$ (= rank S_2'). The next construction is to extend E_2 with its last four rows to produce

$$S_3'' = \begin{bmatrix} E_2 & 0_{82} \\ 0_{46} & E_2' \end{bmatrix}$$

The reader can check that appropriate row operations (excluding permutation) reduce S_3'' to

$$E_3 = \begin{bmatrix} E_2 & 0_{82} \\ 0_{48} & E_3' \end{bmatrix}, \qquad E_3' = \begin{bmatrix} 2 & 3 & 5 & 2 \\ 0 & 1 & 1 & 0 \\ & & 0_{24} & \end{bmatrix}$$

Since rank E_3 = 9 (= rank S_3') we have rank S_3' - rank S_2' = 2 = m, so we conclude from Theorem 4.28 that ν = 2 (see Example 17). Hence by Theorem 4.31, a g.c.r.d. of N(s) and M(s) is given by the last nonzero pair of block rows of E_3, i.e.,

$$D(s) = \begin{bmatrix} 2 & 3 \\ 0 & 1 \end{bmatrix} s + \begin{bmatrix} 5 & 2 \\ 1 & 0 \end{bmatrix}$$

Remarks on Theorem 4.31   (1) When carrying out the reduction procedure any entire row of zeros can be dropped, as for example the last row of E_2 in (4.219).

(2) When N(s) and M(s) are both scalar polynomials (r = m = 1) the result reduces to Theorem 1.9; when r > 1, m = 1 it reduces to Theorem 1.10 in the g.c.d. of a set of r + 1 polynomials. However, the proof for the general case is quite complicated, and is omitted.

(3) In view of the relationship in Theorem 4.29 between the Sylvester- and bezoutian-type matrices, it is not surprising that the latter can also be used to derive a g.c.r.d., but this turns out to be a less attractive approach.

It is possible that the reader was surprised that we gave no method in Section 4.3.3 for testing relative primeness using the block companion matrix C_M in (4.108), associated with the *regular* matrix M(s) in (4.172). This omission will now be remedied, but the delay was deliberate, because it is necessary to make the *additional* assumption that M(s) and N(s) possess a g.c.r.d. D(s) which is *regular*, i.e., which can be written in the form

$$D(s) = I_m s^k + D_1 s^{k-1} + \cdots + D_{k-1} s + D_k \qquad (4.220)$$

for some nonnegative integer $k \leqslant p$ (= δM). This constitutes a significant drawback, since there is no known method for testing a

priori whether $D(s)$ has this property. Subject to this
restriction, however, a result of some interest can be obtained.
Suppose that NM^{-1} is strictly proper, so that in (4.197) $N(s)$ has
degree $p - 1$, i.e., $N_0 = 0$. Let R^{MN} denote the $mpr \times mp$
observability matrix $0(C_M, H_N)$ having block rows H_N, $H_N C_M$, $\ldots$,
$H_N C_M^{mp-1}$, where

$$H_N \triangleq [N_p, N_{p-1}, \ldots, N_1] \tag{4.221}$$

(recall that C_M is $mp \times mp$).

Theorem 4.32   Assume that $N(s)$ and $M(s)$ in (4.197) have $N_0 = 0$,
$M_0 = I_m$, and that they possess a g.c.r.d. in the form (4.220).
(i)   The degree of $D(s)$ is given by

$$k = p - \frac{1}{m} \text{ rank } R^{MN} \tag{4.222}$$

(ii) Consider the matrix consisting of the first $(p - k)rm$ rows of
R^{MN} and let its block columns be R_1, R_2, $\ldots$, R_p (each has
dimensions $(p - k)rm \times m$). Then R_{k+1}, R_{k+2}, $\ldots$, R_p are
linearly independent, and the coefficients of $D(s)$ in (4.220)
are given by solving the linear equations

$$R_i = R_{k+1} D_{k+1-i} + \sum_{j=k+2}^{p} R_j X_{ij}, \quad i = 1, 2, \ldots, k \tag{4.223}$$

Proof.
     (i) It is straightforward to verify (see Problem 4.49) that
$\{C_M, B, H_N\}$ is a c.c. state space realization of $N(s)M^{-1}(s)$, where
B is an $mp \times m$ matrix having all rows zero except the last block,
which consists of I_m. It follows from the dual of Theorem 4.3 that
this realization can be reduced to one which is both c.c. *and* c.o.,
and whose dimension is equal to rank R^{MN}. It was noted in Remark 2
on Theorem 4.19 that this dimension of a minimal realization is
equal to the degree of irreducible m.f.d.'s of NM^{-1}. In order to
evaluate this degree, we use the assumption that the g.c.r.d. $D(s)$
is regular. Thus if $M(s) = \hat{M}(s)D(s)$, then $\hat{M}(s)$ is also regular and

has degree p - k. By definition the degree of irreducible m.f.d.'s
of NM^{-1} is

$$\delta(\det \hat{M}) = m(p - k)$$

We have therefore established that

$$m(p - k) = \text{rank } R^{MN}$$

which is (4.222).

(ii) The proof of (4.223) follows very closely the argument
given for the g.c.d. of two scalar polynomials in Theorem 1.5, and
will be omitted. The reader should compare (4.223) with (1.77). ■

Remarks on Theorem 4.32   (1) The realization of $N(s)M^{-1}(s)$ used in
the proof of part (i) is a direct generalization of the standard
c.c. realization in (2.108) of a scalar transfer function.

(2) An immediate corollary to part (i) is that $N(s)$ and $M(s)$
are r.r.p. if and only if R^{MN} has full rank mp. This result is
applicable only when $M(s)$ is regular, as compared with Theorem
4.28, where $M(s)$ is merely required to be nonsingular. The matrix
R^{MN} can also be related to a block Sylvester matrix which is
essentially the same as S_k in (4.201). However, this is less
useful, since [in addition to the regularity of $M(s)$] it requires
taking k equal to mp, which will in general be larger than ν in
Theorem 4.28.

The final method to be described for determining a g.c.r.d. is
a matrix version of the Routh array. It was seen in Section 1.6.1
that the Routh scheme for scalar polynomials is equivalent to
performing Euclid's algorithm. Assuming that $N(s)$ and $M(s)$ are
both m × m, then Euclid's algorithm for obtaining a g.c.d. still
applies provided that $M(s)$ and each of the remainders is regular.
Let $N(s)$ and $M(s)$ be given by (4.197), with $N_0 = 0$ and M_0 assumed
nonsingular. Then a matrix Routh array of m × m matrices $\{R_{ij}\}$ can
be constructed according to

$$R_{0j} = M_{j-1}, \quad R_{1j} = N_j, \quad j = 1, 2, 3, \ldots \qquad (4.224)$$

$$R_{ij} = R_{i-2,j+1} - (R_{i-2,1}R_{i-1,1}^{-1})R_{i-1,j+1}, \quad i = 2, 3, \ldots;$$
$$j = 1, 2, \ldots \quad (4.225)$$

When m = 1, (4.232) reduces to the standard formula (1.134). As in
the scalar case, if a complete row of zero matrices is obtained
then the last nonvanishing row gives the coefficients of a g.c.r.d.

Example 20   Let

$$N(s) = \begin{bmatrix} 6 & -2 \\ -1 & -3 \end{bmatrix} s^2 + \begin{bmatrix} 6 & -6 \\ -1 & 0 \end{bmatrix} s + \begin{bmatrix} 2 & -4 \\ 0 & -1 \end{bmatrix}$$
$$\underset{N_1}{} \qquad \underset{N_2}{} \qquad \underset{N_3}{}$$

$$M(s) = \begin{bmatrix} 3 & 3 \\ 1 & -3 \end{bmatrix} s^3 + \begin{bmatrix} 7 & 1 \\ 1 & -5 \end{bmatrix} s^2 + \begin{bmatrix} 5 & -3 \\ 0 & -3 \end{bmatrix} s + \begin{bmatrix} 1 & -1 \\ 0 & -1 \end{bmatrix}$$
$$\underset{M_0}{} \qquad \underset{M_1}{} \qquad \underset{M_2}{} \qquad \underset{M_3}{}$$

It is left as an exercise for the reader to construct the array
defined by (4.224) and (4.225), and hence to confirm that the last
two rows are

$$R_{4j}: \begin{bmatrix} 1 & 1 \\ 1 & -1 \end{bmatrix} \quad \begin{bmatrix} 1 & -1 \\ 0 & -1 \end{bmatrix}$$

$$R_{51}: \begin{bmatrix} 0 & 0 \\ 0 & 0 \end{bmatrix}$$

from which it follows that a g.c.r.d. of N(s) and M(s) is

$$D(s) = \begin{bmatrix} 1 & 1 \\ 1 & -1 \end{bmatrix} s + \begin{bmatrix} 1 & -1 \\ 0 & -1 \end{bmatrix}$$

Clearly, the method breaks down if any first block column
matrix $R_{i-1,1}$ is singular, and this restricts its application.
However, in the cases when it can be successfully applied, the
matrix Routh array is attractive because (a) it only involves
matrices of order m, compared with the large matrices needed in
Theorems 4.31 and 4.32; and (b) it can also be employed to obtain
the factors $\hat{N}(s)$ and $\hat{M}(s)$ in (4.170) (see Problem 4.50).

Problems

4.38 Prove that if $N(s)$ and $M(s)$ satisfy the conditions of Theorem
4.22, then $D(s)$ is a common right divisor of $N(s)$, $M(s)$ if and
only if it is a common right divisor of $M(s)$ and the matrix
$R(s)$ in (4.161).

4.39 Let F be an arbitrary $\alpha \times \alpha$ matrix having eigenvalues θ_1, ...,
θ_α and corresponding eigenvectors u_1, ..., u_α. Show that if
μ_{ik}, x_{ik} are corresponding eigenvalue and eigenvector,
respectively, for the constant matrix $N(\theta_i)$, where $N(s)$ is
defined in (4.171), then μ_{ik} and $u_i \otimes x_{ik}$ are a corresponding
eigenvalue and eigenvector for the matrix   .

$$F^q \otimes N_0 + F^{q-1} \otimes N_1 + \cdots + I_\alpha \otimes N_q$$

4.40 Use the method of Problem 3.31 to show that equation (4.182)
has a unique solution F if and only if C_N and C_M have no
eigenvalue in common. Deduce using Theorem 4.23 that this is
equivalent to det $N(s)$ and det $M(s)$ being relatively prime.

4.41 Consider the matrix R'_{MN} defined in (4.178). Show that when
$q = p$, R'_{MN} is equal to the first p block rows of the
observability matrix $C(\tilde{C}_M, Z_p)$, where Z_p and $\tilde{C}_M$ are defined in
(4.192) and (4.193), respectively. When $r = m = 1$, then $\tilde{C}_M$ is
an ordinary companion matrix, and this result reduces to
Theorem 2.6; when $N(s)$ is $r \times m$ and $M(s)$ is 1×1, the result
reduces to that in (4.132).

4.42 With T defined in (4.191), and $\tilde{C}_M$ in (4.193), prove that

(i) $\tilde{C}_M^t T = T\tilde{C}_M$

where the superscript t denotes blockwise transposition.

(ii) $TC_I = I_{mpr}$

where C_I is defined in (4.194).
Hence, using the result in the preceding problem, verify
(4.195).

4.43 Use the result in Problem 4.24(i) to show that $\tilde{C}_M$ in (4.193) satisfies the condition $\tilde{C}_M^t Z = Z\tilde{C}_M$, where Z is the bezoutian matrix in Theorem 4.27, and the superscript t denotes block-wise transposition. Hence deduce that Z^{-1} is block ortho-symmetric (compare with the proof of Theorem 1.14 for the ordinary bezoutian matrix).

4.44 Show that if $N(s)$ and $M(s)$ are both $m \times m$, and have relatively prime determinants, then they are relatively right prime. [Hint: Determine the Smith form of $P(s)$ in (4.141).]

4.45 Show that the block elements Γ_{ij} defined by (4.206), (4.197), and (4.199), are given by

$$\Gamma_{ij} = \sum_{k \geq 0} (M_{L, \pi-i-k} N_{p-j+1+k} - N_{L, \pi-i-k} M_{p-j+1+k})$$

where $M_{Lr} = 0$ for r outside $[0, \pi]$, and similarly for the other terms. [Hint: Use (4.208).]

4.46 Consider a set $N_1(s)$, $N_2(s)$, $\ldots$, $N_k(s)$ of polynomial matrices, each with the same number of columns. Construct a matrix having N_1, $\ldots$, N_k as its block rows. Show that if this matrix is reduced by elementary row transformations to one whose block rows are $D(s)$, 0, 0, $\ldots$, 0, then $D(s)$ is a g.c.r.d. of the set.

4.47 Express the matrix U in (4.212) in terms of controllability/ observability-type matrices, to obtain the analogue of (3.212).

4.48 Show that in (4.216)

$$Q_0 = \begin{bmatrix} P_p & P_{p-1} & \cdots & P_2 & P_1 \\ 0 & P_p & \cdots & P_3 & P_2 \\ \cdot & \cdot & \cdot & \cdot & \cdot \\ 0 & 0 & \cdots & 0 & P_p \end{bmatrix}$$

4.49 Using the notation in the proof of Theorem 4.32 show that the last block column of $(sI_{mp} - C_M)^{-1}$ has entries $M^{-1}(s)$, $sM^{-1}(s)$, ..., $s^{p-1}M^{-1}(s)$ [compare with (2.67)]. Hence confirm that $\{C_M, B, H_N\}$ is a c.c. state space realization of $N(s)M^{-1}(s)$.

4.50 The matrix Routh array defined by (4.224) and (4.225) can be used to obtain the coefficients of $\hat{N}(s)$ and $\hat{M}(s)$ in (4.170) as follows. Define a set of matrices by

$$P_{2k+1,1} = I_m, \quad P_{i1} = H_i P_{i+1,1}, \quad i = 2k, \ 2k-1, \ \dots, \ 2, \ 1$$

$$P_{ij} = P_{i+2,j-1} + H_i P_{i+1,j}, \quad i = 2k-1, \ 2k, \ \dots, \ 1$$

where $H_i \triangleq R_{i-1,1} R_{i1}^{-1}$, these matrices being computed when (4.225) is used, and $2k$ is the number of nonzero H_i. Then

$$\hat{N}(s) = P_{21} s^{k-1} + P_{22} s^{k-2} + \cdots + P_{2k}$$

$$\hat{M}(s) = P_{11} s^k + P_{12} s^{k-1} + \cdots + P_{1,k+1}$$

Apply this scheme to Example 20. Note that the formulas reduce to (1.139) to (1.142) in the scalar case.

   Because of the dimensions of the matrices which arise in their solution, the following problems should preferably be tackled with the aid of a computer program for evaluating rank.

4.51 Apply Theorem 4.28 to the pair of matrices in Problem 4.33(i) to test whether they are r.r.p.

4.52 Consider the matrix $G(s)$ in (4.160). By writing $s = 1/t$, and expanding each element of $G(t)$ in ascending powers of t, obtain $G(s)$ in the form (4.210). Use the statement in Remark 7 on Theorem 4.30, involving a matrix in the form (4.211), to find the McMillan degree of $G(s)$ (see Example 14).

4.53 Apply the result in (4.217) to the matrices $N(s)$ and $M(s)$ in (4.204) and (4.205).

4.54 Apply each of Theorems 4.31 and 4.32 to obtain a g.c.r.d. of

$$N(s) = \begin{bmatrix} s^2 + 8s + 29 & 2s^2 + 4s + 7 \\ s + 1 & 2s + 1 \end{bmatrix}$$

$$M(s) = \begin{bmatrix} s^3 + 8s^2 + 10s - 15 & 3s^2 + 8s - 1 \\ 5s^2 + 46s + 15 & s^3 - 4s^2 + 7s + 3 \end{bmatrix}$$

BIBLIOGRAPHICAL NOTES

An unrivaled source for further details and references on much of the material of this chapter (especially Sections 4.1 and 4.2) is the book by Kailath (1980).

Section 4.1.1. The original proof of Theorem 4.1 was obtained by Wonham (1967) for real systems. The simplified version given, which is valid for systems over an arbitrary field, relies on Lemmas 4.1.2 and 4.1.1, due to Heymann (1968) and Hautus (1977), respectively. Hautus (1970) and Eising (1982) have given alternative proofs of Theorem 4.1. Remark 5 on Theorem 4.1 is part of a series of similar tests devised by Davison et al. (1978); see also the comment on stabilizability, after Example 2, and Paige (1981). Retallack and MacFarlane (1970) gave Theorem 4.2, together with the formula for the case when A has repeated eigenvalues, and further extensions have been made by Kouvaritakis and Cameron (1980), and Cameron and Kouvaritakis (1981). Several other methods for eigenvalue assignment by state feedback can be found in the collections of articles edited by Fallside (1977) and Munro (1979). Bumby et al. (1981) considered the problem over a ring. Our treatment of Theorem 4.3 follows Chen (1970), based on original work of Kalman (1963). For eigenvalue assignment when A and B in (4.1) are polynomial matrices, arising from delay systems, see Lee and Zak (1982).

Section 4.1.2. The development of canonical forms for multivariable systems was initiated by Luenberger (1967). A survey has
been given by Maroulas and Barnett (1979b), and a full and excel-
lent coverage is contained in Kailath's book (1980), with further
insight provided by Antoulas (1981). Theorem 4.4 is due to
Rosenbrock (1970, p. 190), and Remark 3 to Brunovsky (1970), with
further results by Rosenbrock and Hayton (1978). Alternative
proofs of Theorem 4.4 have been obtained by Dickinson (1974) using
state space concepts, and by Flamm (1980). Simple schemes for
computing canonical forms have been devised by Daly (1976) and
Aplevich (1979); the latter includes other applications. For
relationships between controllability indices and so-called minimal
indices, see Koussiouris (1981).

Section 4.1.3. The books by Anderson and Moore (1971), Kwakernaak
and Sivan (1972), and Russell (1979) are good sources here, and
indeed for other parts of Section 4.1. The first use of the
description *matrix* Riccati equation seems to have been by Reid in
1948; see his book (1972) for references and historical remarks.
Kalman (1960) introduced the terminology into control theory. Some
general theorems on these equations are contained in Chapter 5 of
Barnett (1971a). A survey of the algebraic Riccati equation (4.58)
has been given by Martensson (1971) and of the optimal control
problem, by Molinari (1977). The extension of Theorem 4.5, Remark
5, was made by Wonham (1968), with further investigation of the
stabilizing solution carried out by Molinari (1973) and Simaan
(1974). The proof of Theorem 4.6, and the inertia result in
Remark 3, are due to Wimmer (1976). Equation (4.58) with all
matrices complex has been studied by Rodman (1980) and Lancaster
and Rodman (1980), with further development by Wimmer (1982).
Theorem 4.7 is due to Anderson (1966), and a proof of the result in
Remark 1 on Theorem 4.7 was given by Wimmer (1976), based on Coppel
(1974a). The eigenvector solution of the algebraic Riccati
equation originated with MacFarlane (1963b), with an extension to
the case of the matrix Riccati differential equation (Remark 6 on

Theorems 4.5 and 4.6) being made by Walter (1972). Laub (1979) gave the method mentioned in Remark 3 on Theorem 4.8, and Paige and Van Loan (1981) made further improvements. For the discrete-time equation (4.71), see Dorato and Levis (1971) (although this contains some errors), Kucera (1972), and Laub (1979). A procedure involving a generalized eigenvalue problem has been developed by Pappas et al. (1980) and Van Dooren (1981). Roberts (1980) suggested a completely different approach using the matrix sign function. It should be noted that Riccati equations also arise in optimal filtering problems (Anderson and Moore, 1971, 1979; Russell, 1979).

Section 4.1.4. The concept of observers dates back to Luenberger (1964), and is covered in all the standard texts quoted so far for this chapter. The article by Crossley and Porter (1979) is also worth consulting. The statement after Example 6, on the value of q, is from Sridhar and Lindorff (1973). A proof of Theorem 4.11 is contained in a paper by Fletcher (1979). This is part of a series (Porter, 1977; Porter and Bradshaw, 1978; Bradshaw et al., 1978; Fletcher, 1980, 1981a,b,c) in which the method, illustrated in Example 7, has been developed. This provides a considerable extension and improvement of the work of Kimura (1975), who introduced the condition $m + r > n$. We have not had space to describe interesting alternative approaches to output feedback using a generalized inverse matrix (Munro, 1974; Lovass-Nagy et al., 1978, 1981) or a Kronecker product formulation (Power, 1977). Problem 4.12 is based on O'Donnell (1966), and further material on hamiltonian matrices can be found in Laub and Meyer (1974) and Laub (1979). The idea of a prescribed degree of stability in Problem 4.17 is discussed in some detail by Anderson and Moore (1971, p. 53). Problem 4.20 is worked out in full by Bradshaw et al. (1978).

Sections 4.2.1, 4.2.2. Our treatment follows Barnett (1971a, 1975a), Chen (1970), and Kailath (1980), where references can be

found to the original sources for this mainly standard material.
Worthy of particular mention are papers by Kalman (1962, 1963).
Gilbert (1963) gave the scheme of Theorem 4.16, which has been
extended to deal with repeated roots by Shaked and Dixon (1977).
A numerically efficient procedure for state space realization has
been devised by Davison et al. (1978), and Lovass-Nagy et al.
(1977) have suggested using a generalized matrix inverse. The
expressions (4.129) and (4.132) were derived by Barnett (1972b).
For reduction of a polynomial matrix to companion form, see Lee et
al. (1982).

Section 4.2.3. An unequaled source for the whole of this section
is again Kailath (1980), and the monograph by Heymann (1975) is
also most valuable. Basic properties of polynomial and rational
matrices are discussed in Barnett (1971a). The canonical form of
Theorem 4.17 was originally introduced by Smith (1861) for matrices
with integer elements. Theorem 4.18 is also classical (MacDuffee,
1950b, p. 35). The results in Remark 3 on Theorem 4.18 are due,
respectively, to Rosenbrock (1970) and Kucera (1980). A study of
the polynomial equation (4.148) has been made by Emre (1980a).
Theorem 4.20 derives its name from McMillan (1952), whose work in
network theory was first applied to the realization problem by
Kalman (1965b). A thorough treatment of poles and zeros has been
given by MacFarlane and Karcanias (1976), and applications to the
design of multivariable systems were described by MacFarlane
(1975). Some recent ideas on the structure of rational matrices
were summarized by Verghese and Kailath (1981). For a full study
of the approach mentioned in Remark 3 on Theorem 4.19, see Rosenbrock (1970) (also Coppel, 1974b). The concept of a column proper
matrix was introduced by Wolovich (1974), and his book contains a
useful treatment of realization theory. Numerical computation of
matrix fraction descriptions was discussed by Patel (1981). A
proof of the standard form of Theorem 4.22, when M(s) is regular,
is contained in Cullen (1966). Problem 4.21 follows Retallack and
MacFarlane (1970), and is a basic step in proving Theorem 4.2.

Problems 4.30 and 4.31 are taken from Barnett (1968, 1972b), respectively, and Problem 4.32 from Rosenbrock (1970). Kailath (1980) gave further results along the lines of those in Problems 4.34 and 4.35. The division theorem of Flood (1935) in Problem 4.37 was resurrected by Barnett (1971a).

Section 4.3.1. Rosenbrock (1970) pioneered the application of polynomial matrices to the theory of multivariable control systems and a review of applications of relatively prime matrices has been made by Antsaklis (1979). Fuhrmann (1976, 1979, 1981) has done important work on relationships between the polynomial and module theoretic approaches to systems theory.

Section 4.3.2. Theorems 4.24, 4.25, and 4.26 are taken from Barnett (1969a), but the proof given here of the first result is more direct, and was outlined in Barnett (1971a). Remark 1 on Theorem 4.24 was given by Newman (1972) for integer matrices [see also Newman (1982)]. For an application of (4.179) to a control problem, see Bengtsson (1977). The extension in Remark 3 on Theorems 4.25 and 4.26 is due to Feinstein and Bar-Ness (1980), and skew prime matrices, mentioned in Remark 5, were developed by Wolovich (1978). Theorem 4.27 and the expressions (4.195) are due to Barnett and Lancaster (1980b), but the bezoutian matrix in (4.189) originates with Heinig (1977). The relation between the Sylvester and bezoutian matrices in Remark 3 on Theorem 4.27 was given explicitly by Barnett and Lancaster (1980b), and the formula (4.196) is due to Barnett (1969b). The proof of Theorem 1.6, referred to in Remark 4 on Theorem 4.27, can be found in Barnett (1971b).

Section 4.3.3. The generalized Sylvester and bezoutian matrices (4.201) and (4.206), respectively, were introduced by Anderson and Jury (1976). Their paper contains Theorem 4.29 and the remarks thereon, and also essentially presents Theorem 4.28. The actual form of the latter which we have quoted, however, follows Bitmead

et al. (1978), and their paper contains full details of Examples 17
and 19. The Sylvester-type matrix (4.201) generalizes earlier work
of Wolovich (1971) and Rowe (1972). A factorization of Γ defined
by (4.206), corresponding to that for Z in Theorem 4.27, has been
given by Lerer and Tismenetsky (1981). They have also used Γ to
determine the location of the roots of det M(s) = 0. The matrix
Cauchy index and Theorem 4.30 are due to Bitmead and Anderson
(1977). For Remark 7 on Theorems 4.28 and 4.29, see Koussiouris
and Kafiris (1981). The result for the case when M(s) is not
square can be found, together with others, in a paper by Emre and
Silverman (1977). A further development of the Sylvester form of
resultant, including the case when NM^{-1} is not proper, has been
made by Hayton (1980). There are also connections with important
work by Forney (1975) on rational vector spaces.

Section 4.3.4. Theorem 4.31 and the remarks thereon follow Bitmead
et al. (1978). Theorem 4.32 is due to Barnett (1972c), and the
results in Remark 2 to Rowe (1972). For an extension of Theorem
4.32 to more than two polynomial matrices, see Maroulas and
Dascalopoulos (1981). The matrix Routh array was proposed by Shieh
et al. (1978), together with further applications. The inversion
procedure for the scalar Routh array, described in Problem 1.41,
has been extended to the matrix case (Shieh, 1975), and a second
type of matrix Routh array has been devised (Shieh and Tajvari,
1980) which also has applications to control problems. Other
system-theory-based method for g.c.d. determination have been proposed by Emre (1980b), Gohberg et al. (1981, 1983), and Silverman
and Van Dooren (1981). Problem 4.39 is a special case of a result
of Williamson (1931). Problems 4.41 and 4.42 are based on Barnett
and Lancaster (1980b), and the result of Problem 4.44 was noted by
Barnett (1969a). The formula in Problem 4.45 was given by Anderson
and Jury (1976). An algorithm for applying the result in Problem
4.46 was described by Pace and Barnett (1974). The scheme in
Problem 4.50 follows Shieh et al. (1978).

5

Generalized Polynomials and Polynomial Matrices

5.0 INTRODUCTION

Up to now in this book, all polynomials (and polynomial matrices) have been in the familiar "power" form, as initially defined in (1.1). In this chapter we consider the situation when a polynomial $a(\lambda)$, of degree $\delta a = n$, is expressed in a so-called *generalized form*

$$a(\lambda) = \sum_{i=0}^{n} k_i f_{n-i}(\lambda) \qquad (5.1)$$

where the set of real polynomials $f_0(\lambda)$, $f_1(\lambda)$, ..., $f_n(\lambda)$, with $\delta f_i = i$, constitutes an arbitrary basis. In fact, the main thrust of our development concentrates on the case when the basis consists of a set $\{p_i(\lambda)\}$ which is mutually orthogonal with respect to some interval and weighting function. A key idea is to obtain various generalizations of the companion matrix, and this is described in Section 5.1, where the so-called "congenial" matrices (colleagues, comrades, and confederates) are introduced. In Section 5.2.1 we extend the g.c.d. approach of Section 1.3 for scalar polynomials by using the comrade or confederate matrix in place of the companion matrix. This enables the g.c.d. of polynomials to be obtained directly in generalized form. Similarly, in Section 5.2.3, we

369

extend Theorem 4.32 for polynomial matrices relative to a general basis by using block congenial forms to replace the block companion matrix. Another method which we have used extensively in this book, Routh's tabular array, is modified in Section 5.2.2 to deal with generalized polynomials. Section 5.3 presents applications to some standard topics in linear control systems, and a brief mention of some further applications is made in Section 5.4.

It is interesting that quite a few of the results in previous chapters involving companion matrices are seen to be merely special cases of the theorems in this chapter.

5.1 CONGENIAL MATRICES

5.1.1 Orthogonal Basis: Comrade and Colleague

Define a set of real polynomials

$$p_i(\lambda) = \sum_{j=0}^{i} p_{ij}\lambda^j, \quad i = 0, 1, 2, \ldots \tag{5.2}$$

which satisfy the relationships

$$p_0(\lambda) = 1, \quad p_1(\lambda) = \alpha_1\lambda + \beta_1$$

$$p_i(\lambda) = (\alpha_i\lambda + \beta_i)p_{i-1}(\lambda) - \gamma_i p_{i-2}(\lambda), \quad i \geq 2 \tag{5.3}$$

with $\alpha_i > 0$, $\gamma_i > 0$. It can be shown that there exist a weight function $w(\lambda) \geq 0$, and an interval $[c, d]$, such that

$$\int_c^d w(\lambda) \, d\lambda > 0$$

and

$$\int_c^d p_i(\lambda)p_j(\lambda)w(\lambda) \, d\lambda = 0, \quad i \neq j$$
$$\neq 0, \quad i = j \tag{5.4}$$

i.e., the sequence $\{p_i(\lambda)\}$ is orthogonal with respect to $w(\lambda)$ on $c \leq \lambda \leq d$. The converse result is usually presented first, namely

that if the orthogonality condition (5.4) is satisfied, then the
polynomials (5.2) satisfy the recurrence formula (5.3) (assuming
that $p_{ii} > 0$, for all i). In fact, we shall not make use of any
properties of orthogonality as such. Note, however, that if an
arbitrary polynomial $a(\lambda)$ of degree n is expressed relative to the
orthogonal basis (5.2), namely

$$a(\lambda) = a_0 p_n(\lambda) + a_1 p_{n-1}(\lambda) + \cdots + a_n p_0(\lambda) \qquad (5.5)$$

then (5.4) shows that the coefficients in (5.5) are uniquely
determined by

$$a_i = \int_c^d a(\lambda) p_i(\lambda) w(\lambda) \ d(\lambda) \ / \left[\int_c^d p_i^2(\lambda) w(\lambda) \ d\lambda \right] \qquad (5.6)$$

It is trivial to see from (5.2) and (5.3) that

$$p_{ii} = \alpha_1 \alpha_2 \cdots \alpha_i > 0, \quad i \geqslant 1 \qquad (5.7)$$

so in particular, the coefficient of λ^n in (5.5) is $a_0 \alpha_1 \alpha_2 \cdots \alpha_n$.
Setting $\alpha_i = 1$, $\beta_i = 0$, $\gamma_i = 0$ in (5.3) produces the "power" basis
$p_i(\lambda) = \lambda^i$, but it must be stressed that this basis is *not*
orthogonal, since it does not satisfy (5.3) with $\gamma_i > 0$. However,
allowing this substitution for the values of the parameters in
(5.3) enables the familiar forms for "standard" polynomials to be
recovered.

Consider now the *power* form of $a(\lambda)$ in (5.5), assuming for
convenience that $a_0 = 1$. The *monic* form of this polynomial is

$$\tilde{a}(\lambda) = a(\lambda)/(\alpha_1 \alpha_2 \cdots \alpha_n)$$

$$\triangleq \lambda^n + \tilde{a}_1 \lambda^{n-1} + \cdots + \tilde{a}_{n-1}\lambda + \tilde{a}_n \qquad (5.8)$$

It is straightforward to verify, using (5.2), that the relationship
between the coefficients in (5.5) and (5.8) is

$$(\alpha_1 \alpha_2 \cdots \alpha_n)[\tilde{a}_n, \tilde{a}_{n-1}, \ldots, \tilde{a}_1, 1]$$

$$= [a_n, a_{n-1}, \ldots, a_1, 1]P_{n+1} \qquad (5.9)$$

where P_n is the lower triangular $n \times n$ matrix having rows

$$[p_{i0}, p_{i1}, \ldots, p_{ii}, 0, \ldots, 0], \quad i = 0, 1, \ldots, n - 1 \quad (5.10)$$

and the p_{ij} are defined by (5.2) and (5.7). The interest lies in deducing properties of $a(\lambda)$ in the generalized form (5.5), without converting to power form via (5.9). We need a preliminary result:

Lemma 5.1.1 The polynomials defined by (5.3) can be expressed, for $i = 1, 2, 3, \ldots$, as

$$p_i(\lambda) = \det \begin{bmatrix} \alpha_1\lambda + \beta_1 & -1 & & \cdot & \cdot & \cdot \\ -\gamma_2 & \alpha_2\lambda + \beta_2 & -1 & & & \cdot \\ 0 & -\gamma_3 & \alpha_3\lambda + \beta_3 & \cdot & & \vdots \\ & & \cdot & \cdot & \cdot & -1 \\ & 0 & & \cdot & -\gamma_i & \alpha_i\lambda + \beta_i \end{bmatrix}$$

$$(5.11)$$

Proof. We use induction. The result is obvious for $i = 1, 2$. In general, expansion of (5.11) by the last row establishes the recurrence formula in (5.3). ∎

We can now obtain the counterpart to Theorem 1.1, for polynomials in the generalized form (5.5).

Theorem 5.1 The $n \times n$ comrade matrix A in (5.12) has characteristic polynomial $a(\lambda)/(\alpha_1 \cdots \alpha_n)$, where

$$
A \triangleq
\begin{bmatrix}
\dfrac{-\beta_1}{\alpha_1} & \dfrac{1}{\alpha_1} & 0 & 0 & & & \\[2mm]
\dfrac{\gamma_2}{\alpha_2} & \dfrac{-\beta_2}{\alpha_2} & \dfrac{1}{\alpha_2} & 0 & & \mathbf{0} & \\[2mm]
0 & \dfrac{\gamma_3}{\alpha_3} & \dfrac{-\beta_3}{\alpha_3} & \dfrac{1}{\alpha_3} & & & \\[2mm]
 & & \ddots & \ddots & \ddots & & \\[2mm]
\mathbf{0} & & & \ddots & \ddots & \ddots & \\[2mm]
 & & & \dfrac{\gamma_{n-1}}{\alpha_{n-1}} & \dfrac{-\beta_{n-1}}{\alpha_{n-1}} & \dfrac{1}{\alpha_{n-1}} \\[2mm]
\dfrac{-a_n}{\alpha_n} & \dfrac{-a_{n-1}}{\alpha_n} & \cdots & \dfrac{-a_3}{\alpha_n} & \dfrac{-a_2+\gamma_n}{\alpha_n} & \dfrac{-a_1-\beta_n}{\alpha_n}
\end{bmatrix}
\tag{5.12}
$$

and

$$
a(\lambda) = p_n(\lambda) + a_1 p_{n-1}(\lambda) + \cdots + a_n p_0(\lambda)
\tag{5.13}
$$

Proof. Let $D = \text{diag}[\alpha_1, \alpha_2, \ldots, \alpha_n]$, and consider

$$
\det(\lambda D - DA) = \det D \, \det(\lambda I - A)
$$
$$
= (\alpha_1 \alpha_2 \cdots \alpha_n) \, \det(\lambda I - A)
\tag{5.14}
$$

The first $n - 1$ rows of $\det(\lambda D - DA)$ are the same as those of the matrix in (5.11), with $i = n$. Using this fact, it is straightforward to expand $\det(\lambda D - DA)$ by its last row, to produce

$$
\det(\lambda D - DA) = a_n + a_{n-1} p_1(\lambda) + a_{n-2} p_2(\lambda) + \cdots
$$
$$
+ (a_2 - \gamma_n) p_{n-2}(\lambda) + (a_1 + \alpha_n \lambda + \beta_n) p_{n-1}(\lambda)
$$

$$
\tag{5.15}
$$

Using the expression for p_n in terms of p_{n-1} and p_{n-2} in (5.3) shows that (5.15) is identical with $a(\lambda)$ in (5.13). The required result follows at once from (5.14). ∎

Example 1 Consider

$$\tilde{a}(\lambda) = \lambda^4 - \lambda^3 - 4\lambda^2 + 4\lambda \quad [\equiv \lambda(\lambda - 2)(\lambda - 1)(\lambda + 2)] \quad (5.16)$$

and express this relative to a basis of *Legendre polynomials*, defined by (5.3) with

$$\alpha_1 = 1, \quad \beta_1 = 0, \quad \alpha_i = \frac{2i - 1}{i}, \quad \beta_i = 0, \quad \gamma_i = \frac{i - 1}{i}, \quad i \geqslant 2$$
$$(5.17)$$

It is trivial to obtain the explicit expressions

$$p_0(\lambda) = 1, \quad p_1(\lambda) = \lambda, \quad p_2(\lambda) = \tfrac{1}{2}(3\lambda^2 - 1),$$

$$p_3(\lambda) = \tfrac{1}{2}(5\lambda^3 - 3\lambda), \quad p_4(\lambda) = \tfrac{1}{8}(35\lambda^4 - 30\lambda^2 + 3) \quad (5.18)$$

Substituting for powers of λ from (5.18) puts $\tilde{a}(\lambda)$ in (5.16) into the generalized form

$$\frac{8}{35} p_4(\lambda) - \frac{2}{5} p_3(\lambda) - \frac{44}{21} p_2(\lambda) + \frac{17}{5} p_1(\lambda) - \frac{17}{15} p_0(\lambda)$$

Multiplying by (35/8) so as to make the leading coefficient unity yields

$$a(\lambda) = p_4(\lambda) - \frac{7}{4} p_3(\lambda) - \frac{55}{6} p_2(\lambda) + \frac{119}{8} p_1(\lambda) - \frac{119}{24} p_0(\lambda)$$
$$(5.19)$$

which is in the form of (5.13).

    In this chapter we are interested in the situation where a polynomial is *given* in the form (5.19), together with the relationships (5.17), which define the basis. The reader should construct the matrix P_5 for this problem, using the definition (5.10) and the coefficients in (5.18), and verify that (5.9) holds.

    Using (5.17), we can write down the comrade matrix (5.12) for $a(\lambda)$ in (5.19):

$$
A = \begin{bmatrix}
0 & 1 & 0 & 0 \\
\dfrac{1}{3} & 0 & \dfrac{2}{3} & 0 \\
0 & \dfrac{2}{5} & 0 & \dfrac{3}{5} \\
\dfrac{17}{6} & \dfrac{-17}{2} & \dfrac{17}{3} & 1
\end{bmatrix}
\qquad (5.20)
$$

Remarks on Theorem 5.1   (1) The characteristic polynomial of A in
(5.12) is, of course, *monic* in power form, which accounts for the
factor $(\alpha_1 \alpha_2 \cdots \alpha_n)^{-1}$ in the statement of the theorem. If instead
of (5.13) we use (5.5) with $a_0 = (\alpha_1 \cdots \alpha_n)^{-1}$, then $a(\lambda)$ is itself
monic with respect to power form. It is left as a simple exercise
for the reader to show that in this case $a(\lambda)$ in (5.5) is the
characteristic polynomial of a comrade form matrix obtained by
replacing a_i in (5.12) by $a_i(\alpha_1 \cdots \alpha_n)$, i = 1, 2, ..., n.

(2) When $\alpha_i = 1$, $\beta_i = 0$, $\gamma_i = 0$, for all i, then A in (5.12)
reduces to the usual companion form, first defined in Theorem 1.1.
When $\alpha_i = 1$, $\beta_i = 0$, $\gamma_i = 1$, for all i, then A in (5.12) is like
the companion form, but with an extra stripe of 1's below the
principal diagonal. This matrix was called the *colleague* form by
I. J. Good, using a synonymous word beginning with the letter "c."
The term *comrade* thus continues this whimsical tradition.
Incidentally, the colleague matrix corresponds to a basis $p_0 = 1$,
$p_1 = \lambda$, $p_i = \lambda p_{i-1} - p_{i-2}$, $i \geqslant 2$, and these are a set of *Chebyshev*
polynomials denoted by $S_i(\lambda)$.

(3) There are three other possible comrade forms, obtained
from A in the same way that the three alternative companion forms
in (1.30), (1.31), and (1.34) are obtained from the first form
(1.26).

(4) Some other special cases of (5.12) besides the companion
and colleague matrices are worth pointing out. These are the
Schwarz form defined in (3.154), the closely related *Routh* matrix
defined in Problem 3.35, and the *Leslie* matrix

$$
\begin{bmatrix}
f_1 & f_2 & \cdots & f_{n-1} & f_n \\
g_1 & 0 & & & \\
0 & g_2 & & 0 & \\
 & & \ddots & & \\
0 & & & g_{n-1} & 0
\end{bmatrix}
$$

which arises in population models, and is a special case of $J_n A J_n$.

<u>Theorem 5.2</u> The comrade matrix A in (5.12) has the properties:

(i) A is similar to the companion matrix C associated with $\tilde{a}(\lambda)$ in
(5.8), i.e.,

$$
C = \left[\begin{array}{c|c}
0 & I_{n-1} \\
\hline
-\tilde{a}_n & \cdots \quad -\tilde{a}_1
\end{array} \right]
\tag{5.21}
$$

according to

$$
A = P_n C P_n^{-1}
\tag{5.22}
$$

where P_n is defined by (5.10).

(ii) When A has distinct eigenvalues $\lambda_1, \ldots, \lambda_n$, then its right
eigenvectors are

$$
u(\lambda_i) = [1, p_1(\lambda_i), p_2(\lambda_i), \ldots, p_{n-1}(\lambda_i)]^T,
$$

$$
i = 1, 2, \ldots, n \tag{5.23}
$$

and

$$
M^{-1} A M = \text{diag}[\lambda_1, \lambda_2, \ldots, \lambda_n]
\tag{5.24}
$$

where

$$
M \triangleq [u(\lambda_1), u(\lambda_2), \ldots, u(\lambda_n)]
\tag{5.25}
$$

Proof. (i) To establish (5.22) we show that

$$
A P_n v(\lambda) \equiv P_n C v(\lambda)
\tag{5.26}
$$

where $v(\lambda) = [1, \lambda, \lambda^2, \ldots, \lambda^{n-1}]^T$. The first step is to note that (5.2) and (5.9) imply that

$$P_n v(\lambda) = u(\lambda) \tag{5.27}$$

where $u(\lambda)$ is defined by (5.23). It follows directly from (5.12) and (5.27) that the ith element of the vector on the left side of (5.26) is

$$\frac{1}{\alpha_i}(\gamma_i p_{i-2}(\lambda) - \beta_i p_{i-1}(\lambda) + p_i(\lambda)), \quad i = 1, 2, \ldots, n - 1$$

which is equal simply to $\lambda p_{i-1}(\lambda)$, in view of the recurrence formula (5.3). The last element on the left side of (5.26) is

$$\frac{1}{\alpha_n}(-a_n p_0 - a_{n-1} p_1 \cdots -a_1 p_{n-1} + \gamma_n p_{n-2} - \beta_n p_{n-1})$$

$$= \frac{1}{\alpha_n}(-a(\lambda) + p_n + \gamma_n p_{n-2} - \beta_n p_{n-1}) \tag{5.28}$$

by (5.13). A further application of (5.3) reduces (5.28) to $\lambda p_{n-1}(\lambda) - a(\lambda)/\alpha_n$. Consider now the right-hand side of (5.26). It follows from (5.21) that

$$Cv(\lambda) = \lambda v(\lambda) - [0, 0, \ldots, 0, \tilde{a}(\lambda)]^T \tag{5.29}$$

where $\tilde{a}(\lambda)$ is defined in (5.8). Premultiplying (5.29) by P_n and using (5.27) then easily establishes the identity of the elements of $P_n Cv(\lambda)$ with those given above for the left side of (5.26).

(ii) Setting $\lambda = \lambda_1, \lambda_2, \ldots, \lambda_n$ in (5.27) shows that

$$P_n V = M \tag{5.30}$$

where V is the Vandermonde matrix, defined in (1.23), which has columns $v(\lambda_1), v(\lambda_2), \ldots, v(\lambda_n)$. Now [see (1.56)]

$$V^{-1}CV = \text{diag}[\lambda_1, \ldots, \lambda_n] \tag{5.31}$$

where C is the matrix in (5.21), which has the same eigenvalues as A, by the similarity condition (5.22). Substitution of (5.22) and

(5.30) into (5.31) readily produces (5.24). ∎

Remarks on Theorem 5.2   (1) The matrix (5.25) could be called a "generalized" Vandermonde matrix.

(2) A formula corresponding to (5.24) can be derived for the case when A has repeated eigenvalues (see Problem 5.7).

5.1.2 Confederate Matrix

A form of basis more general than $\{p_i(\lambda)\}$ consists of the real polynomials defined by

$$f_0(\lambda) = 1, \quad f_1(\lambda) = \delta_{11}\lambda + \varepsilon_{11},$$

$$f_i(\lambda) = \sum_{k=1}^{i} (\delta_{ik}\lambda + \varepsilon_{ik})f_{k-1}(\lambda), \quad i \geq 2 \tag{5.32}$$

where $\delta_{ii} > 0$, for all i. If we select

$$\delta_{ii} = \alpha_i, \quad \varepsilon_{ii} = \beta_i, \quad \varepsilon_{i,i-1} = -\gamma_i, \quad \text{for all } i$$

$$\delta_{ij} = \varepsilon_{ij} = 0, \quad \text{otherwise} \tag{5.33}$$

then (5.32) reduces to (5.3), i.e., the $f_i(\lambda)$ coincide with the $p_i(\lambda)$.

In general, since $f_i(\lambda)$ has degree i, we may assume from now on without loss of generality that $\delta_{ii} = \alpha_i$, for all i, so that $f_i(\lambda)$ in (5.32) and $p_i(\lambda)$ in (5.3) have the same leading coefficients. Now suppose that each $f_i(\lambda)$ is expressed in terms of the basis (5.3) according to

$$f_i(\lambda) = \sum_{j=0}^{i} c_{ij} p_j(\lambda) \tag{5.34}$$

The c_{ij} are uniquely determined by a formula like (5.6), and our assumption on δ_{ii} implies that $c_{ii} = 1$, for all i. If we define L_n to be the lower triangular n × n matrix having rows

$$[c_{i0}, c_{i1}, \ldots, c_{i,i-1}, 1, 0, \ldots, 0], \quad i = 0, 1, \ldots, n - 1 \tag{5.35}$$

then (5.34) can be written as

$$[1, f_1(\lambda), f_2(\lambda), \ldots, f_{n-1}(\lambda)]^T = L_n u(\lambda) \qquad (5.36)$$

where $u(\lambda)$ is defined by (5.23).

Consider again the polynomial $a(\lambda)$ in (5.13), and suppose that its coefficients relative to the new basis (5.32) are denoted by $\hat{a}_i$, so we have

$$a(\lambda) = f_n(\lambda) + \hat{a}_1 f_{n-1}(\lambda) + \cdots + \hat{a}_n f_0(\lambda) \qquad (5.37)$$

The coefficient of λ^n in (5.37) is $\alpha_1 \alpha_2 \cdots \alpha_n$, which is the *same* as that in (5.13). Hence the relationship between the coefficients in (5.13) and (5.37) is

$$[a_n, a_{n-1}, \ldots, a_1, 1] = [\hat{a}_n, \hat{a}_{n-1}, \ldots, \hat{a}_1, 1] L_{n+1} \qquad (5.38)$$

which corresponds to the formula (5.9). Again, we wish to investigate the properties of $\hat{a}(\lambda)$ in (5.37) directly, without invoking (5.38). This requires yet a further generalization of the companion matrix, and corresponding to Theorem 5.1 we have:

Theorem 5.3   The *confederate matrix*

$$E \triangleq \begin{bmatrix} d_{11} & \frac{1}{\alpha_1} & 0 & 0 & & & \\ d_{21} & d_{22} & \frac{1}{\alpha_2} & 0 & & 0 & \\ d_{31} & d_{32} & d_{33} & \frac{1}{\alpha_3} & & & \\ \cdot & & \cdot & & \cdot & \cdot & \cdot \\ \cdot & & \cdot & & \cdot & d_{n-1,n-1} & \frac{1}{\alpha_{n-1}} \\ d_{n1} - \frac{\hat{a}_n}{\alpha_n} & d_{n2} - \frac{\hat{a}_{n-1}}{\alpha_n} & \cdot & \cdot & \cdot & d_{nn} - \frac{\hat{a}_1}{\alpha_n} \end{bmatrix}$$

$$\qquad (5.39)$$

where

$$d_{ij} = -\frac{1}{\alpha_i}\left(\epsilon_{ij} + \frac{\delta_{i,j-1}}{\alpha_{j-1}} + \sum_{k=j}^{i-1} \delta_{ik}d_{kj}\right), \qquad i, j = 1, \ldots, n,$$

$$i \geqslant j \qquad (5.40)$$

has the properties:

(i)   The characteristic polynomial of E is $a(\lambda)/(\alpha_1 \cdots \alpha_n)$, where
       $a(\lambda)$ is defined in (5.37).

(ii) $E = L_n A L_n^{-1}$ (5.41)

       where L_n is defined by (5.35), and A is the comrade matrix
       (5.12) for the orthogonal basis representation of $a(\lambda)$.

(iii) When F has distinct eigenvalues $\lambda_1, \ldots, \lambda_n$, then its right
       eigenvectors are

$$w(\lambda_i) = [1, f_1(\lambda_i), f_2(\lambda_i), \ldots, f_{n-1}(\lambda_i)]^T,$$

$$i = 1, 2, \ldots, n \quad (5.42)$$

       and

$$N^{-1}EN = diag[\lambda_1, \lambda_2, \ldots, \lambda_n] \qquad (5.43)$$

       where

$$N \triangleq [w(\lambda_1), w(\lambda_2), \ldots, w(\lambda_n)] \qquad (5.44)$$

Proof. We first establish (5.41) by showing that

$$EL_n u(\lambda) \equiv L_n A u(\lambda) \qquad (5.45)$$

where $u(\lambda)$ is defined by (5.23). This parallels the proof of part
(i) of Theorem 5.2, so only an outline of the argument will be
given. First, note that the ith component of the vector on the
left side of (5.45) is, on using (5.36),

$$d_{i1} + d_{i2}f_1(\lambda) + \cdots + d_{ii}f_{i-1}(\lambda) + \frac{1}{\alpha_i}f_i(\lambda),$$

$$i = 1, 2, \ldots, n - 1$$

This can be shown equal to $\lambda f_{i-1}(\lambda)$, after some manipulations using
(5.34) and (5.40). The last element on the left side of (5.45) can
similarly be shown equal to $\lambda f_{n-1}(\lambda) - a(\lambda)/\alpha_n$, where $a(\lambda)$ is the
form in (5.37). On the right-hand side of (5.45) it is easy to
show that

$$Au(\lambda) = \lambda u(\lambda) - [0, 0, \ldots, 0, a(\lambda)/\alpha_n]^T \qquad (5.46)$$

where $a(\lambda)$ has the form in (5.13). Premultiplying both sides of
(5.46) by L_n and using (5.36) then establishes the identity (5.45).

The proof of (i) then follows immediately from (5.41), which
shows that E and A have the same characteristic polynomial.

Finally, it is left as an easy exercise for the reader to
verify (5.42) and (5.43) by setting $\lambda = \lambda_1, \ldots, \lambda_n$ in (5.36) and
using (5.24) and (5.41). ∎

Remarks on Theorem 5.3   (1) The name for (5.39) was chosen in the
spirit of Remark 2 on Theorem 5.1, and this also accounts for the
use of *congenial matrices* as a generic term. The reader can check
that when (5.33) applies, the confederate matrix (5.39) reduces to
the comrade matrix (5.12).

(2) The formula (5.43) can be modified when F has repeated
eigenvalues; N in (5.44) is yet another "generalized" Vandermonde
matrix.

(3) The reader should appreciate that P_n in (5.9) represents
the transformation matrix between power and orthogonal bases, and
similarly L_n in (5.35) is the transformation matrix from the
orthogonal basis to that in (5.32). Hence the similarity relationships (5.22) and (5.41) merely reflect a standard result in linear
algebra on representations relative to different bases. Of
specific interest, however, are the explicit forms of the comrade
and the confederate matrices as displayed in (5.12) and (5.39). In
particular, all congenial matrices exhibit Hessenberg form, which
arises in a variety of applications. Furthermore, a square matrix
is nonderogatory if and only if it is similar to a companion matrix

(see Remark 4 on Theorem 1.1), so (5.22) and (5.41) show that this similarity holds for any congenial matrix.

Problems

5.1 The *Chebyshev polynomials* $\{T_i(\lambda)\}$ of the *first kind* are defined by $T_i(\lambda)$ = cos iθ, λ = cos θ, i = 0, 1, 2, ... Using the trigonometric identity

2cos iθ cos jθ $\equiv$ cos (i + j)θ + cos (i - j)θ,

$$i, j = 0, 1, 2, \ldots$$

prove that

$$\int_0^\pi \cos i\theta \, \cos j\theta \, d\theta = 0, \quad i \neq j$$

and hence prove:

(i) $\{T_i(\lambda)\}$ is an orthogonal polynomial sequence with respect to the weight function $w(\lambda) = (1 - \lambda^2)^{-1/2}$ on the interval $[-1, 1]$.

(ii) $T_i(\lambda) = 2\lambda T_{i-1}(\lambda) - T_{i-2}(\lambda)$, $\quad$ i = 2, 3, ...

5.2 The *Chebyshev polynomials* $\{U_i(\lambda)\}$ of the *second kind* are defined by

$$U_i(\lambda) = \frac{\sin (i + 1)\theta}{\sin \theta}, \quad \lambda = \cos \theta, \quad i = 0, 1, 2, \ldots$$

Prove:

(i) $\{U_i(\lambda)\}$ is an orthogonal polynomial sequence with respect to the weight function $w(\lambda) = (1 - \lambda^2)^{1/2}$ on $[-1, 1]$.

(ii) $U_i(\lambda) = 2\lambda U_{i-1}(\lambda) - U_{i-2}(\lambda)$, $\quad$ i = 2, 3, ...

5.3 The *Hermite polynomials* $\{H_i(\lambda)\}$ are defined by

$$H_i(\lambda) = (-1)^i e^{\lambda^2} D^i (e^{-\lambda^2}), \quad i = 1, 2, \ldots$$

where $D^i \equiv d^i/d\lambda^i$ and $H_0 = 1$. Use Leibnitz's differentiation formula to show that

$$D^i(e^{-\lambda^2}) = -2\lambda D^{i-1}(e^{-\lambda^2}) - 2(i-1)D^{i-2}e^{-\lambda^2}$$

Hence deduce that

$$H_i(\lambda) = 2\lambda H_{i-1}(\lambda) - 2(i-1)H_{i-2}(\lambda), \quad i \geqslant 2$$

5.4 Write down the appropriate comrade matrix for each of the cases when $a(\lambda)$ is expressed in terms of the orthogonal bases of the preceding three problems.

5.5 Consider the matrix M defined in (5.25).

(i) Using the result in (1.24), show that

$$\det M = \alpha_1^{n-1}\alpha_2^{n-2} \cdots \alpha_{n-2}^2\alpha_{n-1} \prod_{1\leqslant i<j\leqslant n}(\lambda_j - \lambda_i)$$

(ii) Use (5.24) to show that if the comrade matrix A in (5.12) has all its eigenvalues distinct, then

$$(MM^T)^{-1}A(MM^T) = A^T$$

Compare this with the companion matrix result in (1.57).

5.6 Consider the matrix N defined in (5.44). Show that

$$(NN^T)^{-1}E(NN^T) = E^T$$

assuming that all the eigenvalues of the confederate matrix E in (5.39) are distinct. This extends the result in Problem 5.5(ii).

5.7 Suppose that an $n \times n$ companion matrix C has the eigenvalue λ_1 repeated n times, and define

$$V_c = \left[v(\lambda), \frac{dv}{d\lambda}, \frac{1}{2!}\frac{d^2v}{d\lambda^2}, \cdots, \frac{1}{(n-1)!}\frac{d^{n-1}v}{d\lambda^{n-1}}\right]_{\lambda=\lambda_1}$$

where $v(\lambda) = [1, \lambda, \lambda^2, \ldots, \lambda^{n-1}]^T$. Show that $CV_c \equiv V_cJ$, where J is an $n \times n$ upper Jordan block for λ_1. Use (5.27) to show that if A is the comrade matrix similar to C via (5.22), then $M_c^{-1}AM_c = J$, where

$$M_c \triangleq \left[u(\lambda), \frac{du}{d\lambda}, \frac{1}{2!}\frac{d^2u}{d\lambda^2}, \cdots, \frac{1}{(n-1)!}\frac{d^{n-1}u}{d\lambda^{n-1}} \right]_{\lambda = \lambda_1}$$

with $u(\lambda)$ defined by (5.23). Deduce the corresponding formula involving E in (5.39).

5.8 Apply Gershgorin's theorem (quoted in Problem 1.15) to the comrade matrix A in (5.12) in order to show that the roots of the generalized polynomial $a(\lambda)$ in (5.13) lie in the union of the disks

$$|z + (\beta_1/\alpha_1)| \leq \frac{1}{\alpha_1}, \quad |z + (\beta_i/\alpha_i)| \leq (1 + \gamma_i)/\alpha_i,$$

$$i = 2, \ldots, n-1$$

$$|z + (\beta_n + a_1)/\alpha_n| \leq (\sum_{j=3}^{n}|a_j| + |\gamma_n - a_2|)/\alpha_n$$

Write down the corresponding result obtained using A^T.

5.2 GREATEST COMMON DIVISORS

5.2.1 G.C.D. Using Congenial Matrices
Let

$$b(\lambda) = b_0 p_m(\lambda) + b_1 p_{m-1}(\lambda) + \cdots + b_m p_0(\lambda) \tag{5.47}$$

be a second polynomial expressed relative to the orthogonal basis (5.2). Our development for the problem of determining a g.c.d. $d(\lambda)$ of $a(\lambda)$ in (5.13) and $b(\lambda)$ in (5.47) will parallel that for the power form case in Sections 1.2 and 1.3. Thus we can assume, without loss of generality, that $m < n$. Consider

$$d(\lambda) = p_k(\lambda) + d_1 p_{k-1}(\lambda) + \cdots + d_k p_0(\lambda), \quad k < n \tag{5.48}$$

relative to the same orthogonal basis (5.2), and having unit leading coefficient. The first step is to construct the following polynomial in the comrade matrix (5.12):

$$b(A) \triangleq b_0 p_m(A) + b_1 p_{m-1}(A) + \cdots + b_m p_0(A) \tag{5.49}$$

where by (5.2), $p_i(A) \triangleq \sum p_{ij} A^j$. The expression (5.49) is the counterpart to (1.37). If we denote the power form of (5.47) by

$$\tilde{b}(\lambda) = \tilde{b}_0 \lambda^m + \tilde{b}_1 \lambda^{m-1} + \cdots + \tilde{b}_m \qquad (5.50)$$

then

$$b(A) \equiv \tilde{b}(A)$$
$$= P_n \tilde{b}(C) P_n^{-1} \qquad (5.51)$$

using the similarity condition (5.22). Since (5.51) shows that $b(A)$ and $b(C)$ have the same rank, it follows that Theorems 1.3 and 1.4, on the degree of the g.c.d., will carry over directly to the generalized case. In fact, we shall show that Theorem 1.5, on the determination of the coefficients of the g.c.d., also applies here. However, we first need to obtain the analogue of Theorem 1.2.

Theorem 5.4 For $b(\lambda)$ in (5.47) with $m < n$, and A in (5.12), the matrix $b(A)$ has rows

$$r_1, \; r_1 p_1(A), \; r_1 p_2(A), \; \ldots, \; r_1 p_{n-1}(A) \qquad (5.52)$$

where

$$r_1 = [b_m, \; b_{m-1}, \; \ldots, \; b_1, \; b_0, \; 0, \; \ldots, \; 0] \qquad (5.53)$$

Proof. As usual, let e_i denote the ith row of I_n, so that the ith row of $b(A)$ is

$$r_i = e_i b(A) \qquad (5.54)$$

Inspection of A in (5.12) shows that its first row is

$$e_1 A = (-\beta_1 e_1 + e_2)/\alpha_1 \qquad (5.55)$$

so that

$$r_2 = e_2 b(A)$$

$$= e_1(\alpha_1 A + \beta_1 I)b(A), \quad \text{by (5.55)}$$

$$= [e_1 b(A)](\alpha_1 A + \beta_1 I)$$

$$= r_1 p_1(A)$$

We prove the subsequent terms in (5.52) by induction. Suppose that

$$r_i = r_1 p_{i-1}(A) \tag{5.56}$$

which we have just established for i = 2, and consider

$$r_{i+1} = e_{i+1} b(A) \tag{5.57}$$

Again from (5.12), the ith row of A is

$$e_i A = (\gamma_i e_{i-1} - \beta_i e_i + e_{i+1})/\alpha_i, \quad i = 2, \ldots, n - 1 \tag{5.58}$$

and substitution of the latter into (5.57) gives

$$r_{i+1} = (e_i \alpha_i A + \beta_i e_i - \gamma_i e_{i-1})b(A)$$

$$= e_i b(A)(\alpha_i A + \beta_i I) - \gamma_i e_{i-1} b(A)$$

$$= r_i(\alpha_i A + \beta_i I) - \gamma_i r_{i-1}, \quad \text{by (5.54)}$$

$$= r_1 p_{i-1}(A)(\alpha_i A + \beta_i I) - \gamma_i r_1 p_{i-2}(A), \quad \text{by (5.56)}$$

$$= r_1[(\alpha_i A + \beta_i I)p_{i-1}(A) - \gamma_i p_{i-2}(A)]$$

$$= r_1 p_i(A)$$

where the last step follows from the recurrence formula (5.3) with λ replaced by A. Since (5.56) holds for i = 2, it must also hold for i = 3, 4, ..., n - 1, which completes the proof of (5.52). Finally, to obtain the expression (5.53) for r_1, note that rearranging (5.55) gives

$$e_1 p_1(A) = e_1(\alpha_1 A + \beta_1 I) = e_2$$

and we again use induction: Suppose that

$$e_1 p_{i-1}(A) = e_i \qquad (5.59)$$

which we have just shown holds for $i = 2$. Then

$$e_1 p_i(A) = e_1(\alpha_i A + \beta_i I) p_{i-1}(A) - e_1 \gamma_i p_{i-2}(A), \qquad \text{by } (5.3)$$

$$= e_1 p_{i-1}(A)(\alpha_i A + \beta_i I) - \gamma_i e_1 p_{i-2}(A)$$

$$= e_i(\alpha_i A + \beta_i I) - \gamma_i e_{i-1}, \qquad \text{by } (5.59)$$

$$= e_{i+1}, \qquad \text{by } (5.58)$$

showing that (5.59) holds for $i = 3, 4, \ldots, n - 1$. Hence from (5.54)

$$r_1 = e_1 b(A) = b_0 e_1 p_m(A) + b_1 e_1 p_{m-1}(A) + \cdots + b_m e_1 p_0(A)$$

$$= b_0 e_{m+1} + b_1 e_m + \cdots + b_m e_1$$

which is precisely (5.53). ∎

We can now state and prove the promised generalizations of Theorems 1.3, 1.4, and 1.5:

Theorem 5.5

(i) The degree k of the g.c.d. of $a(\lambda)$ in (5.13) and $b(\lambda)$ in (5.47) is equal to $n - \text{rank } b(A)$; in particular, $b(A)$ is non-singular if and only if $a(\lambda)$ and $b(\lambda)$ are relatively prime.

(ii) The columns $c_1, c_2, \ldots, c_n$ of $b(A)$ are such that $c_{k+1}, \ldots, c_n$ are linearly independent, and the coefficients $d_1, \ldots, d_k$ in (5.48) are given by

$$c_i = d_{k+1-i} c_{k+1} + \sum_{j=k+2}^{n} x_{ij} c_j, \qquad i = 1, 2, \ldots, k \qquad (5.60)$$

for some x_{ij}.

Proof. Since part (i) follows immediately from (5.51) and Theorems 1.3 and 1.4, it remains to establish part (ii). We can first notice that Theorem 1.5 (in particular, Remark 1) shows that the last n - k columns $\tilde{c}_{k+1}$, ..., $\tilde{c}_n$ (say) of $\tilde{b}(C)$ are linearly independent. The independence of c_{k+1}, ..., c_n then follows from (5.51), since that equation can be written as

$$[c_1, c_2, ..., c_n] = [P_n\tilde{c}_1, P_n\tilde{c}_2, ..., P_n\tilde{c}_n]P_n^{-1} \qquad (5.61)$$

The nonsingularity of P_n implies that $P_n\tilde{c}_{k+1}$, ..., $P_n\tilde{c}_n$ are linearly independent. Since P_n^{-1} is lower triangular with nonzero elements on the principal diagonal [see (5.10)], postmultiplication by P_n^{-1} in (5.61) does not affect the independence of these last n - k columns.

To obtain the final expression (5.60), consider $d(\lambda)$. Since it is a factor of $a(\lambda)$, it follows from what we have proved so far that the last n - k columns δ_{k+1}, ..., δ_n (say) of $d(A)$ are linearly independent. Furthermore, application of (5.53) to $d(A)$ shows that its first row is

$$[d_k, d_{k-1}, ..., d_1, 1, 0, 0, ..., 0]$$

Hence, if we express the first k columns δ_1, δ_2, ..., δ_k of $d(A)$ in terms of the last n - k, we must obtain

$$\delta_i = d_{k+1-i}\delta_{k+1} + \sum_{j=k+2}^{n} y_{ij}\delta_j, \quad i = 1, 2, ..., k \qquad (5.62)$$

for some y_{ij}. Finally, if $b(\lambda) = b_1(\lambda)d(\lambda)$, then

$$b(A) = b_1(A)d(A) \qquad (5.63)$$

where, by definition, $a(\lambda)$ and $b_1(\lambda)$ are relatively prime. Hence by part (i), $b_1(A)$ is nonsingular, and (5.63) shows that $c_i = b_1(A)\delta_i$, i = 1, 2, ..., n. Thus, premultiplication of (5.62) by $b_1(A)$ produces (5.60). ∎

Remarks on Theorems 5.4 and 5.5 (1) The expressions (5.52) reduce
to (1.43) when $p_i(\lambda) = \lambda^i$, $A \equiv C$. Notice, however, that in order
to compute the rows r_1, r_2, $\ldots$, r_n of $b(A)$, we can use (5.56) and
the recurrence formula (5.3) with λ replaced by A, to obtain

$$r_i = r_1 p_{i-1}(A)$$

$$= r_1 [p_{i-2}(A)(\alpha_{i-1}A + \beta_{i-1}I)] - \gamma_{i-1} r_1 p_{i-3}(A)$$

$$= r_{i-1}(\alpha_{i-1}A + \beta_{i-1}I) - \gamma_{i-1} r_{i-2}, \quad i = 3, \ldots, n \quad (5.64)$$

with $r_2 = r_1(\alpha_1 A + \beta_1 I)$, and r_1 given by (5.53). The recurrence
relationship (5.64) between the rows of $b(A)$ is precisely the same
as that between the polynomials in the basis, and provides a convenient method for constructing $b(A)$.

(2) Similarly, when $A \equiv C$, Theorem 5.5 reduces to Theorem 1.5,
in the form using $b(C)$, as in (1.77).

Example 2 Let $a(\lambda)$ be the polynomial in Example 1, equation (5.19),
and take

$$b(\lambda) = 3p_3(\lambda) - 10p_2(\lambda) - 3p_1(\lambda) + 10p_0(\lambda) \quad (5.65)$$

which is expressed relative to the same basis of Legendre polynomials $p_i(\lambda)$ in (5.18). The comrade matrix A associated with $a(\lambda)$
was given in (5.20), so from (5.52) and (5.64) we can write down

$$b(A) = \begin{bmatrix} r_1 \\ r_2 = r_1 A \\ r_3 = r_2\left(\frac{3}{2}A\right) - \frac{1}{2}r_1 \\ r_4 = r_3\left(\frac{5}{3}A\right) - \frac{2}{3}r_2 \end{bmatrix} = \begin{bmatrix} 10 & -3 & -10 & 3 \\ \frac{15}{2} & \frac{-39}{2} & 15 & -3 \\ \frac{-55}{2} & 60 & -40 & \frac{15}{2} \\ \frac{255}{4} & \frac{-663}{4} & \frac{255}{2} & \frac{-51}{2} \end{bmatrix}$$
$$\qquad\qquad\qquad\qquad\qquad\qquad\quad c_1 \quad\;\; c_2 \quad\;\; c_3 \quad\;\; c_4$$

where we have used (5.17) for the values of the parameters in
(5.64). It is readily determined that $\text{rank}[b(A)] = 2$, and that

$$c_1 = \frac{7}{2} c_3 + 15c_4, \quad c_2 = -\frac{9}{2} c_3 - 16c_4$$

so by (5.60) the g.c.d. of $a(\lambda)$ and $b(\lambda)$ is

$$d(\lambda) = p_2(\lambda) - \frac{9}{2} p_1(\lambda) + \frac{7}{2} p_0(\lambda) \tag{5.66}$$

It is instructive to repeat the problem, using the power forms of $a(\lambda)$ and $b(\lambda)$. The former was given in (5.16), and substituting for the $p_i(\lambda)$ from (5.18) into (5.65) gives the corresponding *monic* polynomial

$$\tilde{b}(\lambda) = \lambda^3 - 2\lambda^2 - \lambda + 2 \quad [\equiv (\lambda - 1)(\lambda + 1)(\lambda - 2)] \tag{5.67}$$

The companion matrix for $\tilde{a}(\lambda)$ in (5.16) is

$$C = \begin{bmatrix} 0 & 1 & 0 & 0 \\ 0 & 0 & 1 & 0 \\ 0 & 0 & 0 & 1 \\ 0 & -4 & 4 & 1 \end{bmatrix}$$

and from Theorem 1.2

$$\tilde{b}(C) = \begin{bmatrix} \tilde{r}_1 \\ \tilde{r}_2 = \tilde{r}_1 C \\ \tilde{r}_3 = \tilde{r}_2 C \\ \tilde{r}_4 = \tilde{r}_3 C \end{bmatrix} = \begin{bmatrix} 2 & -1 & -2 & 1 \\ 0 & -2 & 3 & -1 \\ 0 & 4 & -6 & 2 \\ 0 & -8 & 12 & -4 \end{bmatrix}$$
$$\quad\quad \tilde{c}_1 \ \ \tilde{c}_2 \ \ \tilde{c}_3 \ \ \tilde{c}_4$$

Again rank$[\tilde{b}(C)] = 2$, and

$$\tilde{c}_1 = 2\tilde{c}_3 + 6\tilde{c}_4, \quad \tilde{c}_2 = -3\tilde{c}_3 - 7\tilde{c}_4$$

so by Theorem 1.5 [using the expression (1.77)] the g.c.d. in power form is

$$\tilde{d}(\lambda) = \lambda^2 - 3\lambda + 2 \equiv (\lambda - 2)(\lambda - 1)$$

which can be verified from the factorized forms of $\tilde{a}(\lambda)$ and $\tilde{b}(\lambda)$ in (5.16) and (5.67). The reader can also check, using the expressions for the Legendre polynomials in (5.18), that the power form of $d(\lambda)$ in (5.66) is $\frac{3}{2}\tilde{d}(\lambda)$.

(3) Remark 2 on Theorem 1.5 applies equally here; the attractions of the method illustrated above are twofold. First, the g.c.d. problem is reduced to solving a set of linear equations. Second, an extension to deal with a set of polynomials, all relative to the same basis, follows in the same way as does Theorem 1.6 from Theorem 1.5. Specifically, let $d(\lambda)$ in (5.48) now represent the g.c.d. of $a(\lambda)$ in (5.13), and the polynomials

$$b_i(\lambda) = b_{i0}p_{n-1}(\lambda) + b_{i1}p_{n-2}(\lambda) + \cdots + b_{i,n-1}p_0(\lambda),$$

$$i = 1, 2, \ldots, h \quad (5.68)$$

Then the degree of the g.c.d., and its coefficients $d_1, \ldots, d_k$, are given by exactly the same expressions as in Theorem 5.5, with $b(A)$ replaced by the matrix having block rows $b_1(A)$, $b_2(A)$, $\ldots$, $b_h(A)$. We will repeat the procedure adopted in Section 1.3 and consider the more conveniently displayed transpose:

$$[b_1(A^T), b_2(A^T), \ldots, b_h(A^T)] \quad (5.69)$$

By Theorem 5.4, the columns of $b_i(A^T)$ are

$$c_{i1}, p_1(A^T)c_{i1}, p_2(A^T)c_{i1}, \ldots, p_{n-1}(A^T)c_{i1} \quad (5.70)$$

where

$$c_{i1} = [b_{i,n-1}, b_{i,n-2}, \ldots, b_{i0}]^T \quad (5.71)$$

In the expression (5.60), the c_i are now replaced by the *rows* of the matrix in (5.69). The relationships between these rows are unaltered by a permutation of the columns in (5.69). Taking together all the first columns of $b_1(A^T), \ldots, b_h(A^T)$, then all the second columns, and so on, and using (5.70) and (5.71) produces:

Theorem 5.6

(i) The degree k of the g.c.d. $d(\lambda)$ of $a(\lambda)$ in (5.13) and $b_1(\lambda)$,
    ..., $b_h(\lambda)$ in (5.68) is equal to n-rank R_g, where R_g is the
    $n \times nh$ matrix

$$R_g \triangleq [B, p_1(A^T)B, p_2(A^T)B, \ldots, p_{n-1}(A^T)B] \tag{5.72}$$

and

$$B = \begin{bmatrix} b_{1,n-1} & b_{2,n-1} & \cdot & b_{h,n-1} \\ b_{1,n-2} & b_{2,n-2} & \cdot & \cdot \\ \cdot & \cdot & \cdot & \\ b_{10} & b_{20} & \cdot & b_{h0} \end{bmatrix} \tag{5.73}$$

(ii) The rows ρ_1, ρ_2, ..., ρ_n of R_g are such that ρ_{k+1}, ..., ρ_n are
    linearly independent, and the coefficients d_1, ..., d_k in
    (5.48) are given by

$$\rho_i = d_{k+i-1}\rho_{k+1} + \sum_{j=k+2}^{n} x_{ij}\rho_j, \qquad i = 1, \ldots, k \tag{5.74}$$

for some x_{ij}.

Remarks on Theorem 5.6   (1) When $p_i(\lambda) = \lambda^i$, $A \equiv C$, the result
reduces to Theorem 1.6. Again, by using the recurrence formula
(5.3), with λ replaced by A^T, the ith $n \times h$ block in (5.72) is
expressible as

$$R_i \triangleq p_i(A^T)B = [(\alpha_i A^T + \beta_i I)p_{i-1}(A^T) - \gamma_i p_{i-2}(A^T)]B$$

$$= (\alpha_i A^T + \beta_i I)R_{i-1} - \gamma_i R_{i-2}, \qquad i = 3, \ldots, n \tag{5.75}$$

and (5.75) provides a convenient way of constructing R_g.

    (2) The expression (5.72) will be given a control theory
interpretation in Section 5.3 (see Remark 2 on Theorem 5.9).

    (3) The proofs of Theorems 5.5 and 5.6 essentially rely on the
similarity of the comrade and companion matrices. Hence, in view

of Theorem 5.4, and in particular the similarity condition (5.41)
between the confederate and comrade matrices, the reader should
have no difficulty in confirming that Theorems 5.5 and 5.6 still
apply when all the polynomials are expressed relative to the basis
$\{f_i(\lambda)\}$ in (5.32). The sole difference is that A is replaced by
the confederate matrix E in (5.39) (see Problem 5.12 for the
corresponding version of Theorem 5.4).

5.2.2 Routh-Type Tabular Scheme

We saw in Section 1.6.1 that Euclid's algorithm for determining the
g.c.d. of two polynomials in power form can be conveniently performed using a method due to Routh. To develop a corresponding
procedure for the generalized case, it is necessary to appreciate
that division for such polynomials cannot be carried out directly
unless they have equal degrees. This is because there is no way of
expressing the remainder of $p_n(\lambda)$ divided by $p_m(\lambda)$ in generalized
form, without converting each into power form, which we wish to
avoid. For simplicity of exposition, consider two cubics,
expressed relative to the basis in (5.3):

$$a(\lambda) = a_0 p_3(\lambda) + \cdots + a_3 p_0(\lambda), \quad b(\lambda) = b_0 p_3(\lambda) + \cdots$$
$$+ b_3 p_0(\lambda) \quad (5.76)$$

Construct the first three rows of an array $\{r_{ij}\}$ according to the
usual Routh rule:

$$
\begin{array}{ccccc}
r_{0j} & a_0 & a_1 & a_2 & a_3 \\
r_{1j} & b_0 & b_1 & b_2 & b_3 \\
r_{2j} & r_{21} & r_{22} & r_{23} &
\end{array}
$$

where, following (1.136),

$$r_{2j} = \frac{-1}{b_0} \begin{vmatrix} a_0 & a_j \\ b_0 & b_j \end{vmatrix}, \quad j = 1, 2, 3 \quad (5.77)$$

This is equivalent to obtaining the remainder on division of $a(\lambda)$

by $b(\lambda)$, namely

$$c_1(\lambda) = a(\lambda) - \frac{a_0}{b_0} b(\lambda)$$

$$= r_{21}p_2(\lambda) + r_{22}p_1(\lambda) + r_{23}p_0(\lambda)$$

If the polynomials were in power form, the next step in Euclid's algorithm would be to divide $b(\lambda)$ by $c_1(\lambda)$, but this cannot be done since $\delta c_1 < \delta b$. To overcome this difficulty, we multiply $c_1(\lambda)$ by λ, and express this in generalized form:

$$c_1(\lambda) = z_{21}p_3(\lambda) + z_{22}p_2(\lambda) + z_{23}p_1(\lambda) + z_{24}p_0(\lambda) \qquad (5.78)$$

and *then* apply the Routh rule to the rows

$$
\begin{array}{cccc}
b_0 & b_1 & b_2 & b_3 \\
z_{21} & z_{22} & z_{23} & z_{24}
\end{array}
$$

to obtain a new row $\{r_{3j}\}$, with

$$r_{3j} = \frac{-1}{z_{21}} \begin{vmatrix} b_0 & b_j \\ z_{21} & z_{2,j+1} \end{vmatrix}, \quad j = 1, 2, 3 \qquad (5.79)$$

Determination of the coefficients z_{2j} in (5.78) is very easy, but we still postpone explanation of this point for the moment. The new row defined by (5.79) has the same number of elements as $\{r_{2j}\}$ in (5.77) (i.e., the two corresponding polynomials have the same degrees), so we can construct $\{r_{4j}\}$ from

$$r_{4j} = \frac{-1}{r_{31}} \begin{vmatrix} r_{21} & r_{2,j+1} \\ r_{31} & r_{3,j+1} \end{vmatrix}, \quad j = 1, 2 \qquad (5.80)$$

This relates to a remainder

$$c_2(\lambda) = r_{41}p_1(\lambda) + r_{42}p_0(\lambda)$$

from which we construct

$$\lambda c_2(\lambda) = z_{41}p_2(\lambda) + z_{42}p_1(\lambda) + z_{43}p_0(\lambda)$$

We then obtain $\{r_{5j}\}$ from

$$r_{5j} = \frac{-1}{r_{41}} \begin{vmatrix} r_{31} & r_{3,j+1} \\ z_{41} & z_{4,j+1} \end{vmatrix}, \quad j = 1, 2 \tag{5.81}$$

and finally

$$r_{61} = \frac{-1}{r_{51}} \begin{vmatrix} r_{41} & r_{42} \\ r_{51} & r_{52} \end{vmatrix} \tag{5.82}$$

The set of operations described above can be summarized in Table 5.1. In this table, each row immediately below a pair of rows linked by a brace is constructed according to the usual Routh rule, as in (5.77) and (5.79) to (5.82). The coefficients of the polynomials $\lambda c_1(\lambda)$ and $\lambda c_2(\lambda)$ are written in an auxiliary array on the right, as indicated by arrows. The computation alternates from left to right, the rows being calculated in the sequence indicated by the numbers within parentheses. If at some stage, a row in the left half of the array in Table 5.1 consists entirely of zeros, then the preceding row gives the coefficients of a g.c.d. of $a(\lambda)$ and $b(\lambda)$, relative to the same basis $\{p_i(\lambda)\}$. The procedure is

Table 5.1   Generalized Routh Scheme for (5.76)

a_0	a_1	a_2	a_3				
b_0	b_1	b_2	b_3	b_0	b_1	b_2	b_3
(1) r_{21}	r_{22}	$r_{23} \longrightarrow$		z_{21}	z_{22}	z_{23}	z_{24}
r_{31}	r_{32}	r_{33}		r_{31}	r_{32}	r_{33} (2)	
(3) r_{41}	$r_{42} \longrightarrow$			z_{41}	z_{42}	z_{43}	
r_{51}	r_{52}			r_{51}	r_{52} (4)		
(5) r_{61}							

applied in exactly the same way when $a(\lambda)$ and $b(\lambda)$ have equal
degrees n, greater than 3.

It is now appropriate to show how the coefficients z_{ij} are
obtained. We simply arrange the recurrence relationship (5.3) as
follows:

$$\lambda p_{i-1}(\lambda) = \frac{1}{\alpha_i} p_i(\lambda) - \frac{\beta_i}{\alpha_i} p_{i-1}(\lambda) + \frac{\gamma_i}{\alpha_i} p_{i-2}(\lambda), \quad i \geqslant 1 \quad (5.83)$$

which expresses $\lambda p_{i-1}(\lambda)$ in terms of $p_i(\lambda)$, $p_{i-1}(\lambda)$ and $p_{i-2}(\lambda)$
[with $p_{-1}(\lambda) \stackrel{\Delta}{=} 0$]. Notice that in Table 5.1 the rows $\{r_{2j}\}$, $\{r_{4j}\}$
are transformed. In general we have

$$\lambda(r_{2i,1}p_{n-i}(\lambda) + r_{2i,2}p_{n-i-1}(\lambda) + \cdots + r_{2i,n-i+1}p_0(\lambda)$$

$$\equiv z_{2i,1}p_{n-i+1}(\lambda) + z_{2i,2}p_{n-i}(\lambda) + \cdots + z_{2i,n-i+2}p_0(\lambda) \quad (5.84)$$

Application of (5.83) to (5.84) shows that the z's are obtained
directly from the r's according to

$$[z_{2i,1}, z_{2i,2}, \ldots, z_{2i,n-i+2}] = [r_{2i,1}, r_{2i,2}, \ldots$$

$$\ldots, r_{2i,n-i+1}]Q_i,$$

$$i = 1, 2, 3, \ldots \quad (5.85)$$

where Q_i is the $(n - i + 1) \times (n - i + 2)$ matrix

$$Q_i = \begin{bmatrix} \dfrac{1}{\alpha_{n-i+1}} & \dfrac{-\beta_{n-i+1}}{\alpha_{n-i+1}} & \dfrac{\gamma_{n-i+1}}{\alpha_{n-i+1}} & 0 & & & \\ & & & & & & 0 \\ 0 & \dfrac{1}{\alpha_{n-i}} & \dfrac{-\beta_{n-i}}{\alpha_{n-i}} & \dfrac{\gamma_{n-i}}{\alpha_{n-i}} & & & \\ & & \cdot & \cdot & \cdot & & \\ & & & \cdot & \cdot & \cdot & \\ & & & & \dfrac{1}{\alpha_2} & \dfrac{-\beta_2}{\alpha_2} & \dfrac{\gamma_2}{\alpha_2} \\ & 0 & & & & \dfrac{1}{\alpha_1} & \dfrac{-\beta_1}{\alpha_1} \end{bmatrix} \quad (5.86)$$

Notice that Q_{i+1} is obtained from Q_i by deleting the first row and column in (5.86). Also, if A is the comrade matrix in (5.12) associated with a polynomial of degree n + 1, then Q_1 consists of rows 2 to n + 1 of $J_{n+1} A J_{n+1}$.

The formula (5.85) also clears up two further apparent difficulties: first, if $\delta b < \delta a$, then $b(\lambda)$ must be multiplied by the appropriate power of λ before the algorithm can commence. The coefficients can thus be found by repeated application of the corresponding form of (5.85). An identical remedy is applied to a second kind of irregularity, when at any stage some, but not all, of the elements at the beginning of a row are zero.

Example 3   Take as basis the Chebyshev polynomials $\{U_i(\lambda)\}$ of the second kind, defined in Problem 5.2, for which $\alpha_1 = 2$, $\beta_1 = 0$, $\alpha_i = 2$, $\beta_i = 0$, $\gamma_i = 1$, $i \geqslant 2$, and consider

$$a(\lambda) = U_4(\lambda) + 2U_3(\lambda) - 21U_2(\lambda) - 28U_1(\lambda) + 106U_0(\lambda)$$

$$b(\lambda) = U_4(\lambda) - U_2(\lambda) - 2U_0(\lambda) \tag{5.87}$$

The generalized Routh scheme for (5.87) is as follows:

a_i	1	2	-21	-28	106		1	0	-1	0	-2
b_i	1	0	-1	0	-2		1	0	-1	0	-2

(1) r_{2j} : 2   -20   -28   108 $\longrightarrow$ | 1   -10   -13   44   -14 z_{2j}

r_{3j} : 10   12   -44   12 | 10   12   -44   12   (2)

(3) r_{4j} : $\dfrac{-112}{5}$ $\dfrac{-96}{5}$ $\dfrac{528}{5}$ $\longrightarrow$ | $\dfrac{-56}{5}$ $\dfrac{-48}{5}$ $\dfrac{208}{5}$ $\dfrac{-48}{5}$ z_{4j}

r_{5j} : $\dfrac{24}{7}$ $\dfrac{-48}{7}$ $\dfrac{24}{7}$ | $\dfrac{24}{7}$ $\dfrac{-48}{7}$ $\dfrac{24}{7}$ (4)

(5) r_{6j} : -64   128 $\longrightarrow$ | -32   64   -32 z_{6j}

0   0 | 0   0   (6)

As explained for Table 5.1, the row immediately below each linked pair is computed from the Routh rule, and the numbers within

parentheses indicate the order in which this takes place. Each
such labeled row in the auxiliary right-hand side table is reproduced on the left side. For example, the row labeled (2) becomes
$\{r_{3j}\}$, which enables $\{r_{4j}\}$ to be determined. The rows of z's are
constructed from (5.85) and (5.86), the latter giving

$$
Q_1 = \begin{bmatrix}
\frac{1}{2} & 0 & \frac{1}{2} & 0 & 0 \\
0 & \frac{1}{2} & 0 & \frac{1}{2} & 0 \\
0 & 0 & \frac{1}{2} & 0 & \frac{1}{2} \\
0 & 0 & 0 & \frac{1}{2} & 0
\end{bmatrix}
$$

where Q_2 and Q_3 are the matrices below and to the right of the
dashed lines. For example, with i = 2 in (5.85) we obtain

$$
[z_{41}, z_{42}, z_{43}, z_{44}] = \left[\frac{-112}{5}, \frac{-96}{5}, \frac{528}{5}\right]Q_2
$$

$$
= \left[\frac{-56}{5}, \frac{-48}{5}, \frac{208}{5}, \frac{-48}{5}\right]
$$

The final row of zeros on the left side of the array shows that a
g.c.d. of $a(\lambda)$ and $b(\lambda)$ in (5.87) is

$$
-64U_1(\lambda) + 128U_0(\lambda) \equiv -128(\lambda - 1)
$$

The reader should check this result by finding the power forms of
$a(\lambda)$ and $b(\lambda)$.

One final comment: As we saw in Section 1.6.3, the division
by first column elements in formulas such as (5.77) and (5.79) is
unnecessary. However, an optimal fraction-free array like that in
Theorem 1.16 has not yet been produced for the generalized case.

5.2.3 Polynomial Matrices

We briefly consider how Theorem 4.32 on the greatest common right divisor (g.c.r.d.) of two polynomial matrices, extends to the case when

$$M(\lambda) = I_m q_p(\lambda) + M_1 q_{p-1}(\lambda) + \cdots + M_p q_0(\lambda) \qquad (5.88)$$

$$N(\lambda) = N_1 q_{p-1}(\lambda) + N_2 q_{p-2}(\lambda) + \cdots + N_p q_0(\lambda) \qquad (5.89)$$

where $N(\lambda)$ is $r \times m$, and the basis $q_0(\lambda), \ldots, q_p(\lambda)$ is defined by (5.3) [we have used $q_i(\lambda)$ instead of $p_i(\lambda)$ to avoid confusion with the degree p]. Since $M(\lambda)$ is assumed regular, we can quote the result corresponding to Theorem 4.23:

<u>Theorem 5.7</u> The mp × mp *block comrade* matrix

$$A_M \triangleq \begin{bmatrix} \dfrac{-\beta_1}{\alpha_1} I_m & \dfrac{1}{\alpha_1} I_m & 0 & \cdot & \cdot & 0 \\[2ex] \dfrac{\gamma_2}{\alpha_2} I_m & \dfrac{-\beta_2}{\alpha_2} I_m & \dfrac{1}{\alpha_2} I_m & \cdot & \cdot & \cdot \\[2ex] \cdot & \cdot & \cdot & \cdot & \cdot & \cdot \\[1ex] \cdot & \cdot & \cdot & \cdot & \cdot & \cdot \\[2ex] \dfrac{-1}{\alpha_p} M_p & \dfrac{-1}{\alpha_p} M_{p-1} & \cdot & \cdot & \cdot & \dfrac{-M_1 - \beta_p I_m}{\alpha_p} \end{bmatrix} \qquad (5.90)$$

associated with $M(\lambda)$ in (5.88) has the properties

(i) $\det(\lambda I_{mp} - A_M) = \det(M(\lambda))$ \qquad (5.91)

(ii) $A_M = (P_p \otimes I_m) C_M (P_p \otimes I_m)^{-1}$ \qquad (5.92)

where C_M is the block companion matrix in the form (4.173), associated with the monic power form of $M(\lambda)$, and P_p is defined by (5.10).

Remarks on Theorem 5.7   (1) The proof of (5.92) follows the same
lines as the corresponding result (5.22), when m = 1.   The
relationship (5.91) then follows directly from (5.92) and the
property (4.174) of the block companion matrix.

   (2) Clearly, (5.89) is obtained from the ordinary comrade
matrix (5.12) in the same way that the companion matrix generates
its block form.   Similarly, we can now state the generalization of
the block companion result on a g.c.r.d. for (5.88) and (5.89).

Theorem 5.8   Let R_G denote the mpr × mp matrix associated with $M(\lambda)$
and $N(\lambda)$ in (5.88), having block rows

$$H \overset{\Delta}{=} [N_p, N_{p-1}, \ldots, N_1], Hq_1(A_M), Hq_2(A_M), \ldots, Hq_{mp-1}(A_M)$$

(i) $M(\lambda)$ and $N(\lambda)$ are r.r.p. if and only if rank R_G = mp.
(ii) Assume that $M(\lambda)$ and $N(\lambda)$ possess a regular g.c.r.d.

$$D(\lambda) = I_m q_k(\lambda) + D_1 q_{k-1}(\lambda) + \cdots + D_k q_0(\lambda)$$

Then parts (i) and (ii) of Theorem 4.32, on determining k and the
coefficients $D_1, \ldots, D_k$, still hold with R^{MN} replaced by R_G.

   When the basis for the polynomial matrices is $\{f_i(\lambda)\}$ in
(5.32), the reader should by now have no difficulty in writing down
from (5.39) the block confederate matrix E_M.   Theorem 5.8 is then
modified by simply replacing A_M with E_M.

Problems

5.9   Use Theorem 5.5 to determine a g.c.d. of
   (i) $a(\lambda) = T_4 + 2T_3 - 8T_2 - 26T_1 - 17T_0$

         $b(\lambda) = T_3 + 2T_2 - T_1 - 2T_0$

         where $T_i(\lambda)$ are the Chebyshev polynomials defined in
         Problem 5.1.
   (ii) $a(\lambda) = H_4 + 2H_3 - 20H_1 - 28H_0$

         $b(\lambda) = H_3 + 2H_2 + 2H_1 - 4H_0$

where $H_i(\lambda)$ are the Hermite polynomials defined in
Problem 5.3.

In each case, check your result by converting $a(\lambda)$ and $b(\lambda)$
into power form [see Problem 1.21(i)].

5.10 Use Theorem 5.6 to determine a g.c.d. of $a(\lambda)$ in (5.19), $b(\lambda)$
in (5.65), and

$$c(\lambda) = 6p_3 + 50p_2 - 21p_1 - 335p_0$$

relative to the Legendre basis (5.18).

5.11 Use the tabular method of Section 5.2.2 to determine a g.c.d.
for each of the following pairs of polynomials:

(i) $a(\lambda) = T_3 + 12T_2 + 47T_1 + 36T_0$

$b(\lambda) = T_3 + 16T_2 + 71T_1 + 56T_0$

where $T_i(\lambda)$ are the Chebyshev polynomials defined in
Problem 5.1.

(ii) $a(\lambda) = H_3 + 12H_2 + 50H_1 + 72H_0$

$b(\lambda) = H_3 + 16H_2 + 74H_1 + 112H_0$

where $H_i(\lambda)$ are the Hermite polynomials defined in
Problem 5.3.

5.12 Let $\hat{b}(\lambda) = \hat{b}_o f_m(\lambda) + \cdots + \hat{b}_m f_0(\lambda)$, $m < n$, be expressed relative to the basis (5.32). Use the similarity relationship
(5.41) between the confederate matrix E and the comrade
matrix, together with (5.36), to show from Theorem 5.4 that
$\hat{b}(E)$ has rows $\hat{r}_1 = [\hat{b}_m, \ldots, \hat{b}_1, \hat{b}_0, 0, \ldots, 0]$,
$\hat{r}_i = \hat{r}_1 f_{i-1}(E)$, $i = 2, \ldots, n$. Also, use (5.32) to show that

$$\hat{r}_i = \sum_{k=1}^{i-1} \hat{r}_k (\delta_{i-1,k} E + \epsilon_{i-1,k} I), \quad i \geq 2$$

5.13 Use the tabular method of Section 5.2.2 to determine a g.c.d. for

$$a(\lambda) = T_3 + 3T_2 + 7T_1 + 5T_0, \quad b(\lambda) = T_3 + 4T_2 + 11T_1 + 8T_0$$

where $T_i(\lambda)$ are the Chebyshev polynomials defined in Problem 5.1. [Note: An initial zero is obtained at the second step. This requires the corresponding polynomial to be multiplied by λ, using a formula of the type (5.85). The procedure then continues in the manner described.]

5.14 A standard property of the orthogonal polynomials $p_i(\lambda)$ in (5.2) is that all their roots are real, and lie in the interval (c, d). Deduce that the matrix $p_i(A)$ is singular only if the polynomial whose comrade matrix is A has at least one real root in (c, d).

5.15 Consider $\{p_i(\lambda)\}$ defined by (5.3) with $\beta_i = 0$, for all i. Use Theorem 5.5 to prove that $p_{n-1}(\lambda)$ and $p_n(\lambda)$ are relatively prime. In fact, this property holds for all sets of orthogonal polynomials.

5.3 APPLICATIONS TO LINEAR CONTROL SYSTEMS

We now show that some of the results in Chapters 2 and 4 are merely special cases of expressions which arise when the orthogonal polynomial basis (5.2) is used. Consider the usual system description

$$\dot{x}(t) = Fx(t) + Gu(t)$$

$$y(t) = Hx(t) \tag{5.93}$$

where x(t) is the n × 1 state vector, u(t) is the m × 1 control vector, and y(t) is the r × 1 output vector. We define *generalized controllability* and *observability* matrices relative to the basis (5.2) by

$$\tilde{C}(F, G) \triangleq [G, p_1(F)G, p_2(F)G, \ldots, p_{n-1}(F)G] \tag{5.94}$$

$$\tilde{O}(F, H) \text{ has rows } H, Hp_1(F), \ldots, Hp_{n-1}(F) \tag{5.95}$$

These clearly reduce to the standard forms $C(F, G)$ and $O(F, G)$, defined, respectively, in (2.23) and (2.44), when $p_i(\lambda) = \lambda^i$. Although these definitions are perhaps more of mathematical than of practical interest, the following theorem justifies the terminology.

Theorem 5.9   The system (5.93) is c.c. (respectively, c.o.) if and only if $\widetilde{C}(F, G)$ [respectively, $\widetilde{O}(F, H)$] has rank n.

Proof.   Consider first the single input case, when G is a column vector g, say.   It follows that

$$\widetilde{C}(F, g) = [g, p_1(F)g, \ldots, p_{n-1}(F)g]$$

$$= [I, p_1(F), \ldots, p_{n-1}(F)][I_n \otimes g] \qquad (5.96)$$

Take the transpose of the relationship (5.27) between the bases, i.e.,

$$[1, \lambda, \lambda^2, \ldots, \lambda^{n-1}]P_n^T = [1, p_1(\lambda), \ldots, p_{n-1}(\lambda)] \qquad (5.97)$$

where P_n is defined via (5.10).   Replacing λ by F in (5.97) converts it to

$$[I, F, F^2, \ldots, F^{n-1}](P_n^T \otimes I_n) = [I, p_1(F), \ldots, p_{n-1}(F)] \qquad (5.98)$$

Substituting (5.98) into (5.96), and using a property of the Kronecker product [see (A2) in Appendix A], produces

$$\widetilde{C}(F, g) = [I, F, F^2, \ldots, F^{n-1}](P_n^T \otimes g) \qquad (5.99)$$

However, it can also be shown (see Problem 5.19) that $(P_n^T \otimes g) \equiv (I_n \otimes g)P_n^T$, and this reduces (5.99) to

$$\widetilde{C}(F, g) = [I, F, \ldots, F^{n-1}](I_n \otimes g)P_n^T$$

$$= [g, Fg, \ldots, F^{n-1}g]P_n^T$$

$$= C(F, g)P_n^T \qquad (5.100)$$

We now use (5.100) to obtain the relationship between $\widetilde{C}$ and C when G has columns $g_1, \ldots, g_m$. Taking together all the first columns of each block in (5.94), then all the second columns, and so on, shows that

$$\widetilde{C}(F, G) = [\widetilde{C}(F, g_1), \widetilde{C}(F, g_2), \ldots, \widetilde{C}(F, g_m)]T \qquad (5.101)$$

where T is an $mn \times mn$ permutation matrix having columns e_1, e_{m+1}, $e_{2m+1}, \ldots, e_{(n-1)m+1}, e_2, e_{m+2}, \ldots, e_m, \ldots, e_{nm}$, and e_i denotes the ith column of I_{mn}. Substituting (5.100) into (5.101) gives

$$\widetilde{C}(F, G) = [C(F, g_1)P_n^T, \ldots, C(F, g_m)P_n^T]T$$

$$= [C(F, g_1), \ldots, C(F, g_m)](I_m \otimes P_n^T)T \qquad (5.102)$$

Corresponding to (5.101) we also have

$$C(F, G) = [C(F, g_1), C(F, g_2), \ldots, C(F, g_m)]T \qquad (5.103)$$

Combining together (5.102) and (5.103) gives

$$\widetilde{C}(F, G) = C(F, G)T^{-1}(I_m \otimes P_n^T)T \qquad (5.104)$$

Since P_n and T are nonsingular, (5.104) shows that $\widetilde{C}(F, G)$ has the same rank as $C(F, G)$, so the desired result follows immediately from Theorem 2.1.

The observability part of the proof follows similarly. ∎

<u>Remarks on Theorem 5.9</u> (1) The permutation matrix T used in the proof is in fact that which sends a Kronecker product into its reverse [see (A4) in Appendix A], so (5.104) can be written in the neater form

$$\widetilde{C}(F, G) = C(F, G)(P_n^T \otimes I_m) \qquad (5.105)$$

Similarly, the expression involving observability matrices is

$$\widetilde{O}(F, H) = (P_n \otimes I_r)O(F, H)$$

(2) The matrix R_g in (5.72), giving the g.c.d. of a set of generalized polynomials, has precisely the form (5.94) of a generalized controllability matrix. Having realized this point, the results in Section 2.3, are now seen to be special cases of theorems involving generalized polynomials. For example, the counterpart to Theorem 2.8 would read: The $h + 1$ polynomials $a(\lambda)$ in (5.13) and $b_i(\lambda)$ in (5.68) are relatively prime if and only if the pair (A^T, B) is c.c., where A is the comrade matrix (5.12) and B is defined in (5.73). Furthermore, the degree of their g.c.d. is equal to the rank defect of $\widetilde{C}(A^T, B)$, and in this case the coefficients in a g.c.d. are given by (5.74).

(3) By this stage, the reader should appreciate that a corresponding theorem relative to the basis (5.32) can be obtained by replacing $p_i(\lambda)$ by $f_i(\lambda)$ in the definitions (5.94) and (5.95).

An immediate application of (5.94) is to a controllable canonical form for a single-input system, when G is a column vector $g \neq 0$.

Theorem 5.10   There exists a nonsingular transformation $z = \widetilde{P}x$ which transforms $\dot{x} = Fx + gu$ into the canonical form

$$\dot{z} = Az + du \tag{5.106}$$

where A is the comrade matrix (5.12), and $d = [0, 0, \ldots, 0, 1]^T$, if and only if rank $\widetilde{C}(F, g) = n$. In this case

$$\widetilde{P} = \widetilde{C}(A, d)[\widetilde{C}(F, g)]^{-1} \tag{5.107}$$

Proof. The existence part of the result follows from Theorems 2.10 and 5.9. To verify the expression for $\widetilde{P}$, notice that the transformation implies

$$\widetilde{P}F\widetilde{P}^{-1} = A, \quad \widetilde{P}g = d \tag{5.108}$$

Hence, as in the proof of Theorem 2.10,

$$C(A, d) = [\widetilde{P}g, (\widetilde{P}F\widetilde{P}^{-1})\widetilde{P}g, \ldots, (\widetilde{P}F\widetilde{P}^{-1})^{n-1}\widetilde{P}g]$$

$$= \widetilde{P}C(F, g)$$

which gives

$$\widetilde{P} = C(A, d)[C(F, g)]^{-1}$$

Substitution of (5.100) into this last expression produces (5.107). ∎

Remarks on Theorem 5.10   (1) When $p_i(\lambda) = \lambda^i$, (5.106) reduces to the form in Theorem 2.10, namely (2.91), but it is clearly inappropriate to refer to the latter as *the* controllable canonical form. Indeed, the reader should be able to derive the corresponding version of the theorem using the confederate matrix.

(2) For multivariable canonical forms, discussed in Section 4.1.2, the companion matrix blocks, displayed in (4.43), can be replaced by comrade (or congenial) matrices. There are no alterations in the "B" matrices in the canonical forms.

(3) As in the proof of the power form result (Theorem 2.10), a convenient way of calculating $\widetilde{P}$ is to note that it has rows t, $tp_1(A), \ldots, tp_{n-1}(A)$ [i.e., $P \equiv \widetilde{0}(A, t)$]. The elements of t are obtained by solving the n equations obtained from the condition $\widetilde{P}g = d$ in (5.108).

(4) Also parallelling the standard case, there are immediate applications to scalar linear feedback and realization:

Theorem 5.11

(i) If the system $\dot{x} = Fx + gu$ is c.c., then linear state feedback $u = (1/\alpha_n)k\widetilde{P}x$ produces a predetermined closed-loop characteristic polynomial

$$\phi(\lambda) \equiv \det(\lambda I_n - F_c)$$

$$\equiv [p_n(\lambda) + \phi_1 p_{n-1}(\lambda) + \cdots + \phi_n p_0(\lambda)](\alpha_1 \cdots \alpha_n)^{-1} \quad (5.109)$$

where $k = [a_n - \phi_n, \ldots, a_1 - \phi_1]$, $\widetilde{P}$ is defined in (5.107), and F has the characteristic polynomial $a(\lambda)/(\alpha_1 \cdots \alpha_n)$ in Theorem 5.1.

(ii) If $a(\lambda)$ and $b(\lambda)$ are the generalized polynomials in (5.13) and (5.47), with $\delta b < \delta a$, then a c.c. (respectively, c.o.) realization of the transfer function $b(\lambda)/a(\lambda)$ is $\{A, d, h\}(\{A^T, h^T, d^T\})$, where A and d are as in (5.106), and $h = [b_m, \ldots, b_0, 0, \ldots, 0]/\alpha_n$.

Proof. (i) Application of the given feedback expression, together with (5.108), produces the closed-loop matrix

$$F_c \equiv F + gk\widetilde{P}/\alpha_n = \widetilde{P}^{-1}(A + dk/\alpha_n)\widetilde{P} \qquad (5.110)$$

Clearly, dk/α_n is an $n \times n$ matrix having all rows zero except the last, which is k/α_n. Hence $A + dk/\alpha_n$ is the same as A in (5.12), except that its last row is

$$\frac{1}{\alpha_n} [-\phi_n, -\phi_{n-1}, \ldots, (\gamma_n - \phi_2), (-\phi_1 - \beta_n)]$$

By Theorem 5.1, $A + dk/\alpha_n$ has the polynomial on the right in (5.109) as its characteristic polynomial. The similarity relationship (5.110) shows that this is also the characteristic polynomial of F_c.

(ii) Using the similarity condition (5.22) between A and the companion matrix C, we have

$$h(\lambda I - A)^{-1}d = hP_n(\lambda I - C)^{-1}P_n^{-1}d$$

$$= hP_n(\lambda I - C)^{-1}d/(\alpha_1\alpha_2 \cdots \alpha_{n-1})$$

$$= [b_m, \ldots, b_0, 0, \ldots, 0]P_n(\lambda I - C)^{-1}d$$

$$/(\alpha_1 \cdots \alpha_n) \qquad (5.111)$$

We now need the facts: (a) $(\lambda I - C)^{-1}d$ is equal to the last column of $(\lambda I - C)^{-1}$, which by (2.67) is $[1, \lambda, \lambda^2, \ldots, \lambda^{n-1}]^T/\widetilde{a}(\lambda)$, where $\widetilde{a}(\lambda)$ is the characteristic polynomial of C, and from (5.8) $\widetilde{a}(\lambda) = a(\lambda)/(\alpha_1 \cdots \alpha_n)$; (b) using (5.27),

$$b(\lambda) = [b_m, \ldots, b_0, 0, \ldots, 0][1, p_1(\lambda), \ldots, p_{n-1}(\lambda)]^T$$

$$= [b_m, \ldots, b_0, 0, \ldots, 0]P_n[1, \lambda, \ldots, \lambda^{n-1}]^T$$

Amalgamating these results with (5.111), we obtain

$$h(\lambda I - A)^{-1}d = b(\lambda)/a(\lambda) \tag{5.112}$$

as required. The controllability of the realization can be verified directly by observing that $C(A, d)$ has nonzero elements on the
secondary diagonal, and zeros everywhere above it. The c.o.
realization follows immediately by transposing (5.112). ∎

Remarks on Theorem 5.11   (1) Part (i) extends the power form result
in Theorem 2.16, and Remark 5 thereon. It can be shown that the
multi-input result also holds, namely that if (5.93) is c.c., there
exists a matrix K such that F + GK has an arbitrary characteristic
polynomial relative to the basis $\{p_i(\lambda)\}$.

     (2) As in Theorem 2.13, the realizations in part (ii) are
minimal if and only if they are both c.c. and c.o. This is equivalent to $a(\lambda)$ and $b(\lambda)$ being relatively prime.

     (3) Extension of part (ii) to the case of a transfer function
matrix follows in the same way that Theorem 4.15 is derived from
(2.108). Thus

$$\{A \otimes I_m, d \otimes I_m, [B_m, \ldots, B_1, B_0, 0, \ldots, 0]/\alpha_n\} \tag{5.113}$$

is a c.c. realization of the r × m matrix

$$[B_0 p_m(\lambda) + B_1 p_{m-1}(\lambda) + \cdots + B_m p_0(\lambda)]/a(\lambda) \tag{5.114}$$

Problems

5.16 Transform the system in Problem 2.28 into the canonical form
       in Theorem 5.10, relative to the basis of Hermite polynomials
       $H_i(\lambda)$ defined in Problem 5.3. Determine linear feedback such

that the closed-loop polynomial is $(H_3 + 17H_2 + 110H_1 + 228H_0)/8$. Compare your result with Problem 2.41.

5.17 Apply (5.100) to Theorem 2.17 to obtain a corresponding formula for the feedback in Theorem 5.11(i), namely

$$u = -(\alpha_1 \cdots \alpha_{n-1})[\widetilde{C}(F, g)]^{-1}\phi(F)x$$

where $\phi(\lambda)$ is defined in (5.109).

5.18 Using the notation of Theorem 5.11, show that

$$(\lambda I - A)u(\lambda) = a(\lambda)d/\alpha_n$$

where $u(\lambda)$ is defined in (5.23). Hence verify the formula (5.113).

5.19 Prove that if X is an arbitrary $n \times n$ matrix, and y is an arbitrary column n-vector, then

$$X \otimes y = (I_n \otimes y)X$$

5.4   OTHER APPLICATIONS

It is natural to speculate whether other theorems for power form polynomials (or polynomial matrices) carry over to the generalized case. Unfortunately, few results can be reported in this area. For example, the Routh array method of Section 5.2.2 has been applied to the problem of locating the roots of generalized polynomial in the left and right halves of the complex plane. A result just like Theorem 3.3 is obtained, but it requires determination of odd and even parts of the polynomial. This means that in general the polynomial must be converted into power form, which defeats the purpose of the exercise. Similar attempts to generalize the theorems of Hurwitz, Hermite, and Schur-Cohn (Section 3.2) have also proved unsuccessful in obtaining a criterion which can be applied directly to the coefficients of the generalized form.

In view of the attention which we have paid to Sylvester and bezoutian type matrices throughout this book, we close by recording the following analogues, associated with $a(\lambda)$ in (5.13) and $b(\lambda)$ in

(5.47). If we construct

$$p_i(\lambda)a(\lambda) = a_{i0}p_{n+i}(\lambda) + a_{i1}p_{n+i-1}(\lambda) + \cdots + a_{i,n+i}p_0(\lambda),$$

$$i = 1, \ldots, m - 1$$

$$(5.115)$$

$$p_j(\lambda)b(\lambda) = b_{j0}p_{m+j}(\lambda) + b_{j1}p_{m+j-1}(\lambda) + \cdots + b_{j,m+j}p_0(\lambda),$$

$$j = 1, \ldots, n - 1$$

then a Sylvester-type matrix $\tilde{S}$ of order $m + n$ can be defined having rows

$$[0, 0, \ldots, 0, a_{m-i,0}, a_{m-i,1}, \ldots, a_{m-i,m+n-i}],$$

$$\underset{<\!\!-\!(i - 1)\!-\!\!>}{} \qquad\qquad i = 1, 2, \ldots, m$$

$$[0, 0, \ldots, 0, b_{n-j,0}, b_{n-j,1}, \ldots, b_{n-j,m+n-j}],$$

$$\underset{<\!\!-\!(j - 1)\!-\!\!>}{} \qquad\qquad j = 1, 2, \ldots, n$$

with $a_{0k} \equiv a_k$, $b_{0k} \equiv b_k$. This has the property that $a(\lambda)$ and $b(\lambda)$ are relatively prime if and only if $\tilde{S}$ is nonsingular. As for the power form case (see Theorem 1.7), the neatest way of demonstrating this fact is by proving that

$$\begin{bmatrix} I_{n-1} & 0 \\ -\tilde{S}_3\tilde{S}_1^{-1} & I_n \end{bmatrix} \begin{bmatrix} \tilde{S}_1 & \tilde{S}_2 \\ \tilde{S}_3 & \tilde{S}_4 \end{bmatrix} = \begin{bmatrix} \tilde{S}_1 & \tilde{S}_2 \\ 0 & J_n b(A) J_n \end{bmatrix} \qquad (5.116)$$

where we have assumed that $m = n - 1$, and A is the comrade matrix (5.12). Taking determinants of both sides of (5.116), and appealing to Theorem 5.5(i) establishes the desired result, which is a direct generalization of (1.86). Furthermore, an extended Sylvester-type matrix can be constructed, associated with a *set* of generalized polynomials. Exactly as in Theorem 1.10, the coefficients of a g.c.d. are obtained by a reduction to row echelon form. Unfortunately, a major disadvantage lies in the need to compute the coefficients a_{ik}, b_{jk} in (5.115).

Finally, an analogue of the bezoutian matrix Z associated with
$a(\lambda)$ and $b(\lambda)$ can also be obtained by taking as a starting point
the controllability/observability expressions of Theorem 2.11.
This leads to a matrix like Z, expressible as

$$\widetilde{Z} = [\widetilde{C}(A, d)]^{-1}\widetilde{O}(A, b)$$

where $b = [b_m, \ldots, b_0, 0, \ldots, 0]$, $d = [0, \ldots, 0, 1]^T$, and A is
the comrade matrix. It must be admitted that this result is also
of academic interest, because of the lack of a direct way of
expressing the elements of $\widetilde{Z}$ in terms of coefficients of the
generalized polynomials. However, there is some satisfaction, on
reaching the end of the book, in exposing once again the four
strands which have occurred throughout: tabular arrays, and
matrices of companion, Sylvester, and bezoutian type.

Problems

5.20 Associated with the orthogonal polynomials $p_i(\lambda)$ in (5.2),
construct a sequence

$$q_0(\lambda) = b_0, \quad q_1(\lambda) = b_1 + (\alpha_m\lambda + \beta_m)q_0(\lambda)$$

$$q_i(\lambda) = b_i + (\alpha_{m-i+1}\lambda + \beta_{m-i+1})q_{i-1}(\lambda) - \gamma_{m-i+2}q_{i-2}(\lambda),$$
$$i \geq 2$$

Prove that

$$q_m(\lambda) = b_0 p_m(\lambda) + b_1 p_{m-1}(\lambda) + \cdots + b_m p_0(\lambda) \quad [\equiv b(\lambda)]$$

5.21 When $p_i(\lambda) = \lambda^i$, the result in the preceding problem reduces
to Horner's scheme in (1.5), when λ is set equal to a
numerical value. If $b(\lambda)$ is the polynomial in (5.65), use
this method to evaluate $b(3)$, i.e., the remainder on division
of $b(\lambda)$ by $\lambda - 3$.

5.22 Let $a(\lambda)$ and $b(\lambda)$ be the polynomials in (5.13) and (5.47),
respectively, with $\delta b \geq \delta a$. Show that if A is the comrade
matrix in (5.12), then $r(A) \equiv b(A)$, where $r(\lambda)$ is the remainder on division of $b(\lambda)$ by $a(\lambda)$. Using the result of Problem

5.20, show that

$$b(A) = b_0 p_m(A) + \cdots + b_m p_0(A) \equiv Q_m$$

where

$$Q_0 = b_0 I, \quad Q_1 = b_1 I + (\alpha_m A + \beta_m I)Q_0$$

$$Q_i = b_i I + (\alpha_{m-i+1} A + \beta_{m-i+1} I)Q_{i-1} - \gamma_{m-i+2}Q_{i-2}, \quad i \geqslant 2$$

Thus $r(\lambda)$ can be computed relative to the same basis as $a(\lambda)$ and $b(\lambda)$ since (see Theorem 5.4) the first row of $r(A)$ gives the coefficients of $r(\lambda)$.

5.23 Use the method of the preceding problem to obtain the remainder on division of $b(\lambda)$ in (5.65) by $a(\lambda) = p_2 - \frac{1}{2} p_1 + 2p_0$, relative to the Legendre polynomials $p_i(\lambda)$ in (5.18).

BIBLIOGRAPHICAL NOTES

Section 5.0. The term "generalized polynomial" occurs in books on numerical analysis (e.g., Cheney, 1966; Karlin and Studden, 1966). It has been used, however, in quite a different sense by Pernebo (1981), and reflects the fact that the description "generalized" is overworked (see, e.g., Remark 1 on Theorem 5.2, and Remark 2 on Theorem 5.3).

Section 5.1.1. A detailed coverage of properties of orthogonal polynomials, including proofs of the necessity and sufficiency of the recurrence formula (5.3), can be found in Freud (1966) and Chihara (1978). Lemma 5.1.1 is well known, and Theorems 5.1 and 5.2 were presented by Barnett (1975c). It was later found that two other independent discoveries of comrade-type matrices have been made by Specht (1957) and Parodi (1963). Both authors derived bounds like those obtained in Problem 5.8, but neither considered the applications which are developed in Sections 5.2 and 5.3. The term "companion" is a translation from the German "Begleitmatrix" (MacDuffee, 1950b), and the less interesting expression "cyclic" is sometimes used. I. J. Good (1961) coined the label "colleague," from which Barnett (1975c) suggested the name "comrade" (no political connotations!). The relationship of the Schwarz matrix with orthogonal polynomials has been investigated by Genin (1975), and further information and references on Leslie matrices can be found in Usher (1972). The version of (5.24) for the case of repeated eigenvalues was derived by Maroulas (1978), and reproduced in Barnett (1981b).

Section 5.1.2. Theorem 5.3 is taken from Maroulas and Barnett (1979c), where the name "confederate matrix" originated (no link with the Confederacy!). The designation "congenial" for the class of matrices under consideration, was proposed by A. G. J. MacFarlane of Cambridge University, and replaces an earlier suggestion of "friendly" matrices (Barnett, 1981a). Hessenberg matrices are useful in eigenvalue computation (Wilkinson, 1965), solution of linear matrix equations (Golub et al., 1979; Howland and Senez, 1970), frequency response calculations (Laub, 1981), linear state feedback (Miminis and Paige, 1982), and elsewhere. The discrete-time Schwarz matrix of Anderson et al. (1976b), mentioned in the bibliographical notes for Section 3.3.1, has Hessenberg form. Some suggestions for further research on congenial matrices are contained in a survey by Barnett (1981b). The result in Problem 5.5 was given by Barnett (1975d); and that in Problem 5.7 by Barnett (1981b), following Maroulas (1978). Further comments on the method of Problem 5.8 were made by Barnett (1975d).

Section 5.2.1. Theorems 5.4 and 5.5 are taken from Barnett (1975c), although the proof given here of the latter is an improved version. Maroulas and Barnett (1978b) gave Theorem 5.6, together with a result when two different bases are used. For confederate matrix versions, see Maroulas and Barnett (1979c,d).

Section 5.2.2. The Routh-type scheme was presented by Maroulas and Barnett (1978c), and the formula (5.85), together with some other developments, was given in a follow-up paper (Maroulas and Barnett, 1979e).

Section 5.2.3. Theorems 5.7 and 5.8 are taken from Maroulas and Barnett (1978b). Extensions for the block confederate case are contained in Maroulas and Barnett (1979c,d).

Section 5.3. Most of the material can be found in Barnett (1975d), with the multivariable case given by Maroulas and Barnett (1978d). Indications were given by Maroulas and Barnett (1979d) on the modifications involved when the confederate matrix is used. See also a survey by Barnett (1981a).

Section 5.4. The formula (5.116) was derived by Maroulas and Barnett (1978c), and the result on g.c.d. extraction from a generalized Sylvester matrix is due to Barnett (1980). The bezoutian-type matrix was introduced by Maroulas and Barnett (1978d), and Problems 5.20 to 5.22 are based on Maroulas (1978). Details of other work mentioned can be found in Maroulas and Barnett (1978c, 1979a,c-f, 1980) and Barnett and Maroulas (1978). In view of the vast literature on root location for polynomials in power form, there is surprisingly little published for the generalized case, the only additional papers encountered by the author being those of Ostrowski (1965), Zedek (1965), Claessens (1978), Gautschi (1978), and László (1982).

appendix A

Kronecker Product and Matrix Functions

KRONECKER PRODUCT

(A1) If $A = [a_{ij}]$ is $p \times m$, and $B = [b_{ij}]$ is $q \times n$, then $A \otimes B$ is the $(pq) \times (mn)$ matrix having (i, j) block $a_{ij}B$.

  (A2) Basic properties are

$$(A \otimes B)^* = A^* \otimes B^*$$

$$(A \otimes B)(C \otimes D) = AC \otimes BD, \quad \text{where C is } m \times r, \text{ D is } n \times s$$

When $p = m$, $q = n$, then

$$(A \otimes B)^{-1} = A^{-1} \otimes B^{-1}, \quad \text{provided that A and B are nonsingular}$$

The eigenvalues of

(i) $A \otimes B$ are $\lambda_i \mu_j$

(ii) $A \otimes I_n + I_m \otimes B$ are $\lambda_i + \mu_j$

for $i = 1, 2, \ldots, m$, $j = 1, 2, \ldots, n$, where λ_i, μ_j are the eigen-values of A and B, respectively.

  (A3) The *stacking operator* $v(\cdot)$, applied to an $m \times n$ matrix X, is defined as the mn-column vector obtained by stacking together the *rows* of X, i.e.,

$$v(X) = [x_{11}, x_{12}, \ldots, x_{1n}, x_{21}, \ldots, x_{2n}, \ldots, x_{mn}]^T$$

For A and B with dimensions in (A1),

$$v(AXB^T) = (A \otimes B)v(X)$$

This result leads to

(A4) $(A \otimes B) = P_1(B \otimes A)P_2$

where P_1, P_2 are permutation matrices, defined by

$$v(Y) = P_1 v(Y^T), \qquad v(X^T) = P_2 v(X)$$

where X, Y are m × n and p × q. Thus P_1 and P_2 depend *only* upon p, q and m, n, respectively. In particular, when p = m, q = n, then $P_2 = P_1^{-1}$ ($= P_1^T$). (A *permutation* matrix P has a single unit element in each row and column, all other entries being zero, and satisfies $P^T P = I$.)

FUNCTIONS OF MATRICES

(A5) The *minimum polynomial* m(λ) of an n × n matrix A is the unique monic polynomial of least degree such that $m(A) \equiv 0$. Let

$$m(\lambda) = (\lambda - \lambda_1)^{m_1}(\lambda - \lambda_2)^{m_2} \cdots (\lambda - \lambda_s)^{m_s}$$

where λ_1, λ_2, ..., λ_s are the distinct eigenvalues of A, and $\delta m = (m_1 + m_2 + \cdots + m_s) \leqslant n$. A *function* f(A) of the matrix A is defined by f(A) = g(A), where g(λ) is the unique polynomial having $\delta g < \delta m$, such that

$$g(\lambda_k) = f(\lambda_k), \quad g^{(i)}(\lambda_k) = f^{(i)}(\lambda_k), \quad i = 1, 2, ..., m_k - 1$$

$$k = 1, 2, ..., s$$

where

$$f^{(i)}(\lambda_k) = \left[\frac{d^i f(\lambda)}{d\lambda^i} \right]_{\lambda = \lambda_k}$$

In particular, when s = n, m_i = 1, for all i, then *Sylvester's formula* is

$$f(A) = \sum_{i=1}^{n} f(\lambda_i) Z_i$$

where

$$Z_i = \prod_{\substack{j=1 \\ j \neq i}}^{n} (A - \lambda_j I)/(\lambda_i - \lambda_j)$$

(A6) If $f(\lambda)$ has a Taylor series $\sum_{0}^{\infty} c_k \lambda^k$, with radius of convergence r, then $\sum_{0}^{\infty} c_k A^k$ converges to $f(A)$ provided that $|\lambda_i| < r$, for all i.

References: Henderson and Searle (1981); Lancaster (1969).

appendix B

Notation

$\triangleq$	defined equal
$\in$	is a member of
$\blacksquare$	end of proof
$\bar{z}$	conjugate $x - iy$ of a complex number $z = x + iy$
$|z|$	modulus of z, $= (z\bar{z})^{1/2} = (x^2 + y^2)^{1/2}$
$Re(z)$, $Im(z)$	real, imaginary parts of z
$arg(z)$	argument of z
$sgn(\theta)$	$= +1$, $\theta > 0$; -1, $\theta < 0$
$V(\theta_0, \theta_1, \ldots)$	number of variations in sign in θ_0, θ_1, $\ldots$
$\delta a(\lambda)$	degree of a polynomial $a(\lambda)$
$b(\lambda)|a(\lambda)$	polynomial $b(\lambda)$ is a factor of $a(\lambda)$
$a'(\lambda)$, $a''(\lambda)$	derivatives $da(\lambda)/d\lambda$, $d^2a(\lambda)/d\lambda^2$
x	$= [x_1, x_2, \ldots, x_n]$, row n-vector
$|x|$	euclidean norm, $= (\Sigma|x_i|^2)^{1/2}$
$\dot{x}(t)$, $\ddot{x}(t)$	derivatives $dx(t)/dt$, $d^2x(t)/dt^2$
$\bar{x}(s)$	Laplace transform $\mathcal{L}\{x(t)\}$
$\tilde{x}(k)$	z-transform $Z\{x(k)\}$
A	$= [a_{ij}]$, matrix A having a_{ij} in row i, column j
A^T	transpose of A, $= [a_{ji}]$
$\bar{A}$	conjugate of A, $= [\bar{a}_{ij}]$
A^*	conjugate transpose of A, $= (\bar{A})^T$

419

$\text{tr}(A)$	trace of a square matrix A, $= \Sigma\, a_{ii}$
$\|A\|$	euclidean norm, $= (\Sigma\Sigma |a_{ij}|^2)^{1/2} = (\text{tr}(A^*A))^{1/2}$
$\det A$	determinant of a square matrix A
$\text{adj } A$	adjoint of a square matrix A
A^{-1}	inverse of A, $= (\text{adj } A)/(\det A)$
δ_{ij}	Kronecker delta, $\delta_{ii} = 1$, $\delta_{ij} = 0$, $i \neq j$
I_n	unit (identity) matrix of order n, $= [\delta_{ij}]$
J_n	reverse unit matrix of order n, $= [\delta_{i,n-j+1}]$
$\text{diag}[\alpha_1, \ldots, \alpha_n]$	diagonal n × n matrix, $= [\alpha_i \delta_{ij}]$
$\otimes$	Kronecker matrix product
$\pi(A)$, $\nu(A)$, $\delta(A)$	number of eigenvalues of A having positive, negative, and zero real parts
$\text{In}(A)$	inertia of A, $= (\pi(A), \nu(A), \delta(A))$
$\text{In}_c(A)$	inertia of A relative to the unit circle
$\delta A(\lambda)$	degree of a polynomial matrix $A(\lambda)$
$\Phi(t, t_0)$	state transition matrix
$C(A, B)$	controllability matrix for the pair (A, B)
$O(A, H)$	observability matrix for the pair (A, H)
$\delta[G(s)]$	McMillan degree of a rational matrix G(s)
$\{A, B, H\}$	realization of $G(s) = H(sI - A)^{-1}B$
c.(c./o./r./rec.)	completely (controllable/observable/reachable/reconstructible)
g.c.d.	greatest common divisor
g.c.r.d.	greatest common right divisor
m.f.d.	matrix-fraction description
r.r.p.	relatively right prime

References

Ackermann, J. (1972): Der Entwurf linearer Regelungssysteme in Zustandsraum, *Regelungstechnik*, *20*, 297-300.

Ahn, S. M. (1979): Eigenvalues of a matrix inside an ellipse and Chebyshev filter, *IEEE Trans. Autom. Control*, *AC-24*, 986-987.

Anderson, B. D. O. (1966): Solution of quadratic matrix equations, *Electron. Lett.*, *2*, 371-372.

Anderson, B. D. O. (1967): Application of the second method of Lyapunov to the proof of the Markov stability criterion, *Int. J. Control*, *5*, 473-482.

Anderson, B. D. O. (1972a): The reduced Hermite criterion with applications to proof of Liénard-Chipart criterion, *IEEE Trans. Autom. Control*, *AC-17*, 669-672.

Anderson, B. D. O. (1972b): On the computation of the Cauchy index, *Quart. Appl. Math.*, *29*, 577-582.

Anderson, B. D. O., and Jury, E. I. (1973): A simplified Schur-Cohn test, *IEEE Trans. Autom. Control*, *AC-18*, 157-163.

Anderson, B. D. O., and Jury, E. I. (1974): On the reduced Hermite and reduced Schur-Cohn matrix relationships, *Int. J. Control*, *19*, 877-890.

Anderson, B. D. O., and Jury, E. I. (1976): Generalized bezoutian and Sylvester matrices in multivariable linear control, *IEEE Trans. Autom. Control*, *AC-21*, 551-556.

Anderson, B. D. O., and Moore, J. B. (1971): *Linear Optimal Control*, Prentice-Hall, Englewood Cliffs, N.J.

Anderson, B. D. O., and Moore, J. B. (1979): *Optimal Filtering*, Prentice-Hall, Englewood Cliffs, N.J.

421

Anderson, B. D. O., Bose, N. K., and Jury, E. I. (1974): A simple test for zeros of a complex polynomial in a sector, *IEEE Trans. Autom. Control*, *AC-19*, 437-438.

Anderson, B. D. O., Bose, N. K., and Jury, E. I. (1975): On eigenvalues of complex matrices in a sector, *IEEE Trans. Aut. Control*, *AC-20*, 433.

Anderson, B. D. O., Jury, E. I., and Chapparo, L. F. (1976a): On the root distribution of a real polynomial with respect to the unit circle, *Regelungstechnik*, *24*, 101-102.

Anderson, B. D. O., Jury, E. I., and Mansour, M. (1976b): Schwarz matrix properties for continuous and discrete time systems, *Int. J. Control*, *23*, 1-16.

Antoulas, A. C. (1981): On canonical forms for linear constant systems, *Int. J. Control*, *33*, 95-122.

Antsaklis, P. J. (1979): Some relations satisfied by prime polynomial matrices and their role in linear multivariable system theory, *IEEE Trans. Autom. Control*, *AC-24*, 611-616.

Aplevich, J. D. (1979): Tableau methods for analysis and design of linear systems, *Automatica*, *15*, 419-429.

Archbold, A. (1970): *Algebra*, 4th Ed., Pitman, London.

Asner, B. A., Jr. (1970): On the total nonnegativity of the Hurwitz matrix, *SIAM J. Appl. Math.*, *18*, 407-414.

Auslander, D. M., Takahashi, Y., and Rabins, M. J. (1974): *Introducing Systems and Control*, McGraw-Hill, New York.

Ayres, F. (1974): *Theory and Problems of Matrices*, McGraw-Hill, New York.

Balestrino, A. (1979): Circulant matrices and resolvent computation of a matrix in companion form, *Ricerche di Automatica*, *10*, 66-69.

Barnett, S. (1968): Polynomial matrices and a problem in linear system theory, *IEEE Trans. Autom. Control*, *AC-13*, 216-217.

Barnett, S. (1969a): Regular polynomial matrices having relatively prime determinants, *Proc. Camb. Philos. Soc.*, *65*, 585-590.

Barnett, S. (1969b): Degrees of greatest common divisors of invariant factors of two regular polynomial matrices, *Proc. Camb. Philos. Soc.*, *66*, 241-245.

Barnett, S. (1970a): Greatest common divisor of two polynomials, *Linear Algebra Appl.*, *3*, 7-9.

Barnett, S. (1970b): Number of zeros of a complex polynomial inside the unit circle, *Electron. Lett.*, *6*, 164-165.

Barnett, S. (1970c): Qualitative analysis of polynomials using matrices, *IEEE Trans. Autom. Control*, *AC-15*, 380-382.

Barnett, S. (1971a): *Matrices in Control Theory*, Van Nostrand Reinhold, London.

Barnett, S. (1971b): Greatest common divisor of several polynomials, *Proc. Camb. Philos. Soc.*, *70*, 263-268.

Barnett, S. (1971c): A new formulation of the Liénard-Chipart stability criterion, *Proc. Camb. Philos. Soc.*, *70*, 269-274.

Barnett, S. (1971d): A new formulation of the theorems of Hurwitz, Routh and Sturm, *J. Inst. Math. Appl.*, *8*, 240-250.

Barnett, S. (1971e): Location of zeros of a complex polynomial, *Linear Algebra Appl.*, *4*, 71-76.

Barnett, S. (1972a): A note on the bezoutian matrix, *SIAM J. Appl. Math.*, *22*, 84-86.

Barnett, S. (1972b): Relationship between two methods for calculating the least order of a transfer function matrix, *Int. J. Control*, *15*, 509-512.

Barnett, S. (1972c): Regular greatest common divisor of two polynomial matrices, *Proc. Camb. Philos. Soc.*, *72*, 161-165.

Barnett, S. (1973a): New reductions of Hurwitz determinants, *Int. J. Control*, *18*, 977-991.

Barnett, S. (1973b): Interchangeability of the Routh and Jury tabular algorithms for linear system zero location, *Proc. 1973 IEEE Decision and Control Conference*, San Diego, 308-314.

Barnett, S. (1973c): Matrices, polynomials, and linear time-invariant systems, *IEEE Trans. Autom. Control*, *AC-18*, 1-10.

Barnett, S. (1973d): Some applications of matrices to location of zeros of polynomials, *Int. J. Control*, *17*, 823-831.

Barnett, S. (1974a): A new look at classical algorithms for polynomial resultant and g.c.d. calculation, *SIAM Rev.*, *16*, 193-206.

Barnett, S. (1974b): Application of the Routh array to stability of discrete-time linear systems, *Int. J. Control*, *19*, 47-55.

Barnett, S. (1974c): Some topics in algebraic systems theory: a survey, *Int. J. Control*, *19*, 669-688.

Barnett, S. (1974d): Evaluation of discrete linear system performance indices using Kronecker products, *Int. J. Control*, *20*, 849-855.

Barnett, S. (1974e): Simplification of the Liapunov matrix equation $A^TPA - P = -Q$, *IEEE Trans. Autom. Control*, *AC-19*, 446-447.

Barnett, S. (1975a): *Introduction to Mathematical Control Theory*, Clarendon Press, Oxford.

Barnett, S. (1975b): A note on matrix equations and root location, *IEEE Trans. Autom. Control*, *AC-20*, 158-159.

Barnett, S. (1975c): A companion matrix analogue for orthogonal polynomials, *Linear Algebra Appl.*, *12*, 197-208.

Barnett, S. (1975d): Some applications of the comrade matrix, *Int. J. Control*, *21*, 849-855.

Barnett, S. (1976a): Simplification of certain linear matrix equations, *IEEE Trans. Autom. Control*, *AC-21*, 115-116.

Barnett, S. (1976b): Some properties of inners and outers, *Proc. 1976 IEEE Decision and Control Conference*, Clearwater, Fla., 896-900.

Barnett, S. (1976c): A matrix circle in linear control theory, *Bull. Inst. Math. Appl.*, *12*, 173-177.

Barnett, S. (1977): Routh's array and bezoutian matrices, *Int. J. Control*, *26*, 175-181.

Barnett, S. (1979): *Matrix Methods for Engineers and Scientists*, McGraw-Hill, London.

Barnett, S. (1980): Greatest common divisors from generalized Sylvester resultant matrices, *Linear Multilinear Algebra*, *8*, 271-279.

Barnett, S. (1981a): Generalized polynomials and linear systems theory, *Proc. Third IMA Conference on Control Theory* (J. E. Marshall et al., Eds.), Academic Press, London, pp. 3-30.

Barnett, S. (1981b): Congenial matrices, *Linear Algebra Appl.*, *41*, 277-298.

Barnett, S., and Jury, E. I. (1978): Inners and Schur complement, *Linear Algebra Appl.*, *22*, 57-63.

Barnett, S., and Lancaster, P. (1980a): Matrices having striped inverse, *Proc. Fifth European Meeting on Cybernetics and Systems Research*, Vienna; in R. Trappl et al., Eds., *Progress in Cybernetics and Systems Research*, *8*, Hemisphere Publishing Corp., Washington D.C., pp. 333-336.

Barnett, S., and Lancaster, P. (1980b): Some properties of the bezoutian for polynomial matrices, *Linear Multilinear Algebra*, *9*, 99-110.

Barnett, S., and Maroulas, J. (1978): More on Routh's array and bezoutian matrices, *Int. J. Control*, *28*, 657-664.

Barnett, S., and Scraton, R. E. (1982): Location of matrix eigenvalues in the complex plane, *IEEE Trans. Autom. Control*, *AC-27*, 966-967.

Barnett, S., and Siljak, D. D. (1977): Routh's algorithm: a centennial survey, *SIAM Rev.*, *19*, 472-489.

Barnett, S., and Storey, C. (1966): Stability analysis of constant linear systems by Lyapunov's second method, *Electron. Lett.*, *2*, 165-166.

Barnett, S., and Storey, C. (1968): Some applications of the Lyapunov matrix equation, *J. Inst. Math. Appl.*, *4*, 33-42.

Barnett, S., and Storey, C. (1970): *Matrix Methods in Stability Theory*, Nelson, London.

Barroud, A. (1977): A numerical algorithm to solve $A^{T}XA - X = -Q$, *IEEE Trans. Autom. Control*, *AC-22*, 883-885.

Bartels, R. H., and Stewart, G. W. (1972): Solution of the matrix equation AX + XB = C, *Commun. ACM*, *15*, 820-826.

Bashkow, T. R., and Desoer, C. A. (1957): A network proof of a theorem on Hurwitz polynomials and its generalization, *Quart. Appl. Math.*, *14*, 423-426.

Bass, R. W., and Gura, I. (1966): Canonical forms for controllable systems with applications to optimal nonlinear feedback, *Proc. IFAC World Congress*, London.

Belanger, P. R., and McGillivray, T. P. (1976): Computational experience with the solution of the matrix Lyapunov equation, *IEEE Trans. Autom. Control*, *AC-21*, 789-800.

Bellman, R. (1970): *Introduction to Matrix Analysis*, 2nd Ed., McGraw-Hill, New York.

Bengtsson, G. (1977): Output regulation and internal models — a frequency domain approach, *Automatica*, *13*, 333-345.

Bitmead, R. R. (1981): Explicit solutions of the discrete-time Lyapunov matrix equation and Kalman-Yakubovich equations, *IEEE Trans. Autom. Control*, *AC-26*, 1291-1294.

Bitmead, R. R., and Anderson, B. D. O. (1977): The matrix Cauchy index: properties and applications, *SIAM J. Appl. Math.*, *33*, 655-672.

Bitmead, R. R., and Weiss, H. (1979): On the solution of the discrete-time Lyapunov matrix equation in controllable canonical form, *IEEE Trans. Autom. Control*, *AC-24*, 481-482.

Bitmead, R. R., Kung, S.-Y., Anderson, B. D. O., and Kailath, T. (1978): Greatest common divisors via generalized Sylvester and Bezout matrices, *IEEE Trans. Autom. Control*, *AC-23*, 1043-1047.

Bôcher, M. (1964): *Introduction to Higher Algebra*, Dover, New York.

Bradshaw, A., Fletcher, L. R., and Porter, B. (1978): Synthesis of output-feedback control laws for multivariable continuous-time systems, *Int. J. Syst. Sci.*, *9*, 1331-1340.

Brand, L. (1964): The companion matrix and its properties, *Amer. Math. Monthly*, *71*, 629-634.

Brand, L. (1968): Applications of the companion matrix, *Amer. Math. Monthly*, *75*, 146-152.

Braun, M. (1978): *Differential Equations and Their Applications*, 2nd Ed., Springer-Verlag, New York.

Brenner, J. L. (1961): Expanded matrices from matrices with complex elements, *SIAM Rev.*, *3*, 165-166.

Brenner, J. L., and Lyndon, R. C. (1981): Proof of the fundamental theorem of algebra, *Amer. Math. Monthly*, *88*, 253-256.

Brown, W. S. (1978): The subresultant PRS algorithm, *ACM Trans. Math. Software*, *4*, 237-249.

Brunovsky, P. (1970): A classification of linear controllable systems, *Kybernetika*, *3*, 173-187.

Bückner, H. (1952): A formula for an integral occurring in the theory of linear servomechanisms and control systems, *Quart. Appl. Math.*, *10*, 205-213.

Bumby, R., Sontag, E. D., Sussman, H. J., and Vasconcelos, W. (1981): Remarks on the pole-shifting problem over rings, *J. Pure Appl. Algebra*, *20*, 113-127.

Burnside, W. S., and Panton, A. W. (1960): *The Theory of Equations*, Vols. 1 and 2, Dover, New York.

Butman, S., and Sivan, R. (1964): On cancellations, controllability and observability, *IEEE Trans. Autom. Control*, *AC-9*, 317-318.

Cain, B. E. (1980): Inertia theory, *Linear Algebra Appl.*, *30*, 211-240.

Cameron, R., and Kouvaritakis, B. (1981): Minimizing the norm of output feedback controllers used in pole placement: a dyadic approach, *Int. J. Control*, *32*, 759-770.

Carlson, D., and Hill, R. D. (1976): Generalized controllability and inertia theory, *Linear Algebra Appl.*, *15*, 177-187.

Carlson, D., and Hill, R. D. (1977): Controllability and inertia theory for functions of a matrix, *J. Math. Anal. Appl.*, *59*, 260-266.

Carlson, D., and Schneider, H. (1963): Inertia theorems for matrices; the semidefinite case, *J. Math. Anal. Appl.*, *6*, 430-446.

Carlson, D., Datta, B. N., and Johnson, C. R. (1982): A semi-definite Lyapunov theorem, and the characterization of tri-diagonal D-stable matrices, *SIAM J. Alg. Disc. Meth.*, *3*, 293-304.

Chen, C. F. (1974): A new formulation of the Hermite criterion, *Int. J. Control*, *19*, 757-764.

Chen, C.-T. (1970): *Introduction to Linear System Theory*, Holt, Reinhart and Winston, New York.

Chen, C.-T., Desoer, C. A., and Niederlinski, A. (1966): Simplified conditions for controllability and observability of linear time-invariant systems, *IEEE Trans. Autom. Control*, AC-11, 613-614.

Cheney, E. W. (1966): *Introduction to Approximation Theory*, McGraw-Hill, New York.

Chihara, T. S. (1978): *An Introduction to Orthogonal Polynomials*, Gordon and Breach, New York.

Choudhury, D. R. (1973): Algorithm for power of companion matrix and its application, *IEEE Trans. Autom. Control*, AC-18, 179-180.

Claessens, G. (1978): On the location of zeros of a generalized polynomial, *Linear Algebra Appl.*, 22, 79-88.

Coppel, W. A. (1974a): Matrix quadratic equations, *Bull. Austral. Math. Soc.*, 10, 377-401.

Coppel, W. A. (1974b): Matrices of rational functions, *Bull. Austral. Math. Soc.*, 11, 89-113.

Cottle, R. W. (1974): Manifestations of the Schur complement, *Linear Algebra Appl.*, 8, 189-211.

Crossley, T. R., and Porter, B. (1979): State observers and their application, in *Modern Approaches to Control System Design*, N. Munro, Ed.), Peter Peregrinus/IEE, London.

Cullen, C. G. (1965): A note on convergent matrices, *Amer. Math. Monthly*, 72, 1006-1007.

Cullen, C. G. (1966): *Matrices and Linear Transformations*, Addison-Wesley, Reading, Mass.

Cutteridge, O. P. D. (1959): The stability criteria for linear systems, *Proc. IEE*, 106C, 125-132.

Dahlquist, G., and Björck, A. (1974): *Numerical Methods*, Prentice-Hall, Englewood Cliffs, N.J.

Daly, K. C. (1976): The computation of Luenberger canonical forms using elementary similarity transformations, *Int. J. Syst. Sci.*, 7, 1-15.

Datt, B., and Govil, N. K. (1978): On the location of the zeros of a polynomial, *J. Approximation Theory*, 24, 78-82.

Datta, B. N. (1978): On the Routh-Hurwitz-Fujiwara and the Schur-Cohn-Fujiwara theorems for the root-separation problem, *Linear Algebra Appl.*, 22, 235-246.

Davies, A. C. (1974): Bilinear transformation of polynomials, *IEEE Trans. Circuits Syst.*, CAS-21, 792-794.

Davison, E. J., Gesing, W., and Wang, S. H. (1978): An algorithm for obtaining the minimal realization of a linear time-invariant system and determining if a system is stabilizable-detectable, *IEEE Trans. Autom. Control*, AC-23, 1048-1054.

de Souza, E., and Bhattacharyya, S. P. (1981): Controllability and the linear equation Ax = b, *Linear Algebra Appl.*, 36, 97-101.

Dickinson, B. W. (1974): On the fundamental theorem of linear state variable feedback, *IEEE Trans. Autom. Control, AC-19*, 577-579.

Dickinson, B. W. (1980): Analysis of the Lyapunov equation using generalized positive real matrices, *IEEE Trans. Autom. Control, AC-25*, 560-563.

Dobbs, D. E., and Hanks, R. (1980): *A Modern Course on the Theory of Equations*, Polygonal Publishing, Passaic, N.J.

Dorato, P., and Levis, A. H. (1971): Optimal linear regulators: the discrete-time case, *IEEE Trans. Autom. Control, AC-16*, 613-620.

Duffin, R. J. (1969): Algorithms for classical stability problems, *SIAM Rev.*, *11*, 196-213.

Eising, R. (1982): Pole assignment, a new proof and algorithm, *Syst. Control Lett.*, *2*, 6-12.

Elgerd, O. I. (1967): *Control Systems Theory*, McGraw-Hill, New York.

Emre, E. (1980a): The polynomial equation $QQ_c + RP_c = \Phi$ with application to dynamic feedback, *SIAM J. Control Optim.*, *18*, 611-620.

Emre, E. (1980b): Nonsingular factors of polynomial matrices and (A, B)-invariant subspaces, *SIAM J. Control Optim.*, *18*, 288-296.

Emre, E., and Silverman, L. M. (1977): New criteria and system theoretic interpretations for relatively prime polynomial matrices, *IEEE Trans. Autom. Control, AC-22*, 239-242.

Fallside, F., Ed. (1977): *Control System Design by Pole-Zero Assignment*, Academic Press, London.

Feinstein, J., and Bar-Ness, Y. (1980): On the uniqueness of the minimal solution to the matrix polynomial equation $A(\lambda)X(\lambda) + Y(\lambda)B(\lambda) = C(\lambda)$, *J. Franklin Inst.*, *310*, 131-134.

Flamm, D. S. (1980): A new proof of Rosenbrock's theorem on pole assignment, *IEEE Trans. Autom. Control, AC-25*, 1128-1133.

Fletcher, L. R. (1979): Placement des valeurs propres pour les systèmes linéaires multivariables, *C.R. Acad. Sci. Paris*, *289*, 499-501.

Fletcher, L. R. (1980): An intermediate algorithm for pole placement by output feedback in linear multivariable control systems, *Int. J. Control*, *31*, 1121-1136.

Fletcher, L. R. (1981a): On pole placement in linear multivariable systems with direct feedthrough: I. Theoretical considerations, *Int. J. Control*, *33*, 739-749.

Fletcher, L. R. (1981b): On pole placement in linear multivariable systems with direct feedthrough: II. Computational considerations, *Int. J. Control*, *33*, 1147-1154.

Fletcher, L. R. (1981c): An algorithm for pole placement by output feedback, *Proc. Third IMA Conference on Control Theory* (J. E. Marshall et al., Eds.), Academic Press, London, pp. 283-291.

Flood, M. M. (1935): Division by non-singular matrix polynomials, *Ann. Math.*, *36*, 859-869.

Forney, G. D., Jr. (1975): Minimal bases of rational vector spaces with applications to multivariable linear systems, *SIAM J. Control*, *13*, 493-520.

Fortmann, T. E., and Hitz, K. L. (1977): *An Introduction to Linear Control Systems*, Marcel Dekker, New York.

Fossard, A., and Guéguen, C. (1977): *Multivariable System Control*, North-Holland, Amsterdam.

Freud, G. (1966): *Orthogonal Polynomials*, Pergamon Press, Oxford.

Fryer, W. D. (1959): Applications of Routh's algorithm to network theory problems, *IRE Trans. Circuit Theory*, *CT-6*, 144-149.

Fuhrmann, P. A. (1976): Algebraic systems theory: an analyst's point of view, *J. Franklin Inst.*, *301*, 521-540.

Fuhrmann, P. A. (1978): Linear feedback via polynomial models, *Int. J. Control*, *30*, 363-377.

Fuhrmann, P. A. (1981): *Linear Systems and Operators in Hilbert Space*, McGraw-Hill, New York.

Fujiwara, M. (1926): Ueber die algebraische Gleichungen deren Wurzeln in einem Kreise oder in einer Halbebene liegen, *Math. Z.*, *24*, 160-169.

Fuller, A. T., Ed. (1975): *Stability of Motion*, Taylor & Francis, London.

Fuller, A. T. (1977): Edward John Routh, *Int. J. Control*, *26*, 169-173.

Gantmacher, F. R. (1959): *The Theory of Matrices*, Vol. 2, Chelsea, New York.

Gargantini, I. (1971): The numerical stability of the Schur-Cohn criterion, *SIAM J. Numer. Anal.*, *8*, 24-29.

Gautschi, W. (1978): Questions of numerical condition related to polynomials, in *Recent Advances in Numerical Analysis* (C. De Boor and G. H. Golub, Eds.), Academic Press, New York, pp. 45-72.

Genin, Y. (1975): Hurwitz sequences of polynomials, *Philips Res. Rep.*, *30*, 89-102.

Gilbert, E. G. (1963): Controllability and observability in multivariable control systems, *SIAM J. Control*, *1*, 128-151.

Gohberg, I., and Feldman, I. A. (1974): Convolution equations and
projection methods for their solutions, *Transl. Math. Monogr.*, *41*
(Amer. Math. Soc.).

Gohberg, I., Kaashoek, M. A., Lerer, L., and Rodman, L. (1981):
Common multiples and common divisors of matrix polynomials,
I. Spectral method, *Indiana J. Math.*, *30*, 321-356.

Gohberg, I., Kaashoek, M. A., Lerer, L., and Rodman, L. (1983):
Common multiples and common divisors of matrix polynomials, II.
Vandermonde and resultant matrices, *Linear Multilinear Algebra.*

Golub, G., Nash, S., and Van Loan, C. (1979): A Hessenberg-Schur
method for the problem AX + XB = C, *IEEE Trans. Autom. Control*,
AC-24, 909-913.

Good, I. J. (1961): The colleague matrix, a Chebyshev analogue of
the companion matrix, *Quart. J. Math. Oxf. Ser.*, *12*, 61-68.

Grasselli, O. M. (1980): Conditions for controllability and recon-
structability of discrete time linear composite systems, *Int. J.
Control*, *31*, 433-441.

Gutman, S. (1979): Root clustering of a complex matrix in an
algebraic region, *IEEE Trans. Autom. Control*, *AC-24*, 647-650.

Gutman, S. (1982): A test for root-clustering transformability,
IEEE Trans. Autom. Control, *AC-27*, 979-981.

Gutman, S., and Jury, E. I. (1981): A general theory for matrix
root-clustering in subregions of the complex plane, *IEEE Trans.
Autom. Control*, *AC-26*, 853-863.

Hahn, W. (1956): Eine Bemerkung zur zweiten Methode von Ljapunov,
Math. Nachr., *14*, 349-354.

Hahn, W. (1963): *Theory and Application of Liapunov's Direct
Method*, Prentice-Hall, Englewood Cliffs, N.J.

Hahn, W. (1971): Zur stabilitätstheorie linearer autonomer
Differentialgleichungssysteme, *Monatsh. Math.*, *75*, 118-122.

Hamada, N. (1979): Algebraic stability theory, a century from
Routh, *J. Soc. Electron. Commun. (Jap.)*, *62*, 995-1003.

Harn, Y.-P., and Chen, C.-T. (1981): A proof of a discrete
stability test via the Liapunov theorem, *IEEE Trans. Autom.
Control*, *AC-26*, 733-734.

Hartwig, R. E. (1975): Resultants and the solution of
AX - XB = -C, *SIAM J. Appl. Math.*, *23*, 104-117.

Hartwig, R. E. (1982): Applications of the Wronskian and Gram
matrices of $\{t^i e^{\lambda_k t}\}$, *Linear Algebra Appl.*, *43*, 229-241.

Hautus, M. L. J. (1969): Controllability and observability
conditions of linear autonomous systems, *Ned. Akad. Wet. Proc.*,
Ser. A, *72*, 443-448.

Hautus, M. L. J. (1970): Stabilization, controllability and observability of linear autonomous systems, *Ned. Akad. Wet. Proc., Ser. A, 73,* 448-455.

Hautus, M. L. J. (1977): A simple proof of Heymann's lemma, *IEEE Trans. Autom. Control, AC-22,* 885-886.

Haynsworth, E. V. (1968): Determination of the inertia of a partitioned hermitian matrix, *Linear Algebra Appl., 1,* 73-81.

Hayton, G. E. (1980): The generalized resultant matrix, *Int. J. Control, 32,* 567-579.

Heinig, G. (1977): The concepts of a bezoutian and a resolvent for operator bundles, *Funct. Anal. Appl., 11,* 241-243 (transl. from Russian).

Henderson, H. V., and Searle, S. R. (1981): The vec-permutation matrix, the vec operator and Kronecker products: a review, *Linear Multilinear Algebra, 9,* 271-288.

Henrici, P. (1970): Upper bounds for the abscissa of stability of a stable polynomial, *SIAM J. Numer. Anal., 7,* 538-544.

Henrici, P. (1974): *Applied and Computational Complex Analysis,* Vol. 1, Wiley, New York.

Henrici, P. (1977): *Applied and Computational Complex Analysis,* Vol. 2, Wiley, New York.

Heymann, M. (1968): Comments on "On pole assignment in multi-input controllable linear systems," *IEEE Trans. Autom. Control, AC-13,* 748-749.

Heymann, M. (1975): *Structure and Realization Problems in the Theory of Dynamical Systems,* CISM Courses and Lectures No. 204, Springer-Verlag, New York.

Heymann, M. (1979): The pole shifting theorem revisited, *IEEE Trans. Autom. Control, AC-24,* 479-480.

Heymann, M., and Feuer, A. (1974): On a theorem by W. Hahn, *Monatsh. Math., 78,* 391-394.

Hill, R. D. (1977): Eigenvalue location using certain matrix functions and geometric curves, *Linear Algebra Appl., 16,* 83-91.

Householder, A. S. (1968): Bigradients and the problem of Routh and Hurwitz, *SIAM Rev., 10,* 56-66.

Householder, A. S. (1970a): *The Numerical Treatment of a Single Nonlinear Equation,* McGraw-Hill, New York.

Householder, A. S. (1970b): Bezoutiants, elimination and localization, *SIAM Rev., 12,* 73-78.

Howland, J. L. (1978): The resultant iteration for determining the stability of a polynomial, *Math. Comp., 32,* 779-789.

Howland, J. L. (1983): The sign matrix and the separation of matrix eigenvalues, *Linear Algebra Appl.*

Howland, J. L., and Senez, J. A. (1970): A constructive method for the solution of the stability problem, *Numer. Math.*, *16*, 1-7.

Hung, Y. S., and MacFarlane, A. G. J. (1981): On the relationships between the unbounded asymptote behaviour of multivariate root loci, impulse response and infinite zeros, *Int. J. Control*, *34*, 31-69.

Huseyin, O., and Jury, E. I. (1981): Inner formulation of Lyapunov stability test and generalized Schur complement, *Proc. IFAC World Congress*, Kyoto, Japan.

Inouye, Y. (1982): Notes on controllability and constructibility of linear discrete-time systems, *Int. J. Control*, *35*, 1081-1084.

Jacquot, R. G. (1981): *Modern Digital Control Systems*, Marcel Dekker, New York.

Jeltsch, R. (1979): An optimal fraction free Routh array, *Int. J. Control*, *30*, 653-660.

Jones, W. B., and Thron, W. J. (1980): *Continued Fractions: Analytic Theory and Applications*, Addison-Wesley, Reading, Mass.

Jury, E. I. (1964): *Theory and Application of the z-Transform Method*, Wiley, New York.

Jury, E. I. (1971): Inners approach to some problems of system theory, *IEEE Trans. Autom. Control*, *AC-16*, 233-240.

Jury, E. I. (1973): Remarks on "The mechanics of bilinear transformation," *IEEE Trans. Audio Electroacoust.*, *AU-21*, 380-382.

Jury, E. I. (1974): *Inners and Stability of Dynamic Systems*, Wiley, New York; 2nd Ed., Krieger, Fla. (1982).

Jury, E. I. (1975a): The theory and applications of the inners, *Proc. IEEE*, *63*, 1044-1069.

Jury, E. I. (1975b): Comments on "On the Routh-Hurwitz criterion," *IEEE Trans. Autom. Control*, *AC-20*, 292-296.

Jury, E. I. (1978): Stability of multidimensional polynomials, *Proc. IEEE*, *66*, 1018-1047.

Jury, E. I., and Ahn, S. M. (1972): Symmetric and innerwise matrices for the root-clustering and root-distribution of a polynomial, *J. Franklin Inst.*, *293*, 433-450.

Jury, E. I., and Ahn, S. M. (1974): Remarks on the root-clustering of a polynomial in a certain region in the complex plane, *Quart. Appl. Math.*, *32*, 203-205.

Jury, E. I., and Anderson, B. D. O. (1972): Some remarks on simplified stability criteria for continuous linear systems, *IEEE Trans. Autom. Control*, *AC-17*, 371-372.

Jury, E. I., and Anderson, B. D. O. (1981): A note on the "Reduced Schur-Cohn criterion," *IEEE Trans. Autom. Control, AC-26*, 612-614.

Jury, E. I., and Chan, O. W. C. (1973): Combinatorial rules for some useful transformations, *IEEE Trans. Circuit Theory, CT-20*, 476-480.

Kailath, T. (1980): *Linear Systems*, Prentice-Hall, Englewood Cliffs, N.J.

Kalman, R. E. (1960): Contributions to the theory of optimal control, *Bol. Soc. Mat. Mex., 5*, 102-119.

Kalman, R. E. (1962): Canonical structure of linear dynamical systems, *Proc. Nat. Acad. Sci. USA, 48*, 596-600.

Kalman, R. E. (1963): Mathematical description of linear dynamical systems, *SIAM J. Control, 1*, 152-192.

Kalman, R. E. (1965a): On the Hermite-Fujiwara theorem in stability theory, *Quart. Appl. Math., 23*, 279-282.

Kalman, R. E. (1965b): Irreducible realizations and the degree of a rational matrix, *J. Soc. Ind. Appl. Math., 13*, 520-544.

Karlin, S., and Studden, W. J. (1966): *Tchebycheff Systems with Applications in Analysis and Statistics*, Wiley, New York.

Kemperman, J. H. B. (1982): A Hurwitz matrix is totally positive, *SIAM J. Math. Anal., 13*, 331-341.

Kharitonov, V. L. (1981): Distribution of the roots of the characteristic polynomial of an autonomous system, *Autom. and Remote Control (USA), 42*, 589-593.

Kimura, H. (1975): Pole assignment by gain output feedback, *IEEE Trans. Autom. Control, AC-20*, 509-516.

Knuth, D. E. (1969): *The Art of Computer Programming*, Vol. 2, Addison-Wesley, Reading, Mass.

Koussiouris, T. G. (1981): Controllability indices of a system, minimal indices of its transfer function matrix, and their relations, *Int. J. Control, 34*, 613-622.

Koussiouris, T. G., and Kafiris, G. P. (1981): Controllability indices, observability indices and the Hankel matrix, *Int. J. Control, 33*, 773-775.

Kouvaritakis, B., and Cameron, R. (1980): Pole placement with minimized norm controllers, *IEE Proc., 127*, Pt. D., 32-36.

Krein, M. G., and Naimark, M. A. (1981): The method of symmetric and hermitian forms in the theory of the separation of the roots of algebraic equations (transl. by O. Boshko and J. L. Howland), *Linear Multilinear Algebra, 10*, 265-308.

Kucera, V. (1972): The discrete Riccati equation of optimal control, *Kybernetika, 8*, 430-447.

Kucera, V. (1979): *Discrete Linear Control*, Wiley, Chichester.

Kucera, V. (1980): Testing controllability and constructibility in discrete linear systems, *IEEE Trans. Autom. Control, AC-25*, 297-298.

Kuo, B. C. (1977): *Digital Control Systems*, SRL Publishing Co., Champaign, Ill.

Kwakernaak, H., and Sivan, R. (1972): *Linear Optimal Control Systems*, Wiley-Interscience, New York.

Laidacker, M. A. (1969): Another theorem relating Sylvester's matrix and the greatest common divisor, *Math. Mag., 42*, 126-128.

Lancaster, P. (1969): *Theory of Matrices*, Academic Press, New York.

Lancaster, P., and Rodman, L. (1980): Existence and uniqueness theorems for the algebraic Riccati equation, *Int. J. Control, 32*, 285-309.

László, L. (1981): Matrix methods for polynomials, *Linear Algebra Appl., 36*, 129-131.

László, L. (1982): Imaginary part bounds on polynomial zeros, *Linear Algebra Appl., 44*, 173-180.

Laub, A. J. (1979): A Schur method for solving algebraic Riccati equations, *IEEE Trans. Autom. Control, AC-24*, 913-921.

Laub, A. J. (1981): Efficient multivariable frequency response calculations, *IEEE Trans. Autom. Control, AC-26*, 407-408.

Laub, A. J., and Meyer, K. (1974): Canonical forms for symplectic and hamiltonian matrices, *Celestial Mech., 9*, 213-238.

Ledermann, W., and Vajda, S., Eds. (1980): *Handbook of Applicable Mathematics Vol. I: Algebra*, Wiley, Chichester.

Lee, E. B., and Zak, S. H. (1982): On spectrum placement for linear time invariant delay systems, *IEEE Trans. Autom. Control, AC-27*, 446-449.

Lee, E. B., Olbrot, A. W., and Zak, S. H. (1982): Cyclicity of polynomial systems, *IEEE Trans. Autom. Control, AC-27*, 451-453.

Lehnigk, S. H. (1966): *Stability Theorems for Linear Motions*, Prentice-Hall, Englewood Cliffs, N.J.

Lehnigk, S. H. (1967): Liapunov's direct method and the number of zeros with positive real parts of a polynomial with constant complex coefficients, *J. SIAM Control, 5*, 234-244.

Lehnigk, S. H. (1970): On the Hurwitz matrix, *Zeit fur ang. Math. Phys., 21*, 498-500.

Lerer, M., and Tismenetsky, M. (1981): The bezoutian and the eigenvalue-separation problem for matrix polynomials, Preprint MT-497, Israel Institute of Technology, Haifa.

Liénard, A., and Chipart, M. H. (1914): Sur le signe de la partie réelle des racines d'une équation algébrique, *J. Math. Pures Appl.*, *10*, 291-346.

Lipatov, A. V., and Sokolov, N. I. (1979): Some sufficient conditions for stability and instability of continuous linear stationary systems, *Autom. Remote Control (USA)*, *39*, 1285-1291.

Lovass-Nagy, V., Miller, R. J., and Powers, D. L. (1977): On system realization by matrix generalized inverses, *Int. J. Control*, *26*, 745-751.

Lovass-Nagy, V., Miller, R. J., and Powers, D. L. (1978): Further results on output control in the servomechanism sense, *Int. J. Control*, *27*, 133-138.

Lovass-Nagy, V., O'Kennon, M. R., and Rabson, G. (1981): Pole assignment using matrix generalized inverse, *Int. J. Syst. Sci.*, *12*, 383-392.

Luenberger, D. G. (1964): Observing the state of a linear system, *IEEE Trans. Mil. Electron.*, *MIL-8*, 74-80.

Luenberger, D. G. (1967): Canonical forms for linear multivariable systems, *IEEE Trans. Autom. Control*, *AC-12*, 290-295.

McClamroch, N. H. (1980): *State Models of Dynamic Systems*, Springer-Verlag, New York.

MacDuffee, C. C. (1950a): Some applications of matrices in the theory of equations, *Amer. Math. Monthly*, *57*, 154-161.

MacDuffee, C. C. (1950b): *The Theory of Matrices*, Chelsea, New York (reprint of 1933 edition).

McEliece, R. J., and Shearer, J. B. (1978): A property of Euclid's algorithm and an application to Padé approximation, *SIAM J. Appl. Math.*, *34*, 611-615.

MacFarlane, A. G. J. (1963a): The calculation of functionals of the time and frequency response of a linear constant coefficient dynamical system, *Quart. J. Mech. Appl. Math.*, *16*, 259-271.

MacFarlane, A. G. J. (1963b): An eigenvector solution of the optimal linear regulator problem, *J. Electron. Control*, *14*, 643-654.

MacFarlane, A. G. J. (1975): Relationships between recent developments in linear control theory and classical design techniques, *Meas. Control*, *8*, 179-187; 219-223; 279-284; 319-324; 371-375.

MacFarlane, A. G. J., and Karcanias, N. (1976): Poles and zeros of linear multivariable systems: a survey of the algebraic, geometric and complex-variable theory, *Int. J. Control*, *24*, 33-74.

McMillan, B. (1952): Introduction to formal realizability theory II, *Bell Syst. Tech. J.*, *31*, 541-600.

Mansour, M. (1965): Stability criteria of linear systems and the second method of Lyapunov, *Sci. Electr.*, *11*, 87-96.

Mansour, M. (1982): A note on the stability of linear discrete systems and Lyapunov method, *IEEE Trans. Autom. Control*, *AC-27*, 707-708.

Marden, M. (1966): *Geometry of Polynomials*, American Mathematical Society, Providence, R.I.

Maroulas, J. (1978): *Theory of generalized polynomials with applications to linear system theory*, Ph.D. thesis, University of Bradford.

Maroulas, J. (1982): A triangular factorization of bezoutian matrices, *Int. J. Control*, *36*, 1011-1019.

Maroulas, J., and Barnett, S. (1978a): Canonical forms for time-invariant linear control systems: a survey with extensions: Part I. Single-input case, *Int. J. Syst. Sci.*, *9*, 497-514.

Maroulas, J., and Barnett, S. (1978b): Greatest common divisor of generalized polynomials and polynomial matrices, *Linear Algebra Appl.*, *22*, 195-210.

Maroulas, J., and Barnett, S. (1978c): Some new results on the qualitative theory of generalized polynomials, *J. Inst. Math. Appl.*, *22*, 53-70.

Maroulas, J., and Barnett, S. (1978d): Applications of the comrade matrix to linear multivariable systems theory, *Int. J. Control*, *28*, 129-145.

Maroulas, J., and Barnett, S. (1979a): Properties of Hermite's matrices in stability theory, *Int. J. Control*, *29*, 717-732.

Maroulas, J., and Barnett, S. (1979b): Canonical forms for time-invariant linear control systems: a survey with extensions: Part II. Multivariable case, *Int. J. Syst. Sci.*, *10*, 33-50.

Maroulas, J., and Barnett, S. (1979c): Polynomials with respect to a general basis: I. Theory, *J. Math. Anal. Appl.*, *72*, 177-194.

Maroulas, J., and Barnett, S. (1979d): Polynomials with respect to a general basis: II. Applications, *J. Math. Anal. Appl.*, *72*, 599-614.

Maroulas, J., and Barnett, S. (1979e): Further results on the qualitative theory of generalized polynomials, *J. Inst. Math. Appl.*, *23*, 33-42.

Maroulas, J., and Barnett, S. (1979f): Zero location for discrete-time linear system polynomials with respect to an orthogonal basis, *IEEE Trans. Autom. Control*, *AC-24*, 785-787.

Maroulas, J., and Barnett, S. (1980): Continued-fraction expansions for ratios of generalized polynomials, *Appl. Math. Comp.*, *6*, 229-249.

Maroulas, J., and Dascalopoulos, D. (1981): Applications of the generalized Sylvester matrix, *Appl. Math. Comp.*, *8*, 121-135.

Martensson, K. (1971): On the matrix Riccati equation, *Inf. Sci.*, *3*, 17-49.

Mazko, A. G. (1980): The Lyapunov matrix equation for a certain class of regions bounded by algebraic curves, *Sov. Autom. Control (USA)*, *13*, 37-42.

Miller, J. J. H. (1971): On the location of zeros of certain classes of polynomials with applications in numerical analysis, *J. Inst. Math. Appl.*, *8*, 397-406.

Miminis, G. S., and Paige, C. C. (1982): An algorithm for pole assignment of time invariant linear systems, *Int. J. Control*, *35*, 341-354.

Mishina, A. P., and Proskuryakov, I. V. (1965): *Higher Algebra*, Pergamon, Oxford.

Moler, C., and Van Loan, C. (1978): Nineteen dubious ways to compute the exponential of a matrix, *SIAM Rev.*, *20*, 801-836.

Molinari, B. P. (1973): The stabilizing solution of the algebraic Riccati equation, *SIAM J. Control*, *11*, 262-271.

Molinari, B. P. (1977): The time-invariant linear-quadratic optimal control problem, *Automatica*, *13*, 347-357.

Munro, N. (1974): Further results on pole-shifting using output feedback, *Int. J. Control*, *20*, 775-786.

Munro, N., Ed. (1979): *Modern Approaches to Control System Design*, Peter Peregrinus/IEE, London.

Nehari, Z. (1952): *Conformal Mapping*, McGraw-Hill, New York.

Neumann, M. (1981): On the Schur complement and the LU-factorization of a matrix, *Linear Multilinear Algebra*, *9*, 241-254.

Newman, M. (1972): *Integral Matrices*, Academic Press, New York.

Newman, M. (1982): A result about determinantal divisors, *Linear Multilinear Algebra*, *11*, 363-366.

Obreschkoff, N. (1963): *Verteilung und Berechnung der Nullstellen reeler Polynome*, VEB, Berlin.

O'Donnell, J. J. (1966): Asymptotic solution of the matrix Riccati equation of optimal control, *Proc. Fourth Annual Conference on Circuit and System Theory*, University of Illinois, Urbana, Ill., pp. 577-586.

Ortega, J. M. (1973): Stability of difference equations and convergence of iterative processes, *SIAM J. Numer. Anal.*, *10*, 268-282.

Ostrowski, A. M. (1965): On Descartes' rule of signs for certain polynomial developments, *J. Math. Mech.*, *14*, 195-209.

Ostrowski, A., and Schneider, H. (1962): Some theorems on the inertia of general matrices, *J. Math. Anal. Appl.*, *4*, 72-84.

Ouellette, D. V. (1981): Schur complements and statistics, *Linear Algebra Appl.*, *36*, 187-295.

Pace, I. S., and Barnett, S. (1974): Efficient algorithms for linear system calculations: Part I. Smith form and common divisor of polynomial matrices, *Int. J. Syst. Sci.*, *5*, 403-411.

Paige, C. C. (1981): Properties of numerical algorithms related to computing controllability, *IEEE Trans. Autom. Control*, *AC-26*, 130-138.

Paige, C. C., and Van Loan, C. (1981): A Schur decomposition for hamiltonian matrices, *Linear Algebra Appl.*, *41*, 11-32.

Pappas, T., Laub, A. J., and Sandell, N. R., Jr. (1980): On the numerical solution of the discrete-time algebraic Riccati equation, *IEEE Trans. Autom. Control*, *AC-25*, 631-641.

Parks, P. C. (1962): A new proof of the Routh-Hurwitz stability criterion using the second matrix of Lipaunov, *Proc. Camb. Philos. Soc.*, *58*, 694-702.

Parks, P. C. (1963a): Analytic methods for investigating stability — linear and non-linear systems. A survey, *Proc. Inst. Mech. Eng.*, *178*, Pt. 3M, 3-13.

Parks, P. C. (1963b): Further comments on "A symmetric matrix formulation of the Hurwitz-Routh stability criterion," *IEEE Trans. Autom. Control*, *AC-8*, 270-271.

Parks, P. C. (1963c): A new proof of the Hurwitz stability criterion by the second method of Liapunov, with applications to "optimum" transfer functions, *Proc. Joint Automatic Control Conference*, pp. 471-478.

Parks, P. C. (1964): Liapunov and the Schur-Cohn criterion, *IEEE Trans. Autom. Control*, *AC-9*, 121.

Parks, P. C. (1969): Hermite-Hurwitz and Hermite-Bilharz links using matrix multiplication, *Electron. Lett.*, *5*, 55-57.

Parks, P. C. (1977a): A new proof of Hermite's stability criterion and a generalization of Orlando's formula, *Int. J. Control*, *26*, 197-206.

Parks, P. C. (1977b): On the number of roots of an algebraic equation contained between given limits (transl. of C. Hermite), *Int. J. Control*, *26*, 183-195.

Parodi, M. (1959): *La localisation des valeurs caractéristiques des matrices et ses applications*, Gauthier-Villars, Paris.

Parodi, M. (1963): Sur quelques propriétés des zéros des polynomes combinaisons linéaires à coefficients constants de polynomes récurrents, *Bull. Sci. Math.*, *2nd Ser.*, *87*, 43-52.

Patel, R. V. (1981): Computation of matrix fraction descriptions of linear time-invariant systems, *IEEE Trans. Autom. Control*, *AC-26*, 148-161.

Pernebo, L. (1981): An algebraic theory for the design of controllers for linear multivariable systems: Part I. Structure matrices and feedforward design, *IEEE Trans. Autom. Control*, *AC-26*, 171-182.

Policastro, M. (1979): A simple algorithm to perform the bilinear transformation, *Int. J. Control*, *30*, 713-715.

Porter, B. (1967): *Stability Criteria for Linear Dynamical Systems*, Oliver & Boyd, Edinburgh.

Porter, B. (1977): Eigenvalue assignment in linear multivariable systems by output feedback, *Int. J. Control*, *25*, 483-490.

Porter, B., and Bradshaw, A. (1978): Design of linear multivariable continuous-time output-feedback regulators, *Int. J. Syst. Sci.*, *9*, 445-450.

Power, H. M. (1967a): The companion matrix and Liapunov functions for linear, multivariable time-invariant systems, *J. Franklin Inst.*, *283*, 214-234.

Power, H. M. (1967b): The mechanics of the bilinear transformation, *IEEE Trans. Educ.*, *E-10*, 114-116.

Power, H. M. (1969): Canonical form for the matrices of linear discrete-time systems, *Proc. IEE*, *116*, 1245-1252.

Power, H. M. (1970): Note on the Hermite-Fujiwara theorem, *Electron. Lett.*, *6*, 39-40.

Power, H. M. (1977): New approaches to eigenvalue assignment by output feedback and numerator control by state feedback, in *Control System Design by Pole-Zero Assignment* (F. Fallside, Ed.), Academic Press, London.

Power, H. M. (1979): Matrices in control systems, Irish Mathematical Society Conference on Matrix Theory and Its Applications, Dublin.

Pták, V. (1981): The discrete Lyapunov equation in controllable canonical form, *IEEE Trans. Autom. Control*, *AC-26*, 580-581.

Pták, V., and Young, N. J. (1980): A generalization of the zero location theorem of Schur and Cohn, *IEEE Trans. Autom. Control*, *AC-25*, 978-980.

Pták, V., and Young, N. J. (1982): Zero location by hermitian forms: the singular case, *Linear Algebra Appl.*, *43*, 181-196.

Puri, N. N., and Weygandt, C. N. (1964): Calculation of quadratic moments of high-order linear systems via Routh canonical transformation, *IEEE Trans. Appl. Ind.*, *AI-83*, 428-433.

Redheffer, R. (1965): Remarks on a paper by Taussky, *J. Algebra*, *2*, 42-47.

Reid, W. T. (1972): *Riccati Differential Equations*, Academic Press, New York.

Retallack, D. G., and MacFarlane, A. G. J. (1970): Pole shifting techniques for multivariable feedback systems, *Proc. IEE*, *117*, 1037-1038.

Rissanen, J. (1960): Control system synthesis by analogue computer based on the generalized linear feedback concept, *International Seminar on Analogue Computation Applied to the Study of Chemical Processes*, Brussels.

Roberts, J. D. (1980): Linear model reduction and solution of the algebraic Riccati equation by use of the sign function, *Int. J. Control*, *32*, 677-687.

Rodman, L. (1980): On extremal solutions of the algebraic Riccati equations,*Lect. Appl. Math.*, *18*, 311-327.

Roebuck, P. A., and Barnett, S. (1978): A survey of Toeplitz and related matrices, *Int. J. Syst. Sci.*, *9*, 921-934.

Rosenbrock, H. H. (1970): *State-Space and Multivariable Theory*, Nelson, London.

Rosenbrock, H. H., and Hayton, G. E. (1978): The general problem of pole assignment, *Int. J. Control*, *27*, 837-852.

Rowe, A. (1972): The generalized resultant matrix, *J. Inst. Math. Appl.*, *9*, 390-396.

Rugh, W. J. (1975): *Mathematical Desription of Linear Systems*, Marcel Dekker, New York.

Russell, D. L. (1979): *Mathematics of Finite-Dimensional Control Systems*, Marcel Dekker, New York.

Sarma, I. G., and Pai, M. A. (1968): A note on the Lyapunov matrix equation for linear discrete systems, *IEEE Trans. Autom. Control*, *AC-13*, 119-120.

Schwarz, H. R. (1956): Ein Verfahren zur Stabilitätsfrage bei Matrizen-Eigenwort-Problemen, *Z. Angew. Math. Phys.*, *7*, 473-500.

Scott, R. F. (1880): *Theory of Determinants*, Cambridge University Press, Cambridge.

Shaked, U., and Dixon, M. (1977): Generalized minimal realization of transfer-function matrices, *Int. J. Control*, *25*, 785-803.

Shamash, Y. (1980): Comments on the Routh-Hurwitz criterion, *IEEE Trans. Autom. Control*, *AC-25*, 132-133.

Shane, B. A., and Barnett, S. (1974): On the bilinear transformation of companion matrices, *Linear Algebra Appl.*, *9*, 175-184.

Shieh, L. S. (1975): On the inversion of the matrix Routh array, *Int. J. Control*, *22*, 861-867.

Shieh, L. S., and Tajvari, A. (1980): Analysis and synthesis of matrix transfer functions using the new block-state equations in block-tridiagonal forms, *IEE Proc.*, *127D*, 19-31.

Shieh, L. S., Schneider, W. P., and Williams, D. R. (1971): A chain of factored matrices for Routh array inversion and continued fraction inversion, *Int. J. Control*, *13*, 691-703.

Shieh, L. S., Wei, Y. J., and Navarro, J. M. (1978): An algebraic method to determine the common divisor, poles and transmission zeros of matrix transfer functions, *Int. J. Control*, *9*, 949-954.

Sidhu, G. S. (1977): A note on the bezoutian matrix and the controllability and observability of linear systems, *Prob. Control Inf. Theory (Hungary)*, *6*, 129-136.

Silverman, L. M., and Van Dooren, P. (1981): A system theoretic interpretation for g.c.d. extraction, *IEEE Trans. Autom. Control*, *AC-26*, 1273-1276.

Simaan, M. (1974): A note on the stabilizing solution of the algebraic Riccati equation, *Int. J. Control*, *20*, 239-241.

Skala, H. (1971): An application of determinants, *Amer. Math. Monthly*, *78*, 889-890.

Smith, J. H. S. (1861): On systems of linear indeterminate equations and congruences, *Philos. Trans. Roy. Soc. Lond.*, *151*, 293-326.

Specht, W. (1957): Die Lage der Nullstellen eines Polynoms — III, *Math. Nachr.*, *16*, 369-389.

Spiegel, M. R. (1965): *Laplace Transforms*, McGraw-Hill, New York.

Sridhar, B., and Lindorff, D. P. (1973): Pole placement with constant gain output feedback, *Int. J. Control*, *18*, 993-1003.

Stein, P. (1952): Some general theorems on iterants, *J. Res. Natl. Bur. Stand.*, *48*, 82-83.

Stein, P. (1965): On the ranges of two functions of positive definite matrices, *J. Algebra*, *2*, 350-353.

Strejc, V. (1981): *State Space Theory of Discrete Linear Control*, Wiley, Chichester.

Talbot, A. (1960): The number of zeros of a polynomial in a half-plane, *Proc. Camb. Philos. Soc.*, *56*, 132-147.

Taussky, O. (1961a): A generalization of a theorem of Lyapunov, *J. Soc. Ind. Appl. Math.*, *9*, 640-643.

Taussky, O. (1961b): A remark on a theorem of Lyapunov, *J. Math. Anal. Appl.*, *2*, 105-107.

Taussky, O. (1964): Matrices C with $C^n \to 0$, *J. Algebra*, *1*, 5-10.

Taussky-Todd, O. (1968): On stable matrices, in *Programmation en mathématiques numériques*, Centre National de la Recherche Scientifique, Paris.

Tsoi, A. C. (1979): Inverse Routh-Hurwitz array solution to the inverse stability problem, *Electron. Lett.*, *15*, 575-576.

Turnbull, H. W. (1952): *Theory of Equations*, 5th Ed., Oliver & Boyd, Edinburgh.

Usher, M. B. (1972): Developments in the Leslie matrix model, in *Mathematical Methods in Ecology* (J. N. R. Jeffers, Ed.), Blackwell, Oxford.

Van Dooren, P. (1981): A generalized eigenvalue approach for solving Riccati equations, *SIAM J. Sci. Stat. Comput.*, *2*, 121-135.

Van Loan, C. (1979): A note on the evaluation of matrix polynomials, *IEEE Trans. Autom. Control*, *AC-24*, 320-321.

Van Vleck, E. B. (1899): On the determination of a series of Sturm's functions by the calculation of a single determinant, *Ann. Math.*, *2nd Ser.*, *1*, 1-13.

Vardulakis, A. I. G., and Stoyle, P. N. R. (1978): Generalized resultant theorem, *J. Inst. Math. Appl.*, *22*, 331-335.

Varga, A., and Sima, V. (1981): Numerically stable algorithm for transfer function evaluation, *Int. J. Control*, *33*, 1123-1133.

Verghese, G. C., and Kailath, T. (1981): Rational matrix structure, *IEEE Trans. Autom. Control*, *AC-26*, 434-439.

Vieira, A., and Kailath, T. (1977): On another approach to the Schur-Cohn criterion, *IEEE Trans. Circuits Syst.*, *CAS-24*, 218-220.

Walach, E., and Zeheb, E. (1982): Root distribution for the ellipse, *IEEE Trans. Autom. Control*, *AC-27*, 960-963.

Walter, O. H. D. (1972): Real-matrix solutions for the linear optimal regulator, *Proc. IEE*, *119*, 621-624.

Wang, K. (1980): Resultants and group matrices, *Linear Algebra Appl.*, *33*, 111-122.

White, K. P., Jr. (1979): Technique fixation and the Routh-Hurwitz criterion, *IEEE Trans. Autom. Control*, *AC-24*, 987-988.

Wilkinson, J. H. (1965): *The Algebraic Eigenvalue Problem*, Clarendon Press, Oxford.

Wilkinson, J. H., and Reinsch, C. (1971): *Handbook for Automatic Computation*, Vol. II: *Linear Algebra*, Springer-Verlag, Berlin.

Willems, J. C., and Mitter, S. K. (1971): Controllability, observability, pole allocation and state reconstruction, *IEEE Trans. Autom. Control, AC-16*, 582-595.

Willems, J. L. (1970): *Stability Theory of Dynamical Systems*, Nelson, London.

Williamson, J. (1931): The latent roots of a matrix of special type, *Bull. Amer. Math. Soc., 37*, 585-590.

Wimmer, H. K. (1973): On the Ostrowski-Schneider inertia theorem, *J. Math. Anal. Appl., 41*, 164-169.

Wimmer, H. K. (1974a): Inertia theorems for matrices, controllability and linear vibrations, *Linear Algebra Appl., 8*, 337-343.

Wimmer, H. K. (1974b): An inertia theorem for tridiagonal matrices and a criterion of Wall on continued fractions, *Linear Algebra Appl., 9*, 41-44.

Wimmer, H. K. (1976): On the algebraic Riccati equation, *Bull. Austral. Math. Soc., 14*, 457-461.

Wimmer, H. K. (1982): The algebraic Riccati equation without complete controllability, *SIAM J. Alg. Disc. Meth., 3*, 1-12.

Wolovich, W. A. (1971): A frequency domain approach to state feedback and estimation, *Proc. 1971 IEEE Decision and Control Conference*, Miami, Fla.

Wolovich, W. A. (1974): *Linear Multivariable Systems*, Springer-Verlag, New York.

Wolovich, W. A. (1978): Skew prime polynomial matrices, *IEEE Trans. Autom. Control, AC-23*, 880-887.

Wonham, W. M. (1967): On pole assignment in multi-input controllable linear systems, *IEEE Trans. Autom. Control, AC-12*, 660-665.

Wonham, W. M. (1968): On a matrix Riccati equation of stochastic control, *SIAM J. Control, 6*, 681-697.

Young, N. J. (1979a): An identity which implies Cohn's theorem on the zeros of a polynomial, *J. Math. Anal. Appl., 70*, 240-248.

Young, N. J. (1979b): Matrices which maximise any analytic function, *Acta. Math. Acad. Sci. Hung., 34*, 239-243.

Young, N. J. (1979c): Linear fractional transforms of companion matrices, *Glasgow Math. J., 20*, 129-132.

Young, N. J. (1980a): Formulae for the solution of Lyapunov matrix equations, *Int. J. Control, 31*, 159-179.

Young, N. J. (1980b): Functions of canonical matrices, *Linear Multilinear Algebra, 9*, 141-149.

Young, N. J. (1983): A simple proof of Hermite's theorem on the zeros of a polynomial, *Glasgow Math. J.*, *24*.

Zedek, M. (1965): Continuity and location of zeros of linear combinations of polynomials, *Proc. Amer. Math. Soc.*, *16*, 78-84.

Zeheb, E., and Hertz, D. (1982): Complete root distribution with respect to parabolas and some results with respect to hyperbolas and sectors, *Int. J. Control*, *36*, 517-530.

Answers to Problems

Chapter 1

1.2 $(\lambda - 4)^2 \left(\lambda + \dfrac{1}{2} + \dfrac{i\sqrt{3}}{2} \right) \left(\lambda + \dfrac{1}{2} - \dfrac{i\sqrt{3}}{2} \right)$

1.3 $(\lambda - 1)^3 (\lambda + 2)^2$, $(\lambda - 1)(\lambda + 7)(\lambda + 2)^4$, $(\lambda - 1)(\lambda + 2)^2$

1.4 -290

1.7 $(\lambda - 3)^4 + 7(\lambda - 3)^3 + 6(\lambda - 3)^2 - 45(\lambda - 3) - 72$

1.8 $\begin{bmatrix} 1 & 5 & 4 \\ 12 & -39 & -11 \\ -33 & -10 & 53 \end{bmatrix}$ 1.9 $\begin{bmatrix} -7 & 23 & -16 \\ -48 & -39 & 87 \\ 261 & 126 & -387 \end{bmatrix}$

1.10   (i) $\begin{bmatrix} -231 & -100 & 332 \\ 996 & 433 & -1428 \\ -4284 & -1860 & 6145 \end{bmatrix}$ (ii) $-16\lambda^2 + 23\lambda + 7$

1.13 $a_n \lambda^n + a_{n-1} \lambda^{n-1} + \cdots + a_1 \lambda + 1 = 0$

1.21   (i) $\lambda + 1$ (ii) $\lambda^2 + 1$ (iii) $\lambda^2 - 2\lambda + 2$

1.22   (i) $\lambda^2 + 2\lambda - 3$ (ii) $\lambda^2 + \lambda - 2$

1.40   (i) $x = \lambda$, $y = -3\lambda^2 - \lambda + 1$

      (ii) $x = -\frac{1}{6}(2\lambda^2 + 3\lambda)$, $y = \frac{1}{6}(2\lambda^3 - 5\lambda^2 + 6)$

1.41 $\{r_{0j}\} = (1, 4, 4)$, $\{r_{1j}\} = (4, 8, -4)$

Chapter 2

2.6 $[1, -1, 0]^T$ 2.7 $[0.2469, 0.2058]^T$

2.9   Yes   2.10   Yes   2.11   Yes

2.12 $[0, 2, -1]^T$, $[1, 0, 0]^T$ 2.13 $\theta \neq -2, 0, 3$

2.15   c.c. if $u_1 = 0$; if $u_2 = 0$, a basis is $[0, 1, 0, 0]^T$,

      $[0, 0, 1, 0]^T$, $[1, 0, 0, -2\omega]^T$

2.26   (a) $\theta = 10, 12$ (b) $\theta = -2, -\frac{1}{2}, 1$

2.41 $u = (131x_1 - 518x_2 - 137x_3)/13$

2.42 $\theta_1 = \alpha\beta - 1$, $\theta_2 = -\alpha$

Chapter 3

3.6   (i) Yes   (ii) No   (iii) Yes

3.7   (i) 2, 1   (ii) 2, 2   (iii) 2, 4

3.8   (i) $k < 2$ (ii) $0 < k < 2$

3.9 $-120 < k < 540$

3.10   (i) 2 left-half-plane, 2 right   (ii) 2 left, 4 imaginary

      (iii) 3 left, 2 right

3.15   (i) Yes   (ii) No

3.16   (i) 3 outside   (ii) 3 outside, 1 inside

3.17 $-4 < k < 4$ 3.18   Yes

3.19   (i) 3 on unit circle   (ii) 2 inside, 1 outside

3.22   2 left-half-plane, 1 right   3.24   1 inside, 2 outside

3.28 (iii) all roots outside disk

3.46 2 real roots, each of multiplicity 2

3.47 3 distinct roots, one of which is real

3.49 $74\mu^5 - 28\mu^3 + 32\mu^2 + 50\mu$

3.50 $\mu^4 - (14/13)\mu^3 + (16/13)\mu^2 - (2/13)\mu + 3/13$

Chapter 4

4.1 $k = \begin{bmatrix} -a & 2 & 5 \\ -1 & -1 & 0 \end{bmatrix}$

4.2 (i) Yes   (ii) No

4.5 $k = 1$

4.14 $u_1 = -1.71x_1 - 3.16x_2$ 4.15 $u = -2.36\theta - 11.88\dot{\theta}$

4.16 $\begin{bmatrix} 2 - \sqrt{2} & 3 - 2\sqrt{2} \\ 3 - 2\sqrt{2} & 6 - 4\sqrt{2} \end{bmatrix}$ 4.20 $\begin{bmatrix} -47 & 34 & 10 \\ 49 & -35 & -11 \end{bmatrix}$

4.33 (i) Yes   (ii) No

4.47 $[0(A, H)]^T[C(A, B)]^{-1}$

4.54 $\begin{bmatrix} s + 5 & 1 \\ -2 & s \end{bmatrix}$

Chapter 5

5.9 (i) $T_1(\lambda) + T_0(\lambda)$ (ii) $H_1(\lambda) + H_0(\lambda)$

5.10 $p_1(\lambda) - 2p_0(\lambda)$

5.11 (i) $T_2(\lambda) + 6T_1(\lambda) + 5T_0(\lambda)$ (ii) $H_2(\lambda) + 6H_1(\lambda) + 10H_0(\lambda)$

5.13 $T_1(\lambda) + T_0(\lambda)$ 5.16 $u = [131, -518, -137]x/13$

5.21 60   5.23 $(-115p_1(\lambda) + 165p_0(\lambda))/6$

Index